Survival & Rescue Equipment of World War II:

Army Air Forces and U.S. Navy

VOLUME 2

Distress Signals
Personal Aids Kits
Protective Clothing
Survival Training

Dustin Clingenpeel

ECP

ISBN 978-1-943492-61-9

EGP

ELM GROVE PUBLISHING
San Antonio, Texas, USA
www.elmgrovepublishing.com

Contents

Introduction

Historically, air-sea rescue had its inception and greatest advances in emergency equipment, procedures and techniques amid war. What was accomplished in less than five years of conflict probably would not have been accomplished in decades of ordinary development, and retains its relevance in the new millennium. Each article developed had its own specific background or purpose, and was often included as a compliment to other items. Specific requests occurred as a direct result of service activities, an individual's idea or resulted from continuous development.

The US Navy Bureau of Aeronautics' development was carried out at a rapid, high priority rate, when a new item was requested from the fleet. It would start with an unproven idea or a modification of existing equipment and go through stages of development and distribution. Once a piece of equipment was standardized, every effort was made to expedite its production and availability. The process began at the Flight Gear and Safety Equipment Section, Equipment and Materials Branch, Engineering Division, or the BuAer directly associated with the Naval Aircraft Factory. Here the item was taken through the technical stages including the design, preparation of specifications, initiation of test projects, preparation of technical orders, and notes in new or changed equipment. Once the item was ready for manufacture, procurement action was initiated by the Airborne Equipment Section of Maintenance Division NAF, or by the Aviation Supply Office. Quantities were established by the Military Requirements Division of the Bureau of Aeronautics. In some instances, a new type or construction might require manufacturing experience, delaying delivery. No two items followed the same pattern. To serve a practical purpose, limited quantity contracts were awarded to service test, and in some cases procurement was expedited.

Once an item was placed on procurement, delivery schedules were established and quantities were outlined for prescribed supply points. The central supply point was the Naval Aviation Supply Depot, Philadelphia, which housed the Aviation Supply Office. Two primary supply points were maintained, one on each coast. These were Aviation Supply Annex in Oakland, California, and Aviation Supply Annex in Norfolk, Virginia. The supply annexes had the responsibility of maintaining supplies for Naval Air Stations under their jurisdiction. Supply points maintained at an air station could requisition supplies organic to aviation supply from the annexes, or directly from the ASO. Ordnance was typically requisitioned from Naval Supply Depots, and medical items from the Bureau of Medicine & Surgery, however some of these items were maintained at aviation supply points. Each air station had purchasing power through purchasing agents who could make direct procurement locally, or from approved contractors. In some cases, due to urgency, an item would be placed on expedited delivery directly from the contractor to air stations (specifically to NAS Pearl Harbor) to assure it became available as soon as possible.

The USAAF had a more complex system, and when the Wright Field Personal Equipment Laboratory, in Dayton, Ohio, received a request or suggestion for modification, it would undergo a preliminary review before being presented to AAF Headquarters. When the idea was submitted to Operations, Commitments and Requirements Assistant Chief/Air Staff at AAF Headquarters, it was first checked for its potential value. Upon approval from OC&R, Wright Field was authorized funds for development to the point of service testing. The item was then sent to the AAF Proving Ground Command at Eglin Field, Florida, for a series of rigid service tests. When completed, a report and the item would be forwarded to the AAF Board in Orlando, where a decision would be made to recommend the item for standardization, rejection or require further service testing. Their conclusions and recommendations would be submitted to AAF Headquarters for the final decision to approve or disapprove the item, and if approved, standardization and procurement followed. This authority allowed Wright Field to prepare drawings and specifications and to proceed with the negotiation of contracts. The basis of issue and distribution to the various supply points and theaters was determined by AC/AS, Material and Services, under the authority of the Air Technical Service Command. The Emergency Rescue Branch, Personal Equipment Laboratory of the Air Technical Service Command at Wright Field was charged with the responsibility for development of emergency rescue equipment used by the USAAF. The Emergency Rescue Branch was established in August 1943.

The USAAF Air Service Command controlled, handled, and distributed all air forces property. In the continental United States, distribution was coordinated through eleven control area commands and control depots, then through the lowest echelon sub-depot located regionally. ASC depots received a constant stream of material from the factories and channeled it to operational forces. From storage depots or direct from the factory, supplies were moved to control area depots which were responsible for maintaining sub-depots. Sub-depots were the principle suppliers of materials to flying units located at air fields or bases.

Once an item reached a sub-depot or air station responsible for the supply of a squadron or group, they were divided into classification groups being stowed, maintained and distributed accordingly. Individuals were assigned responsibility for the management of those departments. For personal equipment, a Supply Officer was accountable for flying clothing, a Parachute Officer for parachutes and a Material or Safety Officer for life rafts and oxygen equipment.

From experience in early operations a need arose for more efficient maintenance and intelligent use of oxygen equipment. The BuAer were the first to designate the position of an Oxygen Officer authorized under BuAer Circular Letter No. 5-41 dated 19 March 1941. This letter detailed that one officer in each squadron and at each base be designated as Oxygen Officer. This officer would be responsible to the commanding officer for the condition of the oxygen equipment and assure that all flight personnel fully understood its operation prior to use in flight. The USAAF inaugurated the position of Oxygen Officer under AAF Regulation No. 55-7 dated 30 May 1942. It detailed that the first assistant operations officer in each group and squadron be designated as a Unit Oxygen Officer. Both services established the position with existing personnel, but soon found it inadequate for this responsibility. However, it temporarily served the initial demands to prepare pilots and aircrew for high altitude operations. A formal course for the training of oxygen officers for the AAF was established at the School for Aviation Medicine in July 1942 as a five-day course. To meet wartime demands the BuAer established a six-week training course in April 1943 at NAS Jacksonville, and upon graduation, personnel were designated as Oxygen Officers.

The overall training curriculum on safety equipment in 1942 could be best described as deplorable, and survival knowledge severely lacking. It was assumed that most men had some level of outdoor skills, but this proved not to be the case, and by the close of that year it became apparent that there was an urgent need for more training and maintenance of safety equipment. One principle attempt to remedy the situation was by the 8th AF in October 1942, when it was recommended to extend the responsibility of the Oxygen Officer to a new designation of Equipment Officer. This recommendation was based on evidence of numerous reported deficiencies and was to include the proper use and care of flying clothing, adequate knowledge of crash, ditching or oxygen procedures, and life rafts/safety harness procedures. Additionally noted was the improperly equipped crews as a result of theft and hoarding of equipment, ignorance of bail-out procedures, lack of proper crew indoctrination, and lack of cooperation between the officers responsible for supply. Coupled with similar reports from other combat zones regulation 55-7 was revised 4 May 1943 to make the Unit Oxygen Officer an Equipment Officer.

The first Equipment Officers appointed were from group headquarters and each squadron of the 8th Bomber Command. As part of this selection most of those chosen for the positions of both the oxygen and equipment officer were from flying personnel, as it was felt that pilots and aircrews were more likely to respect advice from an individual with actual flying experience. It soon became apparent that officers on flying status were often too preoccupied to properly execute the duties they were delegated to perform. Regulation 55-7 was again revised 28 October 1943, restricting the position to ground officers and assigning a new designation of Personal Equipment Officer.

In November 1943, the course for Unit Oxygen Officers at the School of Aviation Medicine, Randolph Field, Texas, was replaced by a two-week instruction course for the newly created Personal Equipment Officer. This course was eventually transferred to the AAF Tactical Center in Orlando, Florida, increasing the training period to three weeks in all aspects of flying safety equipment and air-sea rescue. Regulation and supervision of the PEO was under the jurisdiction of the Air Surgeon, but due the major expansion of duties it was felt that the program was now out of the scope of Air Surgeon activity. In the new AAF Regulation 55-7A dated 4 March 1944, OC&R AC/AS would be charged with supervising the activities of PEO and made responsible for disseminating information to them. The structure of assigned PEOs falls into four tiers. The top tier was the Command Personal Equipment Officer, who supervised and coordinated all PEOs within the command and principally maintained an administrative roll. On the second tier was the Wing PEO, who recommended the personal and emergency equipment to be used on combat missions and was involved in more hands-on inspection, supervision and coordination of the Group and Squadron PEO, as well as assisting the Command PEO in maintaining records. The Group PEO supervised and coordinated directly with Squadron PEOs and was more directly involved with inspections and continued training. On the bottom tier was the squadron PEO who worked directly with flying personal and assured they were properly trained and equipped for any combat mission.

By 1943 the Pacific fleet was expressing a concern that more training was required in the use and maintenance of emergency rescue equipment. In response, the BuAer went directly to the source of the issue by inaugurating survival training to begin in preflight schools, continuing through to the intermediate and primary courses. Soon after the opening of the Oxygen Officer School at NAS Jacksonville, the syllabus was expanded to include training in emergency equipment such as life rafts, life preservers and other related survival gear. A new designation, Aviation Equipment Officer, was assigned, superseding the position of Oxygen Officer. By January 1944 a new training course was established for the Aviation Equipment Officer at NAS Pensacola. To help compensate for the lack of, and need for, such an officer for fleet activities, the Chief of Naval Operations drafted and distributed Aviation Circular Letter No. 17-44 dated 4 March 1944 which directed squadron commanders to appoint an officer to assume the duties of the Aviation Equipment Officer in units where a trained officer had not yet been assigned. As operations increased in the Pacific, more specialized training was required, in large part due to the increase in specialized emergency equipment, basic survival skills and indoctrination for the health and safety of individuals. As part of the responsibility of the AEO, they were required to disseminate information and keep up with current events conducting survival training in the operational area. It was felt that the survival training aspect was not reflected in their title, and in October 1944 their designation was changed to Aviation Equipment & Survival Officer (AESO). One AEO or AESO would be assigned to a Carrier Air Service Unit (CASU) acting as a maintenance officer in the Assembly & Repair departments. When assigned to a Naval Air Station, they would take on the role, primarily, of instructor. One AESO would be assigned to each squadron who would take on the roles of both maintenance officer and instructor. Prior to the institution of the AEO, the squadron parachute rigger had principally taken on that role. Within a CASU there would be separate facilities that inspected and maintained parachutes, life rafts, flight clothing and other emergency equipment. Aboard a carrier these facilities would be consolidated into one department maintained by the squadron parachute rigger. His responsibility, in large part, was to cater to all the squadron members and their needs regarding personal equipment, which would include training in its use. The AEO would take on this role, but still work in close coordination with the parachute rigger.

For both the AAF and BuAer, the position of Equipment Officer could be divided into custodial, training and inspection duties. Under the custodial duties, he would organize equipment storage and make sure everything needed for a mission was prepared. Training duties consisted of instruction in the care and use of parachutes, oxygen equipment, clothing, flak suits, life rafts and emergency kits. He would also give instruction on emergency procedures relating to bailing out, ditching, fires in flight and survival on land and sea. Before and after each mission, all equipment would be inspected for damage, staining or wear, to assure it was in perfect working order. This included items in storage during mandatory inspection periods. The EO's job was to coordinate with the supply, engineer, parachute and intelligence officers and not assume the responsibility of any one of their jobs, and to coordinate with the flight surgeon sharing the responsibilities in medical and survival training. The Equipment Officer would be responsible for the dissemination of information by any means available to him, the record keeping and reporting of the performance of all this equipment. Most importantly, his duties were not to conflict with other established positions.

The terms MARK-, MK-, Mark- and Mk- (followed by a number) are used interchangeably throughout the text, because they were used interchangeably by the various departments and divisions of the military at different times and in different lists, catalogs and publications, even when referencing the same item. Similarly, Mark- numbers are used both as Roman numerals (I, II, III, etc) and the more familiar Arabic numbers (1, 2, 3, etc). The same item might appear in one reference as Mk-III and as MARK-3 in another. Terms are used according to their source.

"Our aviators and seaman, with confidence that they will fly and sail again tomorrow, dare to face greater odds in the war today. Vision our Services, ingenuity of industry has provided survival and rescue equipment, which has lessened the hazards, improved the safety, and given our men greater courage. May we never be content with present equipment, but constantly improve it with experience, continued study and cooperative effort."

Admiral Russell R. Waesche,
Commandant of the United States Coast Guard

DISTRESS SIGNALS

Part I.
Sea Markers

The fluorescein dye marker was the most successful distress signal for locating survivors during WWII.

When the dye dissolved in water, it would create a fluorescent yellow-green slick 25-50' in diameter that could be visible for 1-3 hours, from distances up to 5 or 6 miles and from heights of 8-10,000', depending on whether the water was rough or not. No other distress signal could equal this performance.

Smoke signals burned for only a few minutes and easily dissipated with wind conditions. Flares burned for an average of 5-10 seconds. Mirrors and signal panels required manual manipulation and direct line of sight, while flashlights and whistles were limited by distance.

(Above) Period color image illustrating the effectiveness of the sea marker.

(Below left) The sea marker only minutes after being deployed. In the ocean a raft is only a tiny speck. From an aircraft at a few thousand feet, the sea marker could be viewed as an extension making the target much larger.

(Below right) The raft was approximately 7' long. With the sea marker it became 50' long and remained visible for hours. (National Archives)

The Bureau of Aeronautics was the first to implement a sea marker, in April 1941, adding it to the compliment of equipment for the M-3 series life raft.

The marker adopted was polished aluminum powder, due to its highly reflective properties. By description, one pound was contained in a cylindrical waterproof cardboard container measuring 3.5" x 7" with friction slip caps on each end, secured with waterproof tape. This unit was in turn stored in a cardboard tube with metal screw caps on each end. When used in a life raft, one of the slip caps would be removed and the powder dumped into the water. The inner container was also designed to break apart when dropped from an aircraft.

This marker was formalized under NavAer Specification M-530, 6 February 1942, as Sea Marker; Aluminum Powder. Due to the use of the critical material, aluminum, this sea marker was only implemented for a very brief period being phased from service by fluorescein dye.

A further revision to Specification M-3 in June 1941 outlined that the sea marker is to be of Federal Specification TT-A-476 titled as Aluminum Powder for paints (aluminum bronze powder) dated 15 November 1932.

Aluminum bronze powder was originally manufactured from a copper alloy mixed with aluminum flakes, and when applied as a coating the object would appear gold in color; bronze was not used in the actual manufacturing process.

The process of manufacture was later revised to compose primarily of aluminum utilized as an additive to paints or dopes commonly used for coating the skin of aircraft, balloons, and dirigibles of the era as a protective coating reflecting UV rays and from other elements.

(Left) Aluminum Bronze Powder, more commonly known simply as Aluminum Powder, was available commercially in one pound cans. In Specification M-3 it details a specific packaging method but also commercial cans were procured and pressed into service being utilized in life rafts, airplanes, and other over water emergency kits.

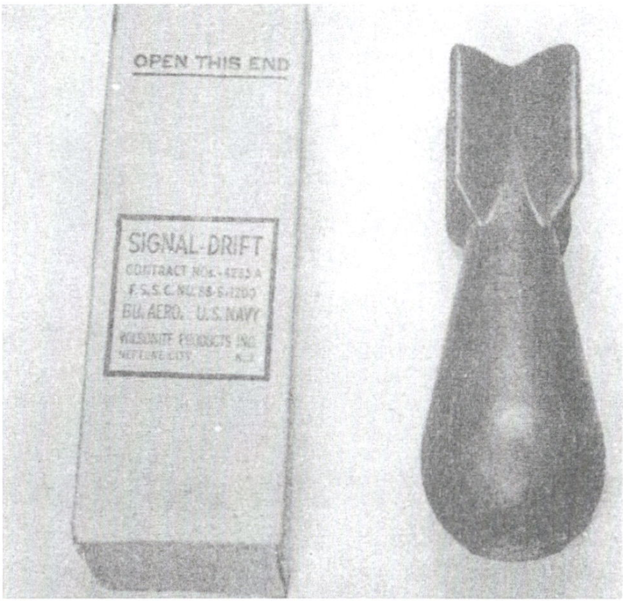

(Above) The bomb shaped signal illustrated above was designated as Signal, Drift, Day, AN-MK-I containing aluminum powder. It consisted of a paper shell designed to break apart upon contact with water when dropped from an aircraft, providing a slick for marking reference points of objects. A similarly designed marker was adopted as Marker, Slick, AN-M59 but contained fluorescein dye.

Prior to the standardization of the aluminum sea marker, the BuAer standardized a fluorescein type under NavAer Specification M-528 13 December 1941 titled as Sea Marker, Fluorescein Dye.

Fluorescein dye was being tested in the summer of 1941, proving to have an equal performance to that of the aluminum powder. Upon entry into WWII, a specification was expedited to eliminate the use of critical war material.

The aluminum sea marker was standardized to meet the immediate requirements for a life raft marking device until suitable quantities of fluorescein dye could be manufactured. Commercial production of fluorescein dye was limited, being utilized primarily for tracing under water streams and locating sewer leaks.

As outlined in Specification M-528, 9.5 ounces of fluorescein soluble sodium salt was to be provided in a cylindrical metal container, with a water soluble salt as a filler, for an overall weight of one pound. The M-528 replaced the M-530 when quantities became available, being standard at a ratio of one can for all life rafts, available under ASO stock number R83-M-160500.

(Above and top center) Fluorescein dye contained in a gray, cone top can measuring 5 1/4" tall with a 3" diameter. This is referenced as a "Crown" top in the can collecting community. It featured nomenclature with simple instructions. The initial production of the M-528 would include a screen in the neck of the can near the cap being eliminated in Amendment-A 21 May 1942. The Calco Chemical Co., a division of the American Cynamid Co., was the only known contractor for the sea marker M-528.

(Above and right) By the summer of 1942, production of the M-528A transitioned to a more cost-effective, seamless cone top can (same as the newly adopted AN-W-5 drinking water unit) measuring 5-5/8" tall with a 2-3/4" diameter.

(Right) Period color image illustrating the brilliant yellow green dye slick being demonstrated by a Naval Aviator utilizing the M-528 can, c.1943. (National Archives)

In August 1942, the USAAF submitted a revision to convert the sea marker M-528 and the AN-W-5 drinking water cans to a rectangular shape, to help with comfort for pararafts and to improve packing characteristics in multi-person life rafts.

By the close of 1942 the BuAer had received complaints of discomfort when sitting on the round cans, and with the development of emergency back pad kits, the proposed rectangular cans proved preferable. Specification M-528 was re-designated under joint Army-Navy Specification AN-S-10, 8 February 1943, with the only change being the style of the can.

The BuAer requested that existing contracts for the round M-528 be converted to the manufacture of the AN-S-10. Following the adoption of the AN-S-10, issue remained one can per all life rafts, and two cans in the M-594 Shipwreck Kit, standardized later that year.

Several issues occurred with the sea marker cans. Firstly, it was found that if not sealed properly, the dye powder would turn to a doughy consistency, rendering it useless. Secondly, the packing characteristics were not optimal. Following field tests, the BuAer issued Technical Order 53-44, dated 17 April 1944, requiring that the M-528 and AN-S-10 be replaced with the life jacket dye marker M-566 at the ratio of three packets per one can in all emergency equipment. In addition, all contracts for the AN-S-10 were to be amended to the production of the M-566 life jacket dye marker, making it nearly impossible to calculate production figures, with consistent revisions to the contracts.

The USAAF did adopt the sea marker but leaned heavily towards the use of pyrotechnics and, with that emphasis, they determined the space consumed by their standard of three AN-S-10 cans per multi-person life raft could be better utilized. Their life rafts were rated for 5 and 7 people, with each having a life preserver which included a dye marker packet. It was determined to eliminate the AN-S-10, and Specification AN-S-10 was officially cancelled 26 October 1944, meaning no further contracts would be awarded. Both the M-528 and AN-S-10 remained in service through the duration of the war.

(Below) The new AN-S-10 was painted gray with red property and nomenclature markings, measuring 4" x 5-1/2" x 1- 3/4". The unit was classed under the same ASO stock number R83-M-160500 as the M-528, available under stock number 7300-379000 for the USAAF. The American Cynamid Co. is the only known contractor.

(Below left) The new sea marker was provided with a plastic screw top cap, which was an improvement over the crimped type provided on the M-528. As part of the original specification, the cap was to be knurled red plastic, and the can domed at the top, to distinguish the sea marker from the water can in low visibility. In the June 1943 revision, these features were amended to a flat top and black plastic cap with improved knurls, making it identical to the water can.

Due to combat experiences, and in line with requirements for personal emergency equipment, it was recommended that a marking device be adopted for attachment to life vests. Following positive results of tests conducted in the summer of 1942, through a coordinated effort between the Calco Chemical Co. and the BuAer, a life jacket dye marker was standardized under NavAer Specification M-566 30 September 1942 titled as Life Jacket Dye Markers.

The outer package consisted of a vinyl coated fabric, which was secured to the life jacket by means of a 1" non-elastic cotton tape. The three ounces of fluorescein dye was a cake contained in a porous bag. An opening was provided at the bottom of the packet with a quick release tab allowing the dye pack to drop out. To prevent loss, the dye pack was secured by a length of a 1" cotton tape to the inside of the outer package.

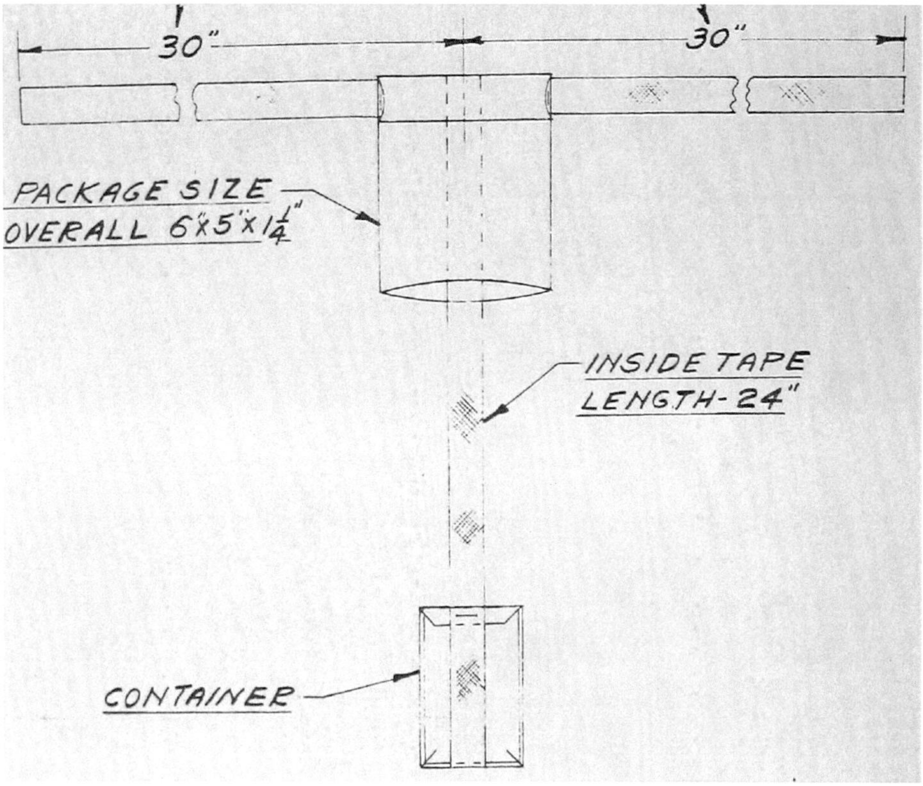

(Left) This drawing, as part of the original specification, illustrates the basic details of the dye marker packet. Of interest is the excessive length of the tie tapes, 30 inches each from the center of the packet. No information has surfaced to explain the intention. The opening at the bottom of the packet is the full width of the packet.

(Below) Three pictures from Technical Note 77-42 dated 27 October 1942, detailing the new life jacket dye marker. One marker was to be issued per life jacket and intended to supplement the M-528 provided in the life raft. The issuance of the technical note was prior to actual availability for service activities. Limited quantities were projected to be available by the end of November 1942 available under ASO stock number R37-P-25. As part of the original specification "TO RELEASE DYE PULL TAB" was printed on the front and "LIFE JACKET DYE MARKER" was to be printed on the back. (National Archives)

(Left and below) On 29 October 1942, Amendment-1 was made effective to Specification M-566 reducing the 1" tie tapes to 8" from center and placing all markings onto the front of the packet. Each marker was individually packed in a Kraft envelope.

(Below) Photograph from August 1943 illustrating the M-566 as per Amendment-1.

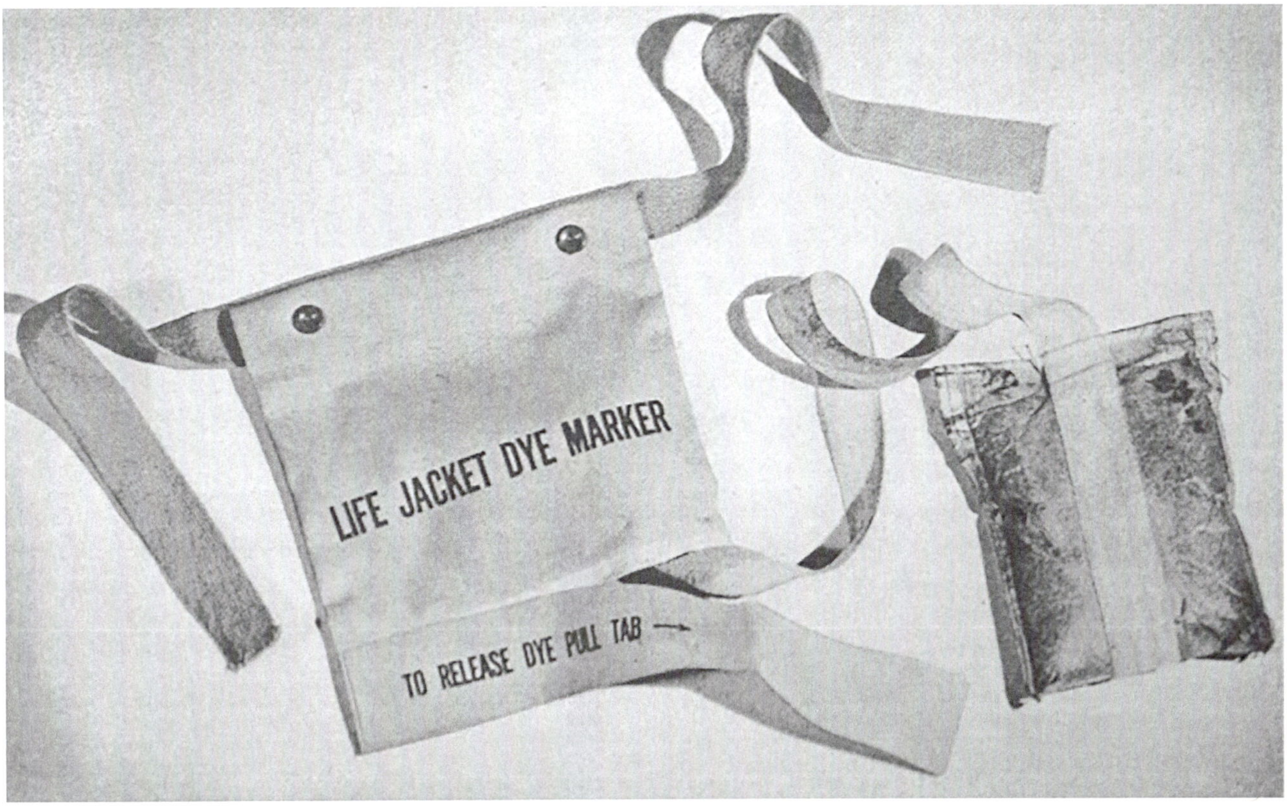

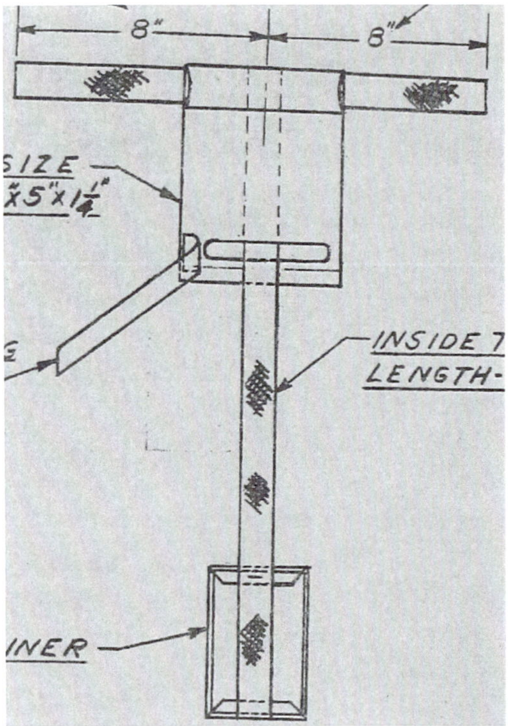

(Left) Amendment-A to Specification M-566 was made effective 30 July 1943, revising the opening at the bottom of the packet to an oval slot, all other construction details remained the same. Since the openings are obscured by the pull tab they are indistinguishable in vintage photographs.

(Above) The American Cynamid Co. (Calco Chemical Co. Division) was the initial contractor for the M-566 life jacket dye marker and in 1943 the American Products Co. were additionally awarded contracts. On the back of the M-566 or M-566A package there are manufacturer codes. It is not known which code belongs to which company.

(Far left) Pilot Lt. jg Raymond F. Kilrain of the USS Yorktown after air strikes on Marcus Island, October 1943. Kilrain wears a single M-566 or M-566a dye marker on his 23P6 life preserver. (National Archives)

(Left) VF-10 pilot Lt. Jg. Donald "Flash" Gordon is sporting two M-566 sea markers on his 23P6 life preserver, March 1944.

By the close of 1943, the M-566 life jacket dye marker was being utilized service wide, with the USAAF recently implementing a life jacket dye marker of their own design.

The feature the USAAF incorporated was to open the packet with a downward pull – tests within the BuAer determined this to be more practical than the side-tab opening. When the user was in the water the packet often required two hands to open, and it was found that the tab had a tendency to rip off leaving the packet sealed.

In early 1944 the BuAer made attempts to place the dye marker under joint Army-Navy form but the USAAF declined, not wanting to defer from their attachment method. The alternative was to adopt the design of a pull-down feature allowing better grip and the use of only one hand to open.

The new packets were standardized under Amendment-B 10 June 1944 titled Life Vest Dye Markers.

To further improve upon the effectiveness of the dye marker, the BuAer distributed Technical Note 44-44 on 25 May 1944, prescribing that two dye markers were to be issued with each life preserver. The intention was that one marker was to be released upon entering the water, and the other to be retained and used when searching aircraft or vessels were sighted.

(Right and far right) Drawing details of the new M-566B dye marker, June 1944.

(Above) Contracts with the Calco Chemical Co. Div. and the American Products Co. for the M-566A were amended to conform to the new Amendment-B. This style of marker would become available in the summer of 1944.

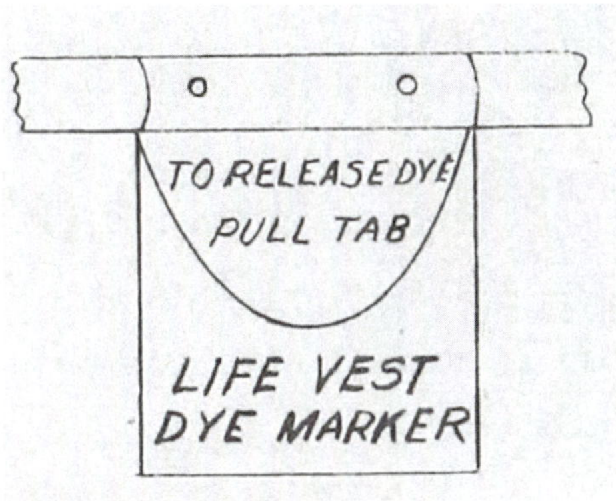

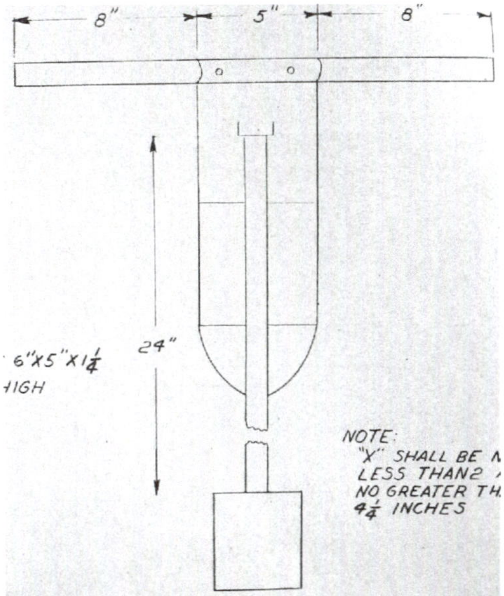

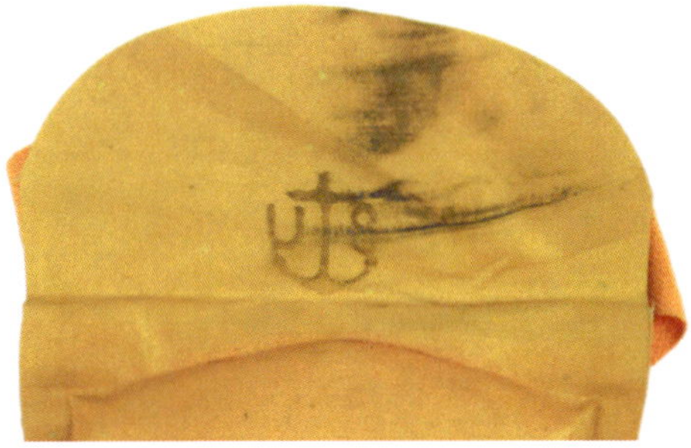

(Far left) Marine aviator Edgar Rafferty of VMF-224 on Yontan Airfield Okinawa c. 1945. Rafferty has the standard compliment of a life preserver and two dye markers, M-566B. (National Archives)

(Left) Ensign Nathan Edmondson wears two M-566B life jacket dye markers, NAS Melbourne Fla, October 1944. Edmondson would later be killed after his F4U exploded in midair, shortly after being catapulted from the USS Hancock on 17 July 1945. (National Archives)

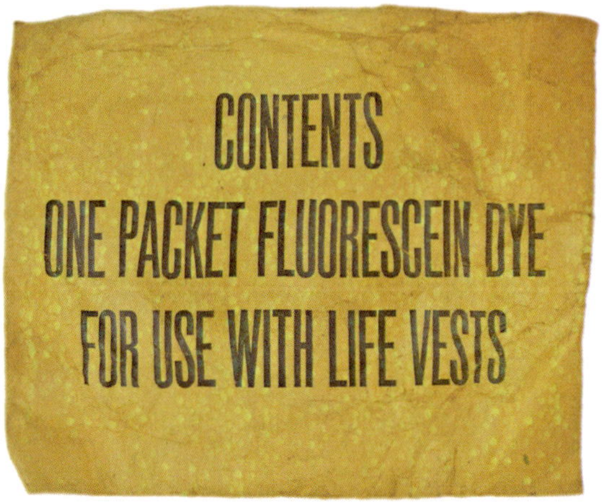

(Above) Due to the increased demand of life jacket dye markers, the Howard Zink Corporation was approved for manufacture. They submitted a deviation from specification, primarily the detail of the tie tapes to be sewn in lieu of folding the packet over and secured with two rivets. Also shown is the Kraft envelope packaging for the dye markers.

(Above left) Lt jg Dottery wears a standard compliment of two sea markers, NAS Melbourne January 1945. (National Archives)

(Above right) Pilot Ens. John Crittenden wears a newly issued Howard Zink sea marker and Type B-4 life preserver after recently coming back aboard the USS Lexington. He spent 45 days on Leyte after being shot down.

(Left) VF-16 pilots Lt jg D.D. Whitmore and Ensign H.R. Gentry, 10 May 1944. The M-566 and M-566A were the standard dye markers in use until late 1944, when the M-566B was introduced. Whitmore still utilizes the 23P6 life preserver, but with three dye markers (up to four dye markers have been observed in period photographs). Gentry utilizes two dye markers with one marker attached to the collar, as prescribed, and the other tied (presumably) to the waist strap D-ring of the life preserver. This photograph pre-dates the distribution of Technical Note 44-44, illustrating that crews fully understood the significance of the sea marker. (National Archives)

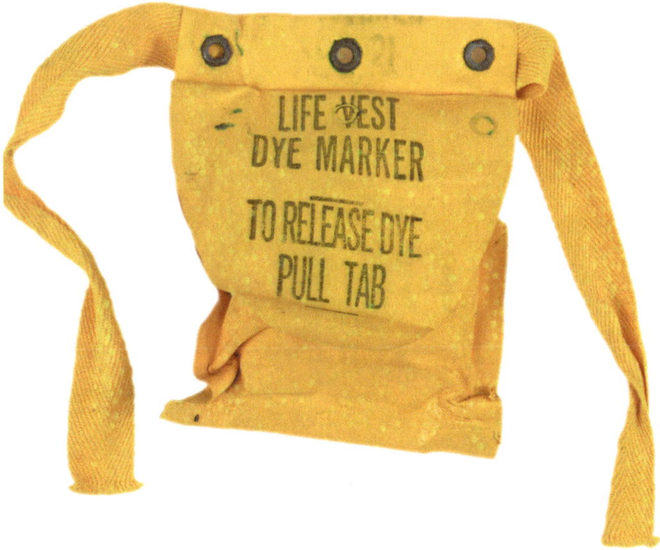

(Above) To help prevent the loss of the dye marker, the tie tapes were reinforced with three brass eyelets, or grommets, effective 10 April 1945. The American Dyewood Co., which principally made dye markers for the USAAF, was awarded a BuAer contract at the time of this amendment, making them the fourth contractor for BuAer dye markers. With only two variations of the April 1945 amendment, it is not perfectly clear which style was supplied by which contractor. These are the last two styles of dye marker that would be utilized prior to VJ-Day. Specification M-566B would be reformatted under Military Specification MIL-P-17980, dated 2 September 1954.

(Left) Lt. jg. Snay wears two grommet type dye markers, July 1945. (National Archives)

(Below) Prior to the revision of Specification M-566A USAAF life jacket dye markers were modified with tie tapes, and field tested at NAS Patuxent River, MD in February and March 1944. Illustrated is one of the modified markers of the standard USAAF specification packet. The dye marker is fully opened detailing the inner assembly. (National Archives)

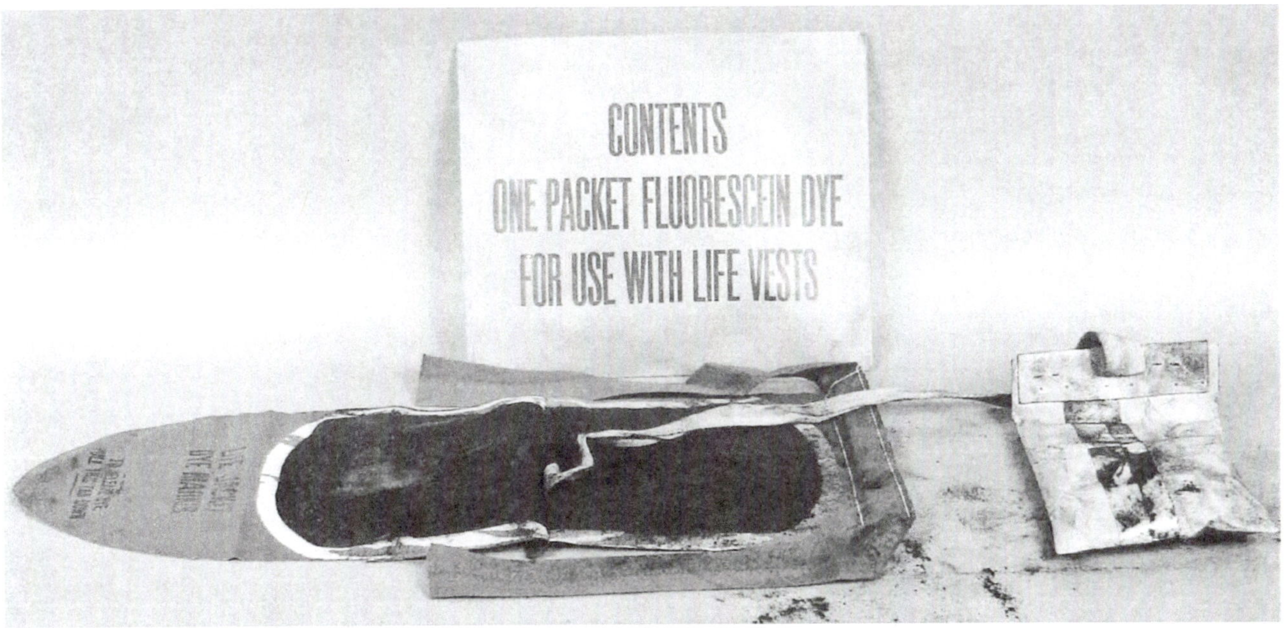

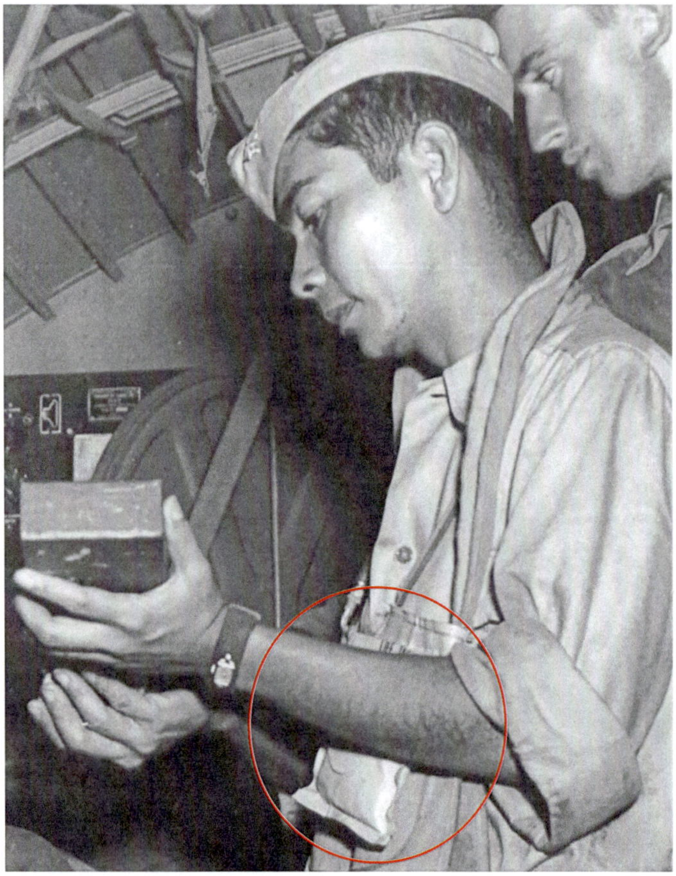

(Above left) Interestingly the modified dye marker is pictured in the December 1944 issue of the Air Sea Rescue Bulletin, illustrating equipment utilized by the US Coast Guard. All equipment listed comprised standard Naval Aeronautical emergency items as listed in standard ASO stock class.

(Above right) A naval medical officer aboard a PBY distributes rations to rescued survivors, c.1944, somewhere in the PTO. Worn on his life preserver is the experimental modified life jacket dye marker. (National Archives)

(Left) The display board of the Liaison Committee for Emergency Rescue Equipment, December 1943, showing the four principle packaging methods used by the aeronautical services. The top two are life jacket dye markers, NavAer Specification M-566 and USAAF specification No. 94-40616. On the bottom is the NavAer Specification M-528 and the Army-Navy Specification AN-S-10. (National Archives)

Like the BuAer, service requirements arose from combat experiences within the USAAF. Influenced by the recent standardization of the life jacket dye marker M-566, Materiel Command recommended development of a packet for use by USAAF personnel in October 1942.

The USAAF was slow in adopting the marker, choosing to field test fluorescein dye from step one, not taking into consideration tests conducted by the BuAer in 1941. Field tests commenced in January 1943 and upon proving satisfactory as a signaling device, the USAAF began the process of engineering the packet and the attachment method.

It was eventually standardized as the Packet, Fluorescein Dye Sea Marker, on 19 June 1943, USAAF Specification No. 94-40616.

Production of the dye marker commenced just prior to the official standardization, with contracts awarded to the American Cynamid Co. and the American Dyewood Co.

(Left) USAAF dye marker Specification 94-40616. The outer package is of the same vinyl coated material as the M-566 and the same three ounces of fluorescein dye in a porous bag. The construction of the packet differs from the BuAer type (at the time of standardization) in not having the addition of the tie-tapes. The side tab opening of the packet would not incorporate into the proposed USAAF attachment method, changing it to a pull-down concept.

(Below and below left) First generation dye marker. Initial production of the packets did not include a cotton backing at the top edge.

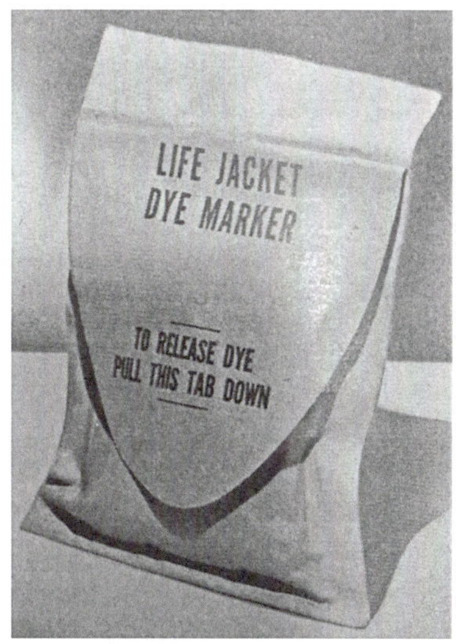

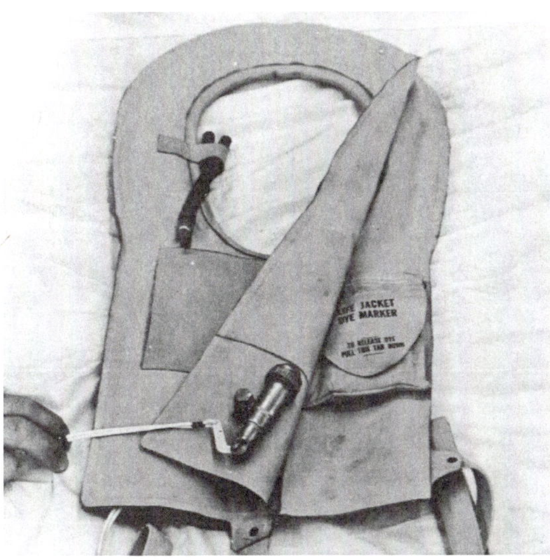

(Left) Attachment method for the USAAF dye marker first outlined in Technical Order 13-1-5, dated 19 April 1943. Packets could be requisitioned utilizing stock number 7300-376000. The marker was to be cemented between the bladders of the life vest on the wearer's left hand side, as shown. The revised technical order of 14 August 1943 stated that all sea markers manufactured prior to June were found to have unsatisfactory heat-sealed seams, and were to be removed from service. Only markers manufactured in July 1943 or later were to be used. It further outlined an improvement to overcome bonding issues to the vest by applying a 2" x 4-1/2" cotton strip sewn to the top edge of the marker. A revised technical order was distributed in November 1943, adding that after modification was completed, each vest was to be stenciled T.O. No. 13-1-5 C.W. C.W. denotes "complied with".

(Right) A dye marker can be seen between the bladders of the Type B-3 life preserver worn by T/SGT Cletus Johnson in Oahu, Hawaii, April 1945. Common practice for attaching the dye marker to the Type B-3 life preserver was to sew it to the outer fabric of the vest, due to the lack of bonding characteristics of the rubber cement. (National Archives)

(Far right) A dye marker is just visible on the Type B-4 life preserver worn by P-61 radar operator, Daniel Hinz on Saipan, c.1945. (National Archives)

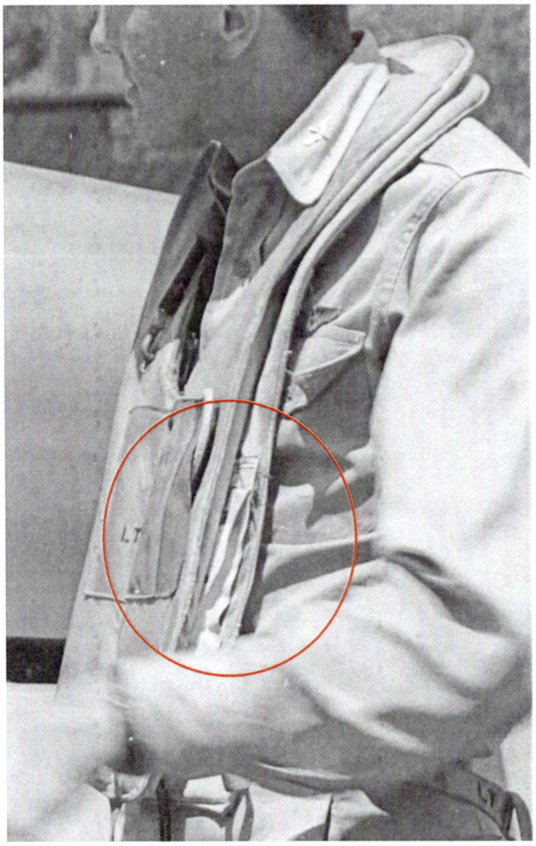

(Below) P-51 pilot of the 20th F.G. Capt. D. Michel is wearing a modified sea marker with tie tapes on his B-4 life preserver, England, 1944.

(Below) Another fighter pilot, Col. Osborn, of the 314th FS wears a similar modified sea marker, March 1944. (National Archives)

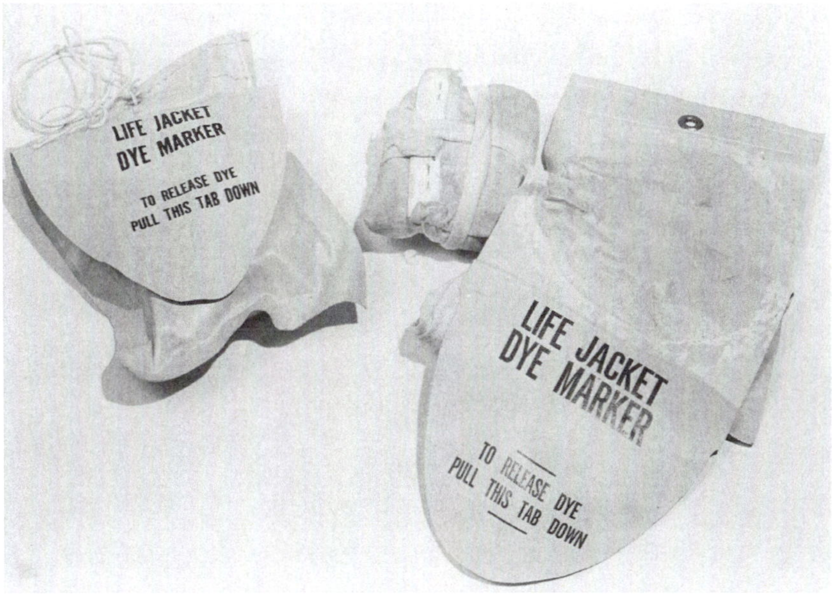

(Above) In 1945, the dye marker 94-40616 was modified to be incorporated with the Type B-5 life preserver by adding a brass grommet. The Type B-5 life preserver included two pockets between the bladders; the pocket on the wearer's left was intended for the dye marker. A length of line would secure the packet to the pocket by means of the eyelets provided on both the marker and the pocket.

In March 1942, Dr. Henry Field, who had recently returned from a public relations mission in Trinidad, and Dr. Harold J. Coolidge of the OSS, began to discuss the psychological factor of military personnel regarding sharks. Dr. Field became interested in this topic after viewing shark activity while flying over the Caribbean and hearing, firsthand, concern from aviators. Together, they drafted a memorandum to President Roosevelt on the feasibility of development of a shark deterrent. After much deliberation between military leaders, it was concluded that development of a shark repellent would greatly benefit morale and decrease anxiety among survivors. The Bureau of Aeronautics issued a directive in June 1942 to initiate research on a shark deterrent.

In July 1942, the BuAer contracted personnel with the Committee on Medical Research of the Office of Scientific Research & Development and Marine Studios Inc. for preliminary testing. After receiving reports of shark attacks on shipwreck victims, the Bureau of Ships wanted to develop a unit for use by all naval forces operating in forward areas, and in April 1943 took over the project, assigning the Naval Research Laboratory the task, working in cooperation with the OSRD. On September 1, 1943 the NRL assumed full responsibility of the investigation.

Testing began in July 1942, conducted at Woods Hole Oceanographic Institute, Massachusetts. Live dog sharks were collected from the institute docks and placed in laboratory tanks. From July-September a series of controlled elimination experiments were conducted, using 150 possible repellents in the form of fish poisons, irritants, stenches, poisonous gases, inks and a variety of other chemicals. For these tests, groups of 3-5 dog sharks were observed and individually given a score card. Strikes were then recorded on treated and untreated bait with each test about one to two hours in duration. The overall results were inconclusive. Irritants and stenches showed no results. Ink clouds only inhibited sight. Sharks would eat the poisoned meat only to die later, making it unsuitable as a repellent for protecting humans. Maleic acid, copper sulfate and decomposing shark meat were the only three materials found to have some effect. The results were discouraging and the project was almost abandoned.

Senior investigator and professional shark catcher, Stewart Springer, introduced the idea of using decomposing shark meat, based on his knowledge of fishing lore. It was recognized by fishermen that rotting shark flesh rendered the immediate area unfishable and drove sharks away. Woods Hole investigators, as a last resort, concentrated efforts on decomposing shark meat. The Lab tests found that decomposing shark meat exudes a chemical called ammonium acetate, which when dissolved in water leaves acetic acid. Acetic acid was found to be the repelling agent. Since copper sulfate and acetic acid proved to be the most effective repellent, it was decided to combine both the copper ion and acetate ion into one compound, creating what was called copper acetate, for maximum effect. Initial tank tests were promising and field tests were now in order.

The first field tests were conducted in the gulf of Guayaquil near Ecuador, South America, from December 6 1942 to February 10 1943, by the OSRD. Four agents were evaluated: maleic acid, copper sulfate, copper acetate and extract from decomposing shark meat. A series of fishing tests was conducted using three lines at various depths, with one control recording strikes on each line throughout the day. In progress report no.5, submitted to the committee of OSRD dated 28 February 1943, William Burden wrote, "The tests for copper acetate were 100% satisfactory." All others were found to be unfavorable in this report. Further studies were concentrated on the repellent quality of copper acetate.

Additional field tests were conducted off the coast of Florida, the Keys and in the Gulf of Mexico, from May 1943 to January 1944, by the NRL with cooperation from the OSRD. Copper acetate was tested under a variety of conditions and depths. The tests included various fishing techniques, chumming, and even using trash to manipulate shark behavior. Cakes were secured to life jackets for simulation purposes, placed on the surface and up to 12' deep. Many shark species were encountered during these evaluations, giving overall mixed results. Under low shark activity copper acetate was found to be up to 80% effective. With increased shark activity or frenzy feeding, effectiveness dropped to under 50%. The two methods used to test copper acetate were in the form of cakes for both surface and depth, and sprays for surface tests.

Copper acetate is a colorless solution, and to better see the dispersion for test purposes from the surface, dye was added in tests conducted off Mayport, Florida in June 1943. Supported by tests off Biloxi, Mississippi in August 1943, it was concluded that dark dye possessed repellent qualities affecting the visual senses of sharks on the surface. It was realized that dark dye could be used as a camouflage or concealment, much like an octopus or squid. Another important factor was on the psychological aspect; with a dark dye, the user could see the weapon in action. The NRL consulted the Calco Chemical Division of the American Cynamid Co. to develop a water-soluble dye, compatible with copper acetate. After a series of lab tests Calco presented a dark dyestuff of the nigrosine type. When dissolved in water, nigrosine would color the water black or a very dark blue. In NRL report no. P-2230, dated 25 February 1944, entitled "First Partial Report on the use of Chemical Materials as shark Repellents," it was concluded that the Calco dyestuff provided an effective visual stimulus against sharks.

For maximum effect, the NRL concluded that a compound of copper acetate and Calco dyestuff should be prepared. The properties of both substances could affect multiple senses increasing the repellent effectiveness. The final mixture included 80% nigrosine dyestuff and 20% copper acetate. A series of three tests was scheduled off Mayport, Florida, on May 27 and June 3 1944. For these tests, an experimental packet for individual use was developed in the form of a chemical cake contained in a waterproof container. These final fishing tests on the shark repellent proved the mixture to be effective. Based on the results, an interim specification was drafted until further data from service tests could be gathered.

The Bureau of Ships drafted Interim Specification R51S48 (INT), Shark Chaser (Life Jacket) on 15 June 1944. The Packet designed for individual use consisted of a 5.5 oz. cake of repellent in a cotton bag, protected by a blue outer envelope made from a vinyl-copolymer-coated fabric. The unit measured 8.5" x 4.5" x .75" with cotton tie tapes to attach to inflatable life preservers, and a lanyard with safety pin for use with kapok type life vests. The unit functioned in the same manner as the dye markers, but the chemical bag could be returned and secured by a snap fastener on the outer envelope. The chemical cake was composed of 76% dyestuff, 19% copper acetate, and 5% soluble wax binder. Specification R51S48 Shark Chaser (Life Jacket) was standardized 1 October 1945. Until this time the shark chaser was considered experimental.

The USAAF drafted Specification No. 94-40828 titled as Packet, Shark Deterrent on 25 July 1944. The packet differed slightly from the US Navy version in that it did not have tie tapes and was not reusable. The overall unit was constructed from the same material and chemical mixture. On 23 September 1944, the NRL released the final report on the use of chemical materials as shark repellents. Based on the evidence, the USAAF was the first to procure the shark deterrent in large quantities. Contract 33038-AC-6701 was awarded in December 1944, under Specification 94-40828, for 85,000 packets, to American Products Co. Deliveries were being made to general supply depots in February 1945. Official nomenclature: Packet, Shark Deterrent, Class 13, Stock no. 8300-623748. The original stock number issued upon contract was 8300-623738, though this was later changed for reasons unknown. Additional reading includes ADTIC Informational Bulletin No.4 titled "Sharks," dated July 1944. The four-page, fold-out bulletin provides basic information on types of shark and protective measures, illustrating regions of shark activity across the world.

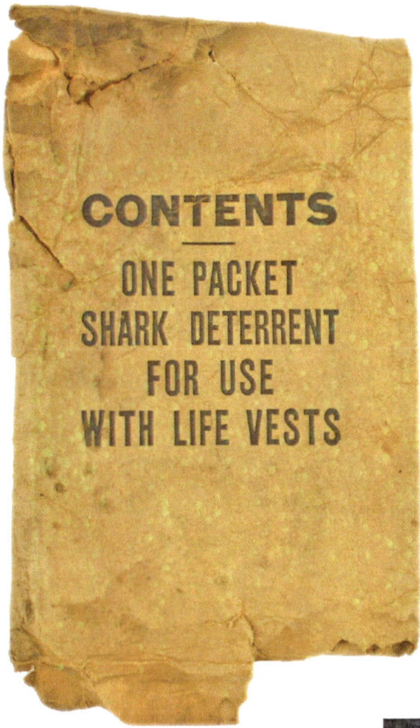

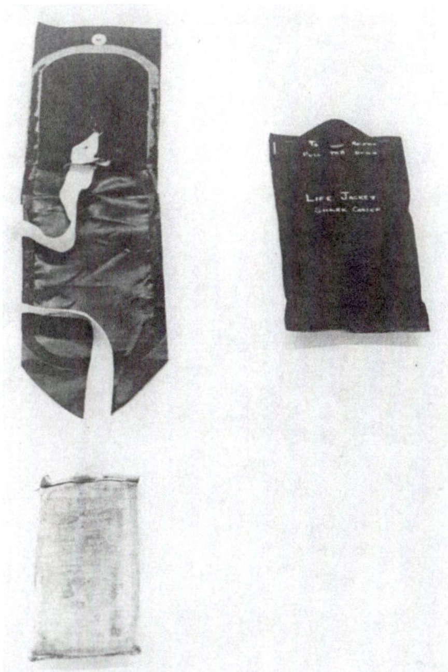

(Above) Shark chaser and its packing envelope as procured by the USAAF.

(Above right) A wartime image shows the package fully-opened illustrating the inner bag, together with a sealed example. (USAF)

(Below left and right) Attachment method for the shark chaser as outlined in Technical Order 13-1-27 dated 26 March 1945. The method is identical to that of the dye marker but on the wearer's right hand side.

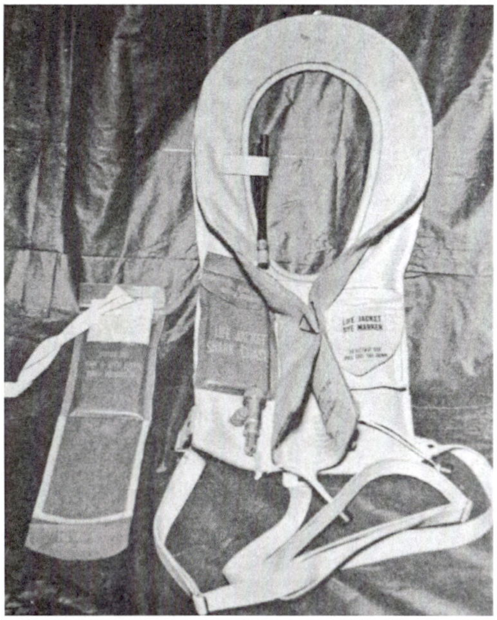

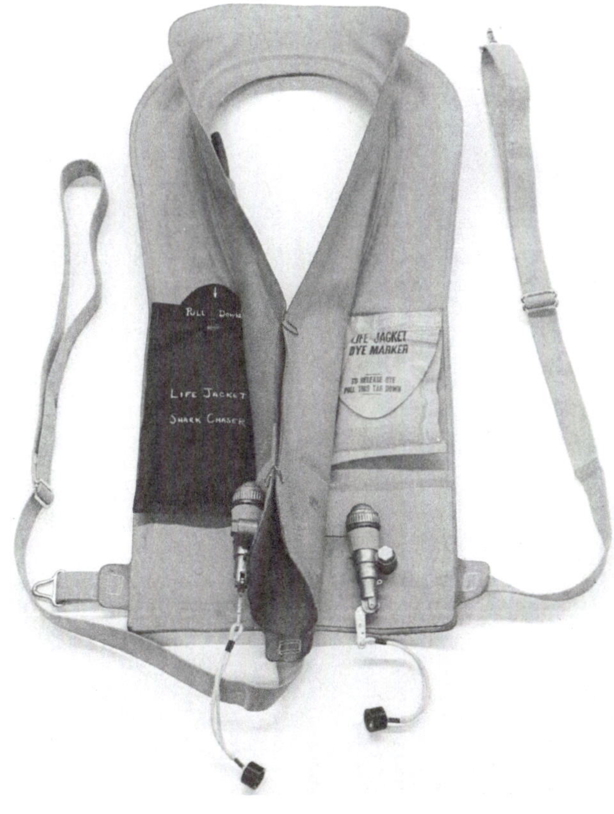

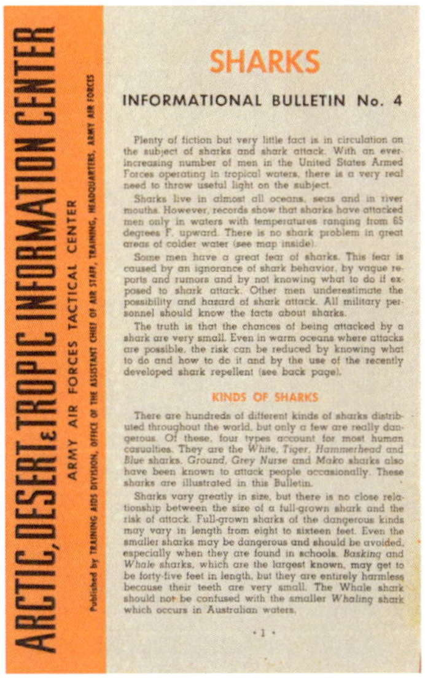

ARCTIC, DESERT & TROPIC INFORMATION CENTER

ARMY AIR FORCES TACTICAL CENTER

Published by TRAINING AIDS DIVISION, OFFICE OF THE ASSISTANT CHIEF OF AIR STAFF, TRAINING, HEADQUARTERS, ARMY AIR FORCES

SHARKS

INFORMATIONAL BULLETIN No. 4

Plenty of fiction but very little fact is in circulation on the subject of sharks and shark attack. With an ever-increasing number of men in the United States Armed Forces operating in tropical waters, there is a very real need to throw useful light on the subject.

Sharks live in almost all oceans, seas and in river mouths. However, records show that sharks have attacked men only in waters with temperatures ranging from 65 degrees F. upward. There is no shark problem in great areas of colder water (see map inside).

Some men have a great fear of sharks. This fear is caused by an ignorance of shark behavior, by vague reports and rumors and by not knowing what to do if exposed to shark attack. Other men underestimate the possibility and hazard of shark attack. All military personnel should know the facts about sharks.

The truth is that the chances of being attacked by a shark are very small. Even in warm oceans where attacks are possible, the risk can be reduced by knowing what to do and how to do it and by the use of the recently developed shark repellent (see back page).

KINDS OF SHARKS

There are hundreds of different kinds of sharks distributed throughout the world, but only a few are really dangerous. Of these, four types account for most human casualties. They are the *White, Tiger, Hammerhead* and *Blue* sharks. *Ground, Grey Nurse* and *Mako* sharks also have been known to attack people occasionally. These sharks are illustrated in this Bulletin.

Sharks vary greatly in size, but there is no close relationship between the size of a full-grown shark and the risk of attack. Full-grown sharks of the dangerous kinds may vary in length from eight to sixteen feet. Even the smaller sharks may be dangerous and should be avoided, especially when they are found in schools. *Basking* and *Whale* sharks, which are the largest known, may get to be forty-five feet in length, but they are entirely harmless because their teeth are very small. The Whale shark should not be confused with the smaller *Whaling* shark which occurs in Australian waters.

• 1 •

(Left) Four page Pamphlet printed and distributed by the ADTIC titled "Sharks." It gives a basic description of the types of shark and theory habits followed by defense against attack. The pamphlet folds out into a world map highlighting areas of shark activity. (USAF)

In June 1944 the BuAer procured 2,500 packets for field testing. The experimental packets were the same as those used in the tests from May and June. These packets had no lanyard with safety pin and the chemical cake was contained in a paper bag protected by a waterproof envelope labeled Life Jacket Shark Chaser.

In January 1945, the ASO awarded contract 288s-28906 under BuShips Specification R51S48 (INT) for 160,000 packets to Zinc Howard Corp.

Deliveries to major supply points were expected by March 1945. The official nomenclature described it as Shark Chaser (Life Jacket) ASO stock number R37-S-75.

Technical Note 87-45 was distributed 17 September 1945, describing the shark chaser, attachment and its use. Technical Note 87-45 stated, "Until further data has been accumulated from service areas to confirm the experimental results of the Shark Chaser, BuAer considers that the primary shark defense should continue to be the conventional technique outlined in 'Shark Sense' and other publications." It went on to say "...activities using the Shark Chaser are requested to furnish the Bureau of Aeronautics with any data available from actual survival experience."

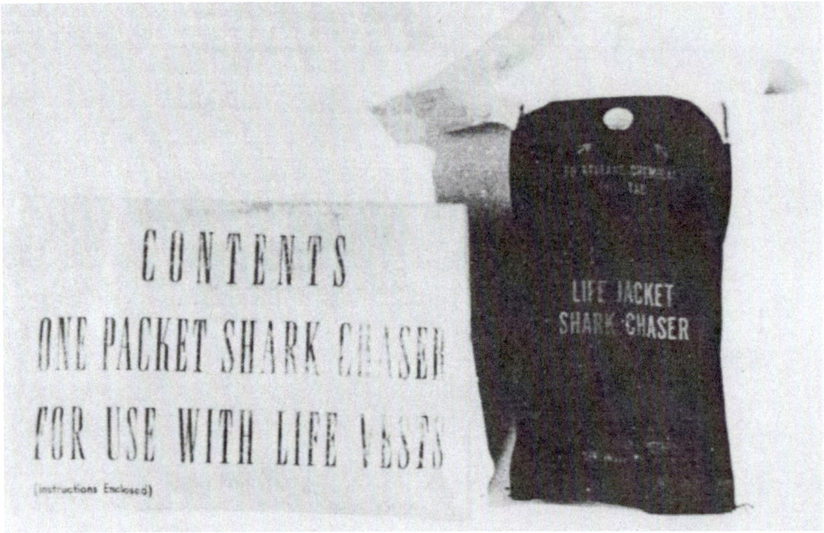

(Above) Detail of the shark chaser and its paper envelope provided with Technical Note 87-45.

(Above right) Shark Sense was a 23-page informational booklet about sharks, discussing facts about the species and precautionary measures, printed in March 1944 by the Aviation Training Division.

(Right) Shark chaser as procured by the BuAer in January 1945.

(Below) Two variations of the outer package. The lower chaser has a sewn tie tape, like that of the dye marker in production at the time of procurement. The upper packet has the tie tape secured with the addition of two eyelets, much like the April 1945 amendment to the M-566 dye marker. (Courtesy Paul Faulkner)

The project investigators understood that, based on existing knowledge, a 100% effective deterrent was not realistic. The objective was to find the most effective repellent, as quickly as possible. It was agreed that 67% effectiveness was satisfactory. The primary focus was orientated towards morale, and if the repellent actually worked that was a bonus. Shark deterrent was considered a medical project until it was found that it was possible to repel sharks. Calculations were determined by the following formula: number on control minus number on repellent {divided by} number on control X 100 = % effectiveness.

Ultimately the tests conducted in the labs and in the field were inconsistent. Further studies or amendments would rely on actual service use in the combat theatres.

Survival reports and other literature mention the use of chlorine capsules or tablets for use against sharks within naval aviation. These documents date from 1942-1944 and are not isolated to one unit, suggesting common usage amongst the fleet, so cannot be easily dismissed. In tests conducted at Woods Hole in 1942, results showed that chlorine had no effect. It is not clear where the idea of using chlorine as a repellent originated. No documentation has surfaced to suggest this as an official recommendation. The chlorine tablets or capsules are described as being stored in a vial or match safe.

(Above) 2nd Lt. John Artel is wearing a shark chaser in addition to the standard compliment of two dye markers on his Type B-4 life preserver at Cherry Point, NC, July 1945.

(Above) Three of the four pilots of VF(N)-55 are wearing shark chaser packets at NAS Barbers Pt. in the summer of 1945.

(Right) Two pilots of VMF-911 wearing shark chaser packets on B-4 life vests, c.1945. (National Archives)

(Below) Like that of the USAAF dye marker 94-40616, the shark chaser was revised to better integrate with the Type B-5 life preserver by reducing its length and adding a single eyelet to the top flap. The shark chaser was to be stored in the right hand pocket of the life preserver, secured by a lanyard from the packet to the pocket. It is not clear when the revision took place and, as can be seen, it is void of any nomenclature. The shark chaser specification 94-40828 was reformatted under MILSPEC MIL-S-4627 sometime after 1947, which is often seen on surviving packets of this type. Bureau of Ships Specification R51S48 was also reformatted under MILSPEC MIL-S-2785.

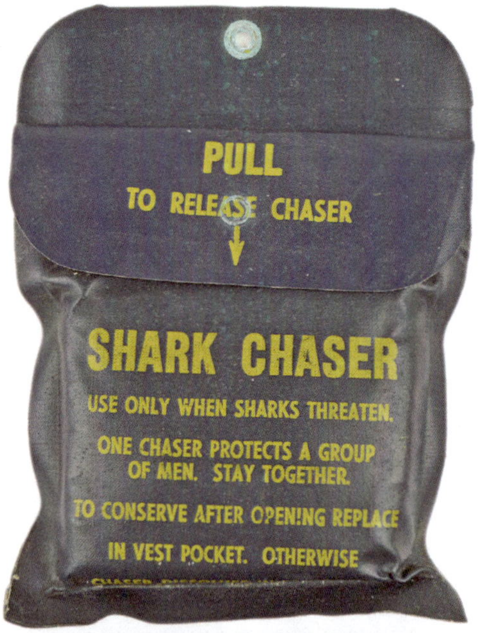

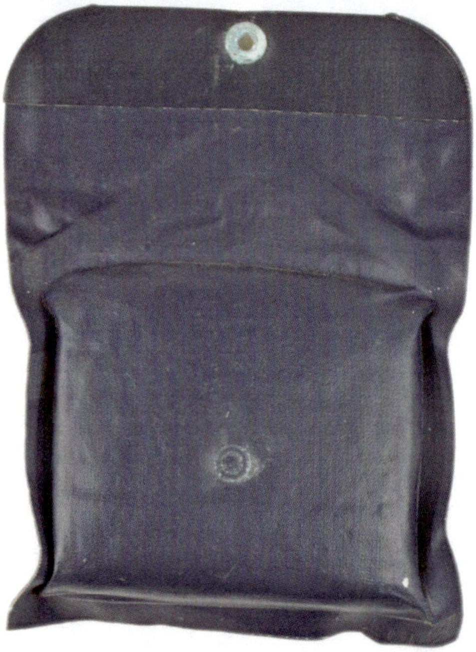

(Above) Period illustration showing the shark chaser packet both sealed and open, complete with its cord for attaching to the pocket of the B-5 life vest.

DISTRESS SIGNALS

Part II.

Pyrotechnic Smoke Signals

This is the first of two segments covering pyrotechnics, used extensively for air sea rescue. Each aeronautical service had differing emphasis towards their preferred types, with the US Navy over water and the USAAF over land.

In the prewar years, there was little that could be considered "emergency equipment" aside from a basic compliment of a life raft, life preserver and a parachute. The standard was to supply the aircraft with signaling devices to include both smoke and flares being used for communication, targeting, marking or distress.

The standard rescue procedure of the BuAer, during the late 1930s and into WWII, was first to let survivors know that they had been spotted and to mark their location by dropping drift signals – either the AN-MK-I or Signal, Drift, Night, AN-MK-4

(Left) After the Mark-4 float light was released from an aircraft, the impact with the water drove a firing pin on a timing device. Positive buoyancy then caused the device to reach the surface before ignition of the pyrotechnic. Once ignited, the light emitted a bright 12" flame together with white smoke, burning for three to four minutes. The combination of the flame and smoke made this device an effective day and night signal, though it was specified for night use in its nomenclature. (National Archives)

(Below) The body of the Mark-4 was constructed of wood for buoyancy. It measured 3" in diameter and 13" in length. Float lights were carried aboard all patrol and scouting aircraft to include VS, VB and VT carrier planes.

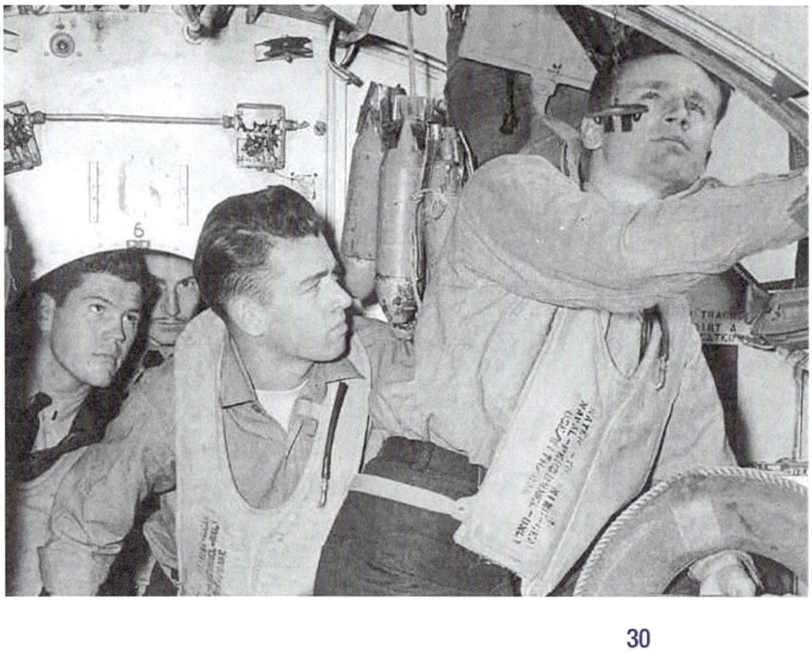

(Left) Mark-4 lights can be seen hanging on the fuselage of a US Navy PBY c.1944. These were standard equipment for all patrol type aircraft. The light was very versatile, being used to mark contact with submarines, determine wind direction, location of life rafts, and as markers for night landings. (NASM)

(Right) In 1943 the Mark-4 was approved for joint Army-Navy use, giving it the new designation as AN-Mark-4. This signal became part of the typical compliment for USAAF bombardment aircraft operating in the PTO and CBI. The AN-Mark-4 light can be seen hanging on the fuselage of a B-24 of the 436th B.S., March 1943 in the CBI. (National Archives)

(Below) Prior to joint Army-Navy adoption of the Mark-4 drift signal, the USAAF utilized the Signal, Drift M25 for the same applications. A cylinder with fins for stabilization, it measured 16" long with a 2-1/2" diameter. Its operation was the same as the Mark-4, emitting a 6" flame and plume of smoke.

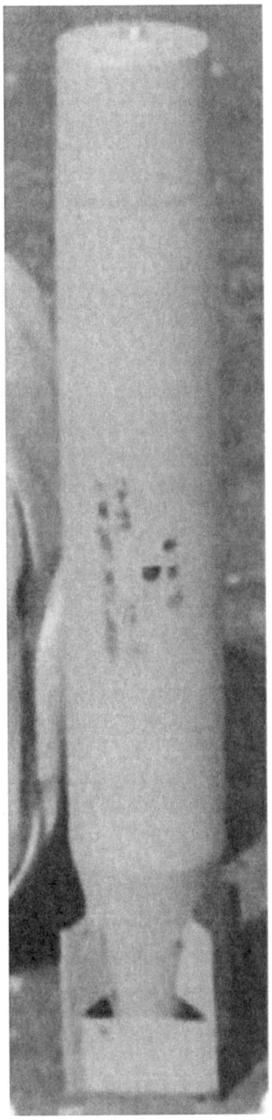

(Above) Standard operating procedure after survivors were spotted by a patrolling aircraft. Once these two aviators had been spotted, a drift signal was dropped to mark their location. A fluorescein dye marker was also dropped, and the slick can be seen to the right of the image. These signals assist an approaching surface vessel or seaplane to locate the survivors.

(Left and below) A new Signal, Drift, Night AN-MK-5 Mod-I superseded the Mark-4, sharing the same characteristics. The main body was made from wood but with a flat nose. Upon contact with water, a firing pin engaged the pyrotechnic, burning for 15-17 minutes. The Mark-5 was 19" in length and 3" in diameter. These three signals were originally designed to be stationary reference points for air navigation or drift. They were quickly adapted for air sea rescue work, marking locations of survivors.

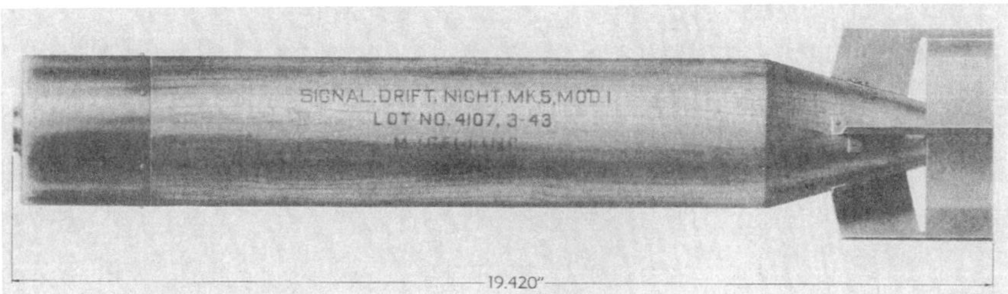

The Mark-4 float light was optimal for marking survivors adrift at sea. Over land, the hand grenade H.C. M-8, White, was utilized, categorized strictly as a distress signal by WWII.

The HC M8 was an Army Ordnance item dating back to at least 1935 and used aboard USAAF aircraft to mark locations such as targets, friendly areas, vehicles and tanks, as well as being used as distress signals. The inclusion of the M8 in life raft equipment was not contemplated in the 1930s, in large part because they were found to be sensitive to moisture and the grenade was found to ignite prematurely when wet. In 1937, the M-8 was revised to remedy this defect, but even with the new design it was still found to have moisture sensitivity issues which caused it to require frequent replacement in aircraft that operated at high altitudes. This was the principle reason the USAAF did not include them in life raft equipment.

By the close of the 1930s, the BuAer was slowly integrating signaling devices as standard into life raft equipment, officially adding the HC M8 smoke grenade in April 1941 at a ratio of two per multi-person life raft, available under ASO stock number R83-G-770100.

At this time, the M-3 series life raft included the signaling mirror, whistle, sea marker and smoke grenade, providing a variety of means for attracting searching aircraft and vessels. The USAAF would never adopt the grenade for life raft equipment, preferring flares instead.

(Left) The Hand Grenade, HC White Smoke, M8, re-designated under joint Army-Navy standardization in June 1942, consisted of a grey metal container, 2.5" in diameter and 5" long, with yellow markings. When activated, the grenade would emit a greyish white smoke for three to four minutes.

(Above right) Smoke grenades were standard issue for patrol aircraft. The stowage bag and a grenade can be seen near the gunner's blister in a PBY, April 1944 over Kiska Island. (Robin Bartel Collection)

(Below) Period color photograph illustrating the ignited grenade.

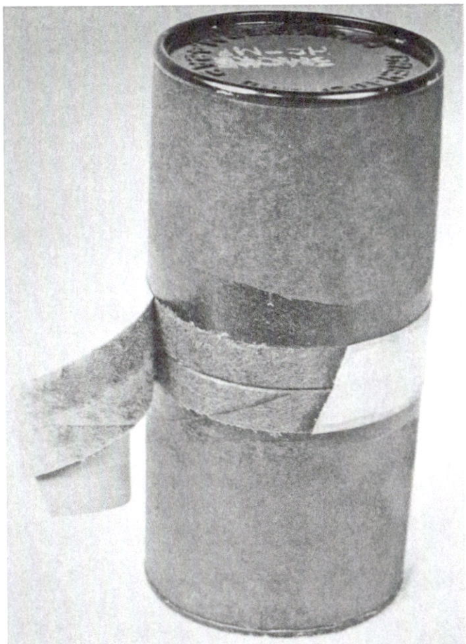

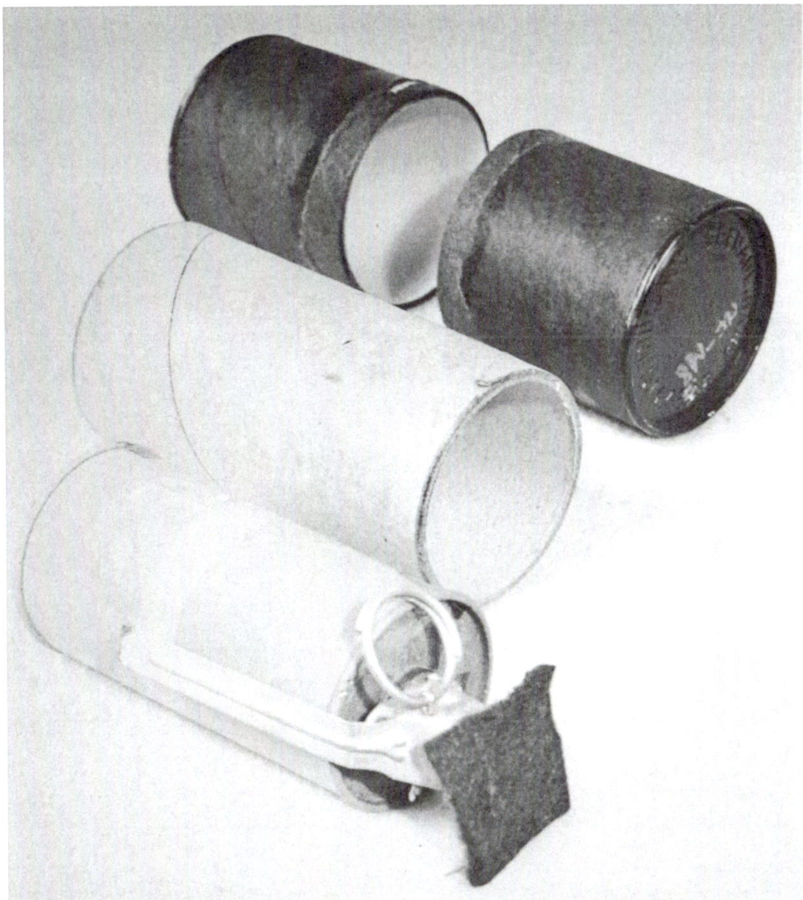

(Above) The grenade was contained in a tarred cardboard tube with metal end caps, sealed by waterproof tape at the center of the two halves.

(Right) It was further protected by an inner cardboard sleeve. (National Archives)

(Left) Cardboard tube for the HC M8 or AN-M8. Note white paper label with nomenclature with a yellow band, indicating it was packed in November 1942.

(Above) A variation of the cardboard tube with the nomenclature stamped on one of the metal end caps and sealed with a strip of water proof tape.

(Left) 1945 Color images of the AN-M8 and AN-Mark-4 smoke signals. (National Archives)

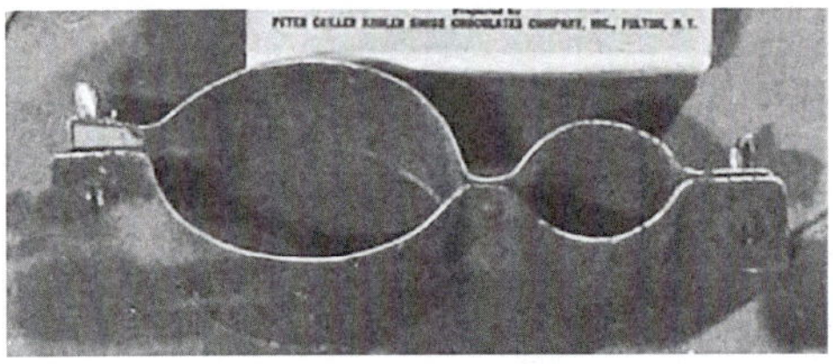

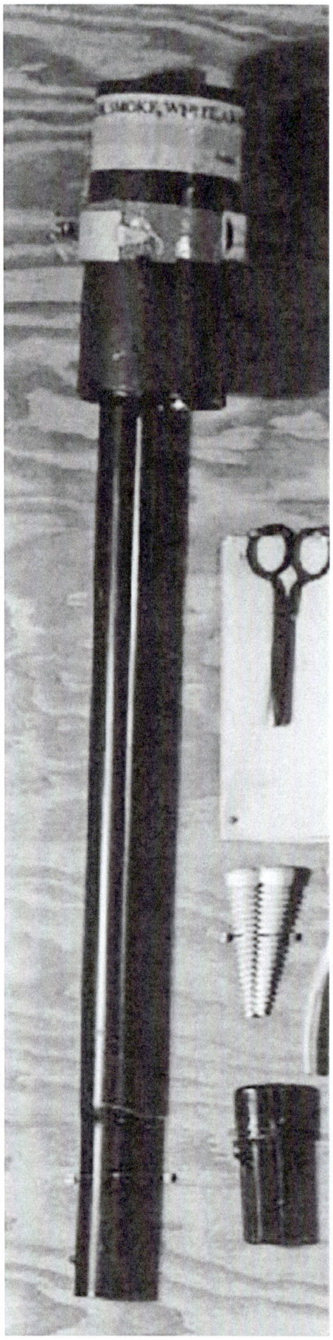

(Above) When ignited, the M8 burned at an extreme temperature and could not be hand-held. To make the grenade suitable as a signal device for use in life rafts, a clamp was devised that would attach to a section of an oar and allow the user to extend the grenade away from the raft. Several variations of the clamp exist but all operated under the same principle.

(Right) The clamp resting on the standard two smoke grenades as part of an emergency equipment display.

(Far right) The clamp as fitted on an oar section. The clamp was available under ASO stock number R83-C-44600. (National Archives)

At the time the USAAF entered the war, it was satisfied with pyrotechnic flares as part of life raft equipment being universal for both day and night use. The BuAer was just utilizing the M8 pyrotechnic grenade.

As the war progressed, the recurring theme of service requirements arose. In July 1943 the USAAF field tested the recently-adopted M16 smoke grenade series in hopes of better increasing the signaling and communication capabilities over the standard white smoke.

The M16 series consisted of the colors orange, red, yellow, blue, green, violet and black. Though the white smoke grenade had been used primarily for aircraft, with other colors available, a more refined means of communication could be established. This field test had very little to do with actual distress signaling, but would influence future naval testing. By this date, the BuAer had received numerous reports on the service status regarding the M8 smoke grenade, and the list of complaints included the following: when submerged in water for any length of time, the grenade would not function; that the cardboard container was not suitably waterproof; that sparks would burn holes in the raft; and that often the clamp was missing or lost, rendering the grenade useless. Air Sea Rescue personnel often commented that the white smoke blended in with low hanging clouds, and was difficult to distinguish when white caps were present.

The BuAer established a three-part service requirement to determine first of all the most effective color for attracting attention; secondly, that the signal needed to be waterproof; and thirdly, it needed to be a hand signal, eliminating the use of the clamp.

In November 1943, the M-16 series smoke grenade was tested at NAS Patuxent River under a series of weather and lighting conditions to determine the most suitable color. In addition, two companies submitted designs for a compact hand signal, both in orange.

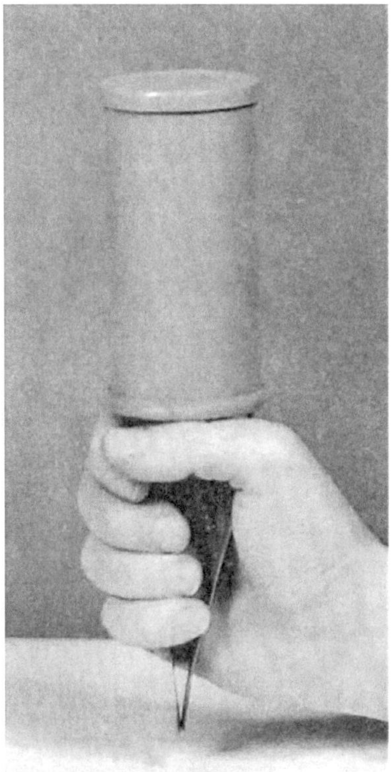

(Above) The Chemurgic Corporation submitted a hand-held smoke device. As shown above, a cap conceals a pull ring to ignite the signal emitting an orange smoke. Two tabs fold down and were used as a handle. A sample was submitted to the BuAer in August 1943, was field tested in conjunction with the M8 for comparison, and immediately found superior to size and the contrast the orange smoke provided. At this time, it was recommended that this signal replace the M8 as soon as practical. (National Archives)

(Left) Due to the positive results from the field test in August, the BuAer initiated a more in-depth field trial of all available colors utilizing the M-16 smoke grenade series for evaluation to be conducted at NAS Patuxent River in November 1943. In this series of field tests Aerial Products Inc. submitted two sizes of smoke signals to be evaluated which they labeled as the "Mae West" grenade. In their design emphasis the smaller signal was to be carried by an individual and the larger unit to be used in life rafts. The signal functions under the same concept as the Chemurgic Corp. model, with a pull ring and fold down handle. The final conclusions of the tests proved that the color orange provided the most superior contrast to back grounds in all types of conditions. (National Archives)

(Above) A color photograph of the Aerial Products "Mae West" grenade being demonstrated at NAS Patuxent River, November 1943. (National Archives)

(Right) The refined signal submitted by Aerial Products Inc., which they now called the A-P Life Jacket Smoke Signal Mark 1 Orange. It still included the pull-down handle. The Bureau of Ordnance designates equipment with a "Mark" and subsequent revisions are noted with "Modification" followed by a number. Due to confusion, the BuOrd instituted a new system in March 1944, stating that all first model types will be designated with a Mark and Modification 0 (zero).

During the field tests, the Chemurgic Corp. signal would not ignite after being submerged, and if it ignited it would only burn for an average of 30 seconds, receiving a fail rating upon conclusion of the tests. No further inquiries with this company were conducted by the BuAer.

The Aerial Products signal proved satisfactory and the company continued to improve the signal in coordination with the BuAer. Further tests were conducted in the early months of 1944, improving design characteristics, namely waterproofing, and dropping the concept of the large model.

The final product met a rigid series of requirements of being completely waterproof, compact, durable and simple to use. The adopted signal was designated Distress Smoke Hand Signal, Mark-1 Mod-0 having a 1-5/8" diameter and a length of 3-7/8." The outer metal container would be colored orange with printed instructions on its use. Upon removal of the paper cap, the pull ring is exposed and when ignited the signal would burn for 18 seconds emitting a brilliant orange smoke.

Distribution of the signal was to be a ratio of two per individual and 4 signals to one HC AN-M8 in emergency equipment. Initial procurement was for 600,000 signals, awarded in August 1944, with expected deliveries in November under BuOrd contract 6941. The USAAF field tested the signal in October and November 1944, concluding that they should also adopt it for life raft equipment. Following negotiations in January 1945 the signal was approved for joint Army-Navy standardization 28 February 1945.

In February 1945, an additional 3,000,000 signals were procured under then new designation Signal, Distress, Smoke, Hand, AN-Mark-1 Mod-1 under BuOrd contract 8296. The contract was split with 1.5 million signals being allocated to the USAAF and, upon availability, they were to be supplied in their multi-person life rafts at a ratio of 8 signals and 2 per pararaft. No physical changes were made to the signal but, because it was placed under Army-Navy format, the nomenclature was changed noted by Mod-1 (Modifciation-1).

(Above and right) The two types of hand-held smoke signal under BuOrd contracts 6941 and 8296 BuOrd drawing number 398780. Each signal was manufactured at a cost of $1.58. (National Archives)

(Below and below right) Pilot Lt. Littlefield of AG 87 wears a Mark-1 smoke signal secured to his life preserver aboard the USS Ticonderoga, July 1945. Below, an unknown pilot of VF-34 has a hand-held smoke signal attached to his life preserver. (National Archives)

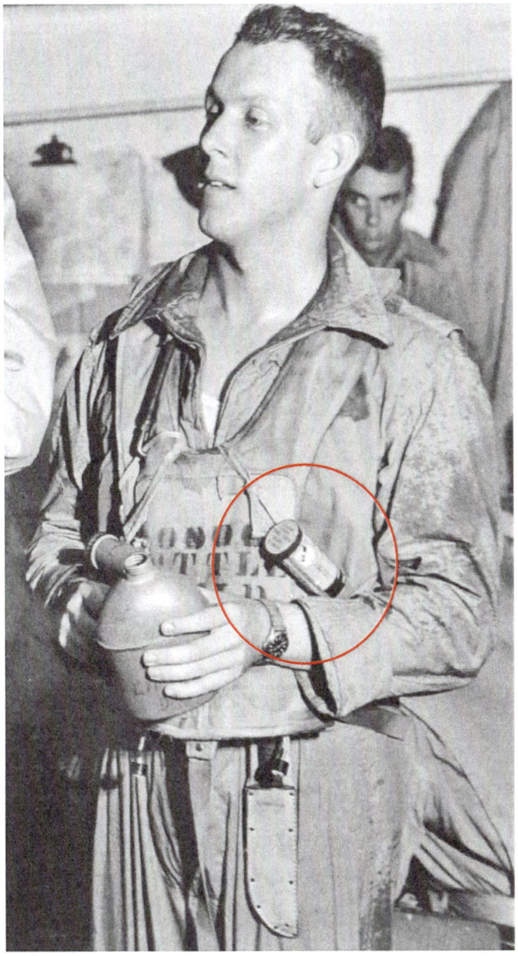

(Above) The Mark-I hand-held smoke signal. With the typical practice of carrying personal emergency signals a survivor has the means to broadcast his location without assistance of the provisions provided in life raft or other segregated kits. (National Archives)

As part of the original concept and design, the Mark-1 smoke signals were intended to be carried in a suitable location for individuals, such as in the back-pad kit or flight suit pockets.

In December 1944, a method to carry two signals was devised at NAS Jacksonville by means of pockets on the life preserver. At this time, the signals were available in very limited quantities, issued only to pilots operating over water, and had to be turned in after each flight during stateside operations. The bulk of what was available was distributed to units preparing for overseas embarkation.

A recommendations for the modification was submitted to the BuAer engineering division in February 1945, with field test evaluations conducted by personnel having overseas combat experience, who approved of the concept. Technical Note 15-45 was distributed 23 February 1945, to modify life preservers for the attachment of the smoke signal pockets.

It was additionally recommended that this modification be applied to production of the Type B-4 life preserver, but it was never initiated due to the development of the Mark-2 life vest.

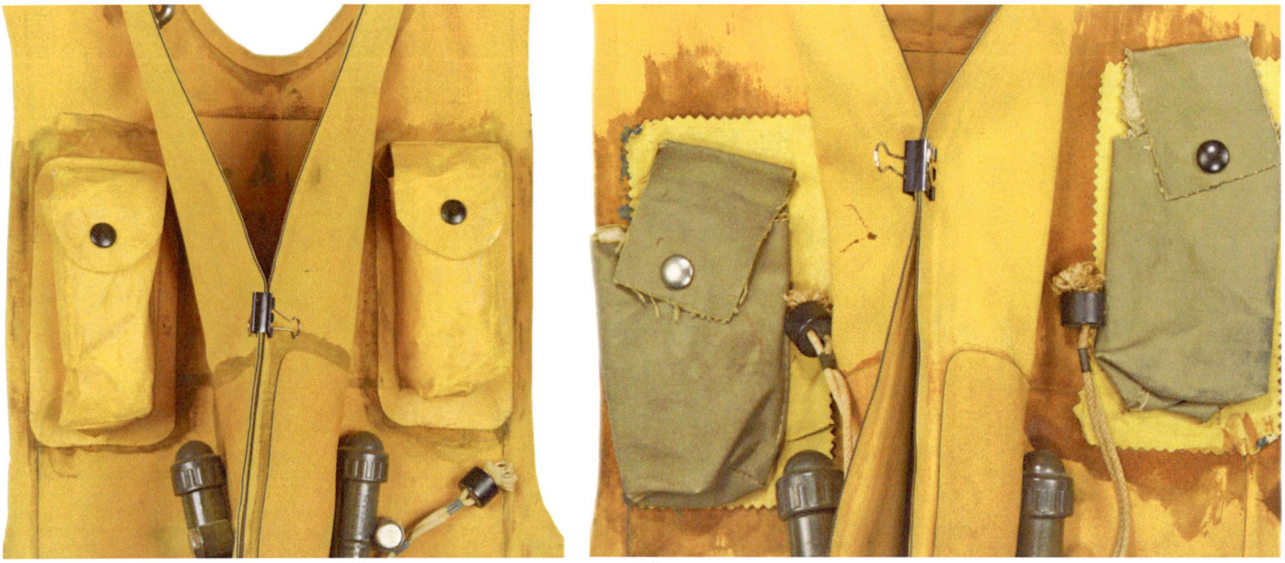

(Above) Examples of the smoke signal pockets applied to a US Navy Contract Type B-4 life preserver, as per Technical Note 15-45. There were variations in material utilized for the construction of the pockets, such as a grey rubberized fabric. USAAF life preservers having the same application have also been found, but it is difficult to determine if these were in Naval use or by USAAF technical order, without alternate identifying markings. Approval by the USAAF for the modification was authorized through a Technical Instruction TI-252 by the Air Technical Service Command 29 March 1945, detailing that until suitable technical orders could be drafted, BuAer Technical Note 15-45 was to be distributed. Formal USAAF Technical Order 13-1-31, dated 21 July 1945, was released for the attachment of two pockets for either the Mark-I Mod-0 or T-49 distress signal.

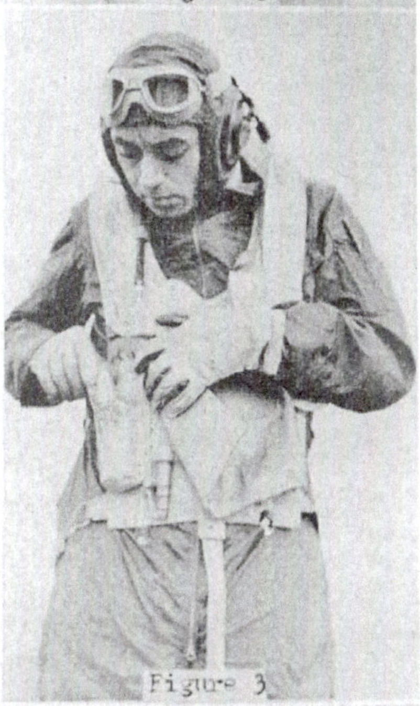

(Above) 1945 color image of the hand-held smoke signal in use showing the bright orange colored smoke. (National Archives)

(Left) Series of illustrations which accompanied Technical Note 15-45.

In December 1944, a formal request to develop a combined day and night distress signal was initiated by fleet personnel. The principal concern was for a waterproof night distress signal, if nothing else, to replace the MK-IV hand projector and Very's signals, which were standard at that time, for all BuAer emergency equipment. If a combined unit could not be devised, a night signal designed like that of the Mark-1 hand-held smoke signal would be of great benefit.

The idea was geared towards the new emphasis of "Rescue," considering the amount of combat operations being conducted at night. To immediately address the issue, the BuAer adopted the US Army Ordnance T-49 distress signal.

Building off the Mark-1 hand-held smoke signal, Aerial Products Inc. successfully incorporated a night distress signal for a combined hand held unit. Samples were manufactured and tested at NAS San Juan through the months of March to May 1945. Upon a successful evaluation, the new unit was given the designation Signal, Distress, Day and Night, Mark-13 Mod-0 in July 1945 as with the approval for procurement.

The existing BuOrd contract 8296 for the AN Mark-1 Mod-1 hand-held smoke was amended in August 1945, the new order was for 1.5 million day and night distress signals. It is not known how many hand-held smoke signals from the original 3 million on order were manufactured prior to the change order to the Mark-13.

Due to production delays, the new signal was not available until January 1946. This signal would officially replace the MK-IV hand projector, T-49 (M-75) distress signal, and the Mark-1 hand-held smoke, with the USAAF adopting it as well in 1946 replacing all subsequent distress signals utilized in their emergency equipment.

(Above) USAAF Technical Order 13-1-31 provided information for the attachment of two pockets that were to be applied to either the Type B-3 or B-4 life preservers. Each pocket was to accommodate one signal. The physical application was identical to that as prescribed in the BuAer technical note but different in dimensions. The drawing in Technical Note 15-45 required the pocket to be 4.5" in height to accommodate the Mark-1 Mod-0 smoke signal. For the USAAF, the pocket was to accommodate either the Mark-1 Mod-1 or T-49 hand held two-star red signal. The prescribed application required the pockets to be 5-1/8" in height, since the T-49 was a taller signal. This image accompanied 13-1-31 illustrating the prescribed application for the USAAF.

(Above) The development of the Mark-13 Mod-0 and the Mark-2 life vest paralleled each other, and from the success of the pockets outlined in Technical Note 15-45, this design feature was incorporated into the standardized vest but made specifically for the Mark-13 MoD-0 day and night signal. After the use of either the star or smoke, the signal could be placed back into the pockets for further use. Being demonstrated is the first model Mark-2 life vest c.1946.

(Right) The Mark-13 Mod-0 measured 5' in length x 1-5/8' diameter. It was split, with one end emitting an orange smoke for 18 seconds and on the other a two-star signal. Paper caps were provided just like that of the Mark-1 hand-held smoke. On the red star side, small bumps encompassed the rim to identify it in the dark.

DISTRESS SIGNALS

Part III.

Pyrotechnic Flares

The lineage of signal pistols and flares utilized by the aeronautical services during WWII begins in 1877. The significance of this date was the refinement of the propelling distress signal, flare or star. Prior to that date the signal pistol, better described as a flare holder, did not project a signal. The design resembled a pistol with a grip, short but oversized barrel and percussion hammer. The pyrotechnic material was muzzle loaded, activated by a percussion cap and burned in place.

In 1877 US naval officer Edward Wilson Very developed a single shot, breech loading pistol that utilized a centerfire cartridge that propelled a signal star into the air. The US Navy adopted the concept as the Model 1882, titled as the (Very) Signal Pistol with a four-gauge centerfire cartridge. The Very Signal Pistol was then revised and adopted as the Model 1894, changing the design of the pistol and decreasing the size of the cartridge to a ten-gauge.

The Model 1896 overhauled the pistol to the design that is more readily recognized as the Very's Signal Pistol, also known as the Mark-I. The Mark-I was quickly revised in 1900 to the Mark-II. In 1915 the Mark-III Very's Signal Pistol went into production.

(Above and right) The Mark-III Very's Signal Pistol would be the standard for the US Navy from 1915 to 1942. The pistol frame was constructed of brass with a blued steel barrel and walnut grips, and fired a ten-gauge cartridge weighing 2.42 pounds. This device was used for communication and distress signaling, typically stored on surface vessels in a "Service Box" containing the pistol, cartridges and a holster. (Robin Bartel Collection)

(Left) Signal Pistol Holder, MK-V and Cartridge Belt, MK-I, for the ten-gauge signals. (Robin Bartel Collection)

(Below) In August 1941, a new lightweight Very's signal pistol evolved, constructed of new materials to reduce its weight. It was designated Mark-V Very Signal Pistol. In September 1942, the Mark-V was to replace and supersede the Mark-III as supply permitted. The Mark-V Very's Pistol was made of an aluminum die casting, utilizing a frame and handle of a phenolic material. (Robin Bartel Collection)

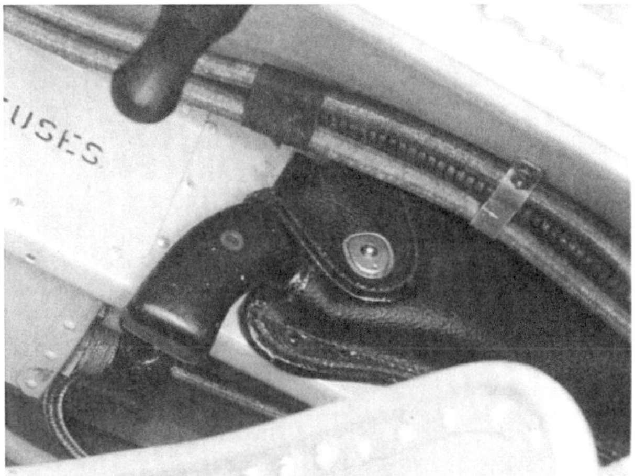

(Above left) Signals were standard in all Naval aircraft, for communication and recognition purposes. Allotments depended on the type of airplane and its delegated duties. The standard compliment is shown in this image. Included are MK-4 float lights, smoke grenades and the Mark-III Very's Pistol with ten-gauge signals. This is part of display of equipment provided on a PBY patrol aircraft of VP-42, NAS Seattle, December 1939. (National Archives)

(Above right) By the 1930s, the Mark-III Very's pistol was integrated as standard equipment in all naval aircraft, including VF class and single seaters. For VF class aircraft, the pistol was typically placed to the pilot's right in an easily accessible location. The primary purpose was to provide basic communication from aircraft to aircraft, or aircraft to ground, but could also be used for distress signaling. (National Archives)

(Above) Two members of Escort Scouting Squadron VGS-12 pose with an excessive number of Mark-III Pistols! Kodiak Alaska, June 1942. (National Archives)

Just prior to the entry of the United States in WWI, the US Army adopted the US Navy Mark-III Very's Pistol for infantry ground use. Flare pistols were a vital communication device in the trench warfare that ensued. A large procurement was placed by the US Army for the Mark-III, but was soon cancelled after the Allied Expeditionary Forces began to arrive in Europe.

The AEF required that the cartridge be interchangeable with the French 25mm model, however, a small quantity of Mark-III Pistols with the ten-gauge signals did find service use within the AEF. The US Army found the ten-gauge signals lacking in visibility, and that the 25mm (1") allowed for a larger amount of charge and size of star.

Both the ten-gauge and French 25mm signals were available in red, white and green, but the 25mm provided greater variety, including parachute types. In addition, the 25mm had multiple star arrangements of either three or six in a single cartridge for the colors red and green.

The 25mm was adopted for AEF ground forces and the Army Air Service. The Air Service also adopted the French 35mm for use in aircraft.

The 35mm cartridges available included the colors red, white and green, in both one and six star configurations, as well as cartridges that changed colors in combinations of red to green, green to red, red to white, green to white, white to red and white to green.

(Left) In March 1918, General Pershing directed that the French system of pyrotechnics be adopted standardizing the French Very's Pistol 25mm Mark-IV. The initial pistols were requisitioned from the French forces until manufacturing could commence in the United States. The pistols fabricated in the USA were manufactured to French specifications. Additionally, illustrated is the casing of the 25mm brass signal cartridge. (Robin Bartel collection)

(Right) In 1918, the Air Service developed and adopted the 35mm Signal Pistol, Mark-I (Aviation). 29,699 Mark-I pistols were procured in August 1918, being almost a direct copy of the French design. Both the 25mm and 35mm pistols were utilized by the Air Service, often commercial models, until being superseded by the M2 Pyrotechnic Pistol. (National Archives)

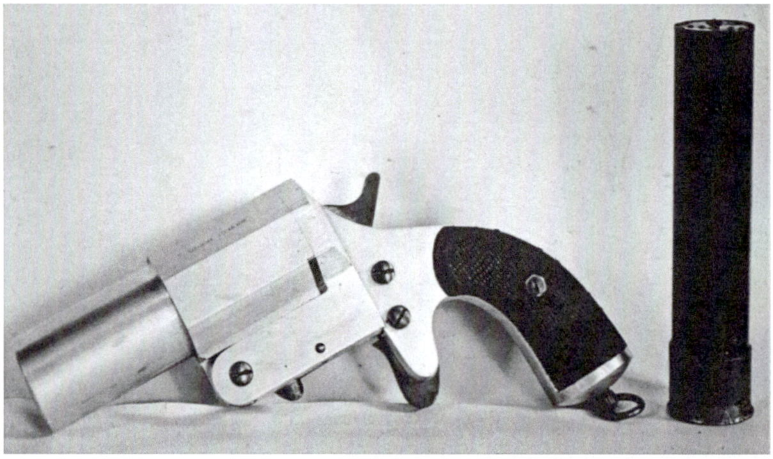

During WWI, the Very Signal Pistol, Mark-III and Mark-IV were the standard types in use by the different services.

Both the ground forces and Air Corps had found those pistols inadequate to meet their needs. As soon as the Ordnance Department could be reorganized on a peacetime basis, plans for a development program were outlined. The first two principle guidelines established were to improve pyrotechnics of the Very pistol and increase the range of grenades fired from service rifles.

A board of officers, known as the Army Pyrotechnic Board, convened in July 1919 to investigate the matter. They laid down general specifications for a satisfactory pyrotechnic projector along the lines of the Very pistol. To maximize on the development process steps were taken to procure pyrotechnic weapons and materials from the British and French. After evaluating the possibilities, the board recommended the development of a projector which should be separate from, but not exceed the weight of the service rifle, and capable of projecting signals to a minimum height of 600'.

In 1922, the Infantry Board received four different types of projectors for testing, three of which were sufficiently promising. However, because of the shortage of funds it was found necessary to limit development to the one considered closest to meeting the desired characteristics. Contrary to the recommendations of the Army Pyrotechnic Board, it was the decision of the Adjutant General to pursue a bayonet *tromblon* type, like that of the VB grenade launcher.

Both the Infantry and Cavalry Boards tested the new projector, finding it satisfactory for the discharging of pyrotechnic material and superior to the Very pistol. The projector was standardized as the Signal Projector, M1, with Signal Projector Support, M1 23 August 1928. The new ground projector was expected to meet the requirements expressed ever since WWI for a universal projector that could be used both for projecting signals and self-propelled grenades.

The M1 Projector was essentially a small mortar provided with its own firing mechanism, being supported by the service rifle, or M1 Support. Upon standardization, the Discharger, Mark IV for Rifles, M1903 and M1917, was declared obsolete. The Pistol, Very (25mm) Mark IV, became limited standard. To assist airplanes in fog and hazy weather, landing fields were authorized the M1 Projector as a standard allowance, 22 March 1934.

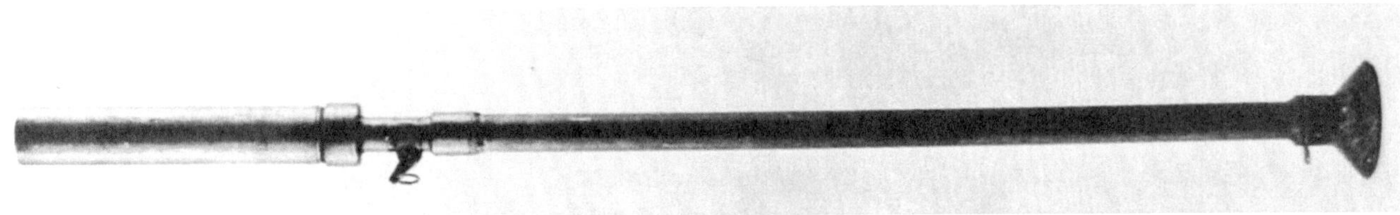

(Above) Upon standardization of the M3 Ground Projector, the M1 Projector was placed on limited standard status. In 1941 the Field Artillery Board found they had a need for a projector for training purposes in the use of high burst ranging signals, and recommended that the M1 Projector be re-standardized. With slight modifications, the M1 was standardized as the M1A1 Projector 24 December 1941. The principle change was that the barrel was no longer adaptable for the service rifle being threaded onto the mount. The barrel diameter was also increased to accommodate the M27 ground signal. Production began in March 1942 by the Knapp-Monarch Company, with 287 in use by the summer of 1943. Altogether there were 1,269 M1A1 Projectors produced with no further procurements made after June 1943. Illustrated is the M1A1 Projector w/mount which retains much of the same appearance as the M1 that it superseded.

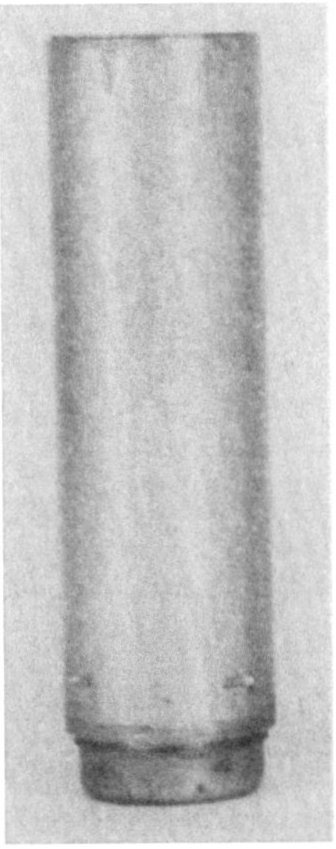

(Left) A ground signal, predecessor to the M17 thru M22, used in conjunction with the M1 Projector.

(Below) The High-Burst Ranging Ground Signal M27. When fired it produced a puff of smoke about 550' in the air. This signal was used for training purposes, to simulate the high burst of artillery shells.

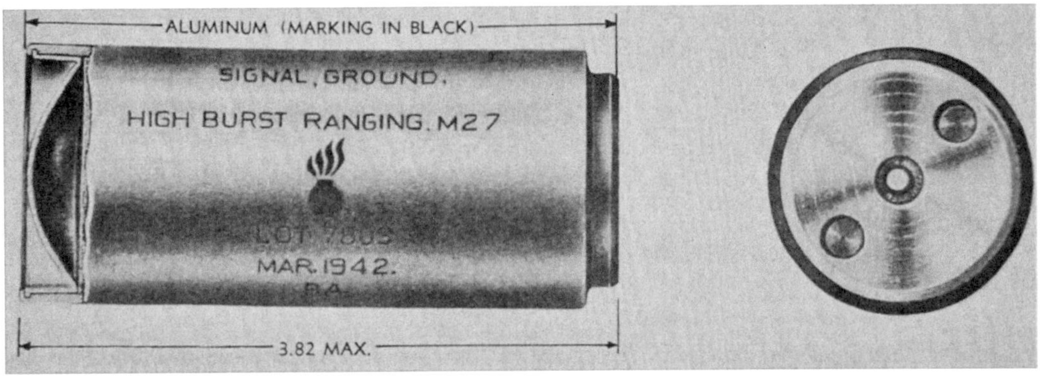

An Air Corps Pyrotechnic Board was appointed, 14 May 1926, to develop pyrotechnic material for their needs. On 10 April 1928, the Subcommittee on Pyrotechnics for Air Corps outlined performance requirements for pyrotechnics, directing that the projector should be along the general lines of the Very Pistol but with improvements in design. Advertisements of the type of projector desired brought only one bid from the Kilgore Manufacturing Company. They presented a discharger known as the "Driggs-Faber System."

Louis Driggs and Henry Faber were awarded a patent, in May 1929, for a flare pistol described as "...having a detachable barrel carrying the features of a typical pistol having a pistol grip, trigger with guard, hammer and a barrel, better described as a socket." The casing of the pyrotechnic was the actual barrel, secured in place by the pistol and muzzle loaded. In 1929, a series of tests was conducted utilizing the Driggs-Faber pistol along with British, French and other American types. These tests were conducted at Pickatinny Arsenal and Langley Field, with the Ordnance Department receiving reports on the various tests through 1930.

The Driggs-Faber model proved to be the most favorable, and the Air Corps urged its standardization. On 10 March 1932, a limited, expedited procurement was authorized, providing that the necessary refinements could be executed. The parameters were that the pistol could be loaded and fired, and the casing extracted, with one hand.

The Air Corps required that the design be reconfigured to meet these additional service requirements. At the time of release of procurement, the discharger was designated as Projector, Pyrotechnic, M2, later changed to Pistol, Pyrotechnic, M2, 25 May 1933. After further tests and modifications, the Pistol, Pyrotechnic, M2 was approved as standard, 14 September 1933, referenced under Ordnance Dwg. No. 42-26. In order for Kilgore to fulfill its contract with the government, a subsidiary was established under the entity of the International Flare-Signal Company.

The M2 consisted of a bronze frame and barrel with Bakelite grips, designed to be muzzle loaded. The pistols that preceded it were breech loaded. It incorporated features such as a short barrel, oversized handle for operation while wearing heavy flying gloves and an ejector switch operated by the thumb. It also eliminated the process off cocking a hammer. It weighed 3.09 pounds. A quantity contract was awarded 24 January 1935 for the production of 19,437 to meet service requirements. Royalties were to be paid for the use of the Driggs-Faber System, with the patent retained by Kilgore. The M2 pistol was placed on the requirement program for new and old aircraft 1 August 1936. With quantities available, all aircraft types except pursuit and training were authorized an allowance of one M2 Pistol, superseding all previous types of Very's pistols, in May 1937. In the revised allowance list of May 1941, pursuit aircraft were authorized an allotment of one M2 Pistol and 4 parachute signals, one of each M11, M14, M15 and M16.

Prior to the adoption of the M2, service activities reported that red was the only cartridge available for the Mark-IV, which limited communication capabilities. The Mark-III (ten-gauge) provided three different colors, but there was a widely-based complaint that the star burned too quickly and could only be seen at very short distances. At this time, the concept was orientated towards observer or reconnaissance aircraft to communicate with ground troops. As part of the development process, leading to the adoption of the M2, a new series of signals was introduced for use with the M2 Pistol.

New, 37mm (1.56") signals were adopted, with the authorized allowance list for aircraft calling for five types, from 1937 through to 1941. These signals include the Flare, Aircraft, Parachute M9; Signal, Aircraft, Red Star, Parachute M11; Signal, Aircraft, Red Star Cluster, Parachute M14; Signal, Aircraft, White Star Blinker, Parachute M15 and the Signal, Aircraft, Green Star Blinker, Parachute M16. A signal, White Star, Parachute M10 was additionally developed, and though not included in aircraft allowances, was utilized.

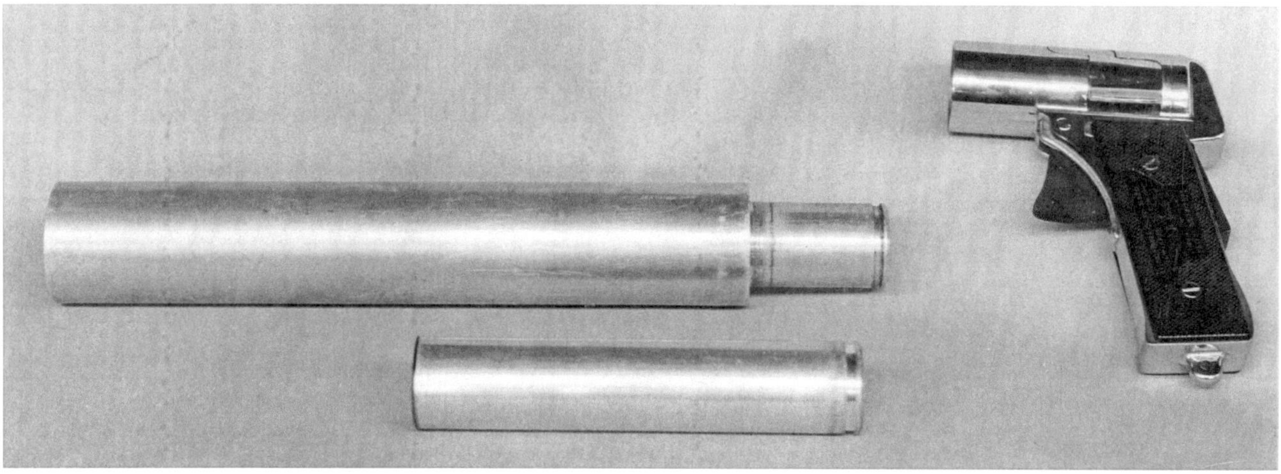

(Above) The new improved Driggs-Faber signal pistol, shown with the new type of signal flares upon adoption. Both signals were parachute types. The top signal in the picture is the M9; the smaller version carried the designations M10, M11, M14, M15 and M16.(National Archives)

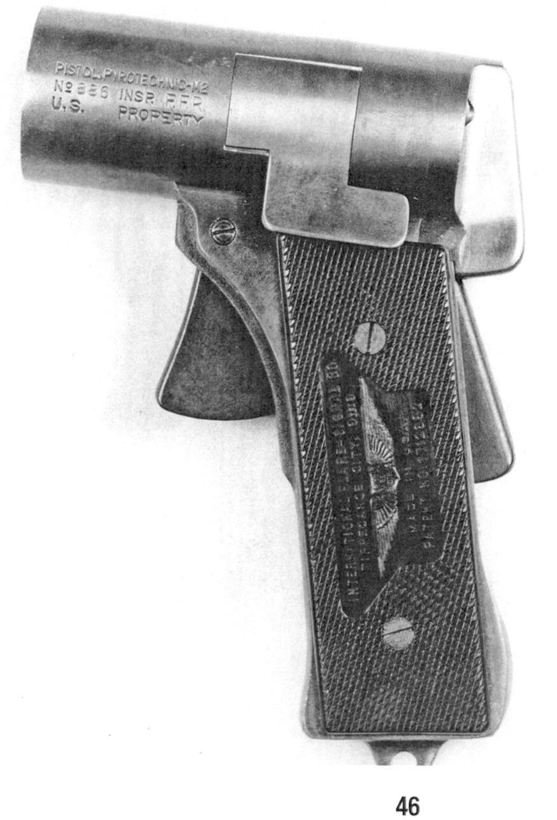

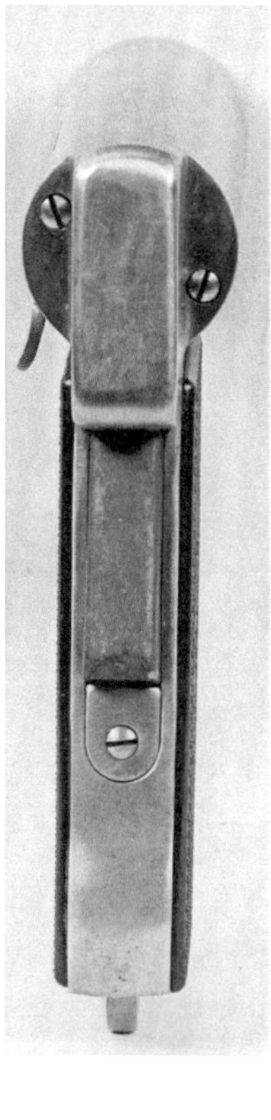

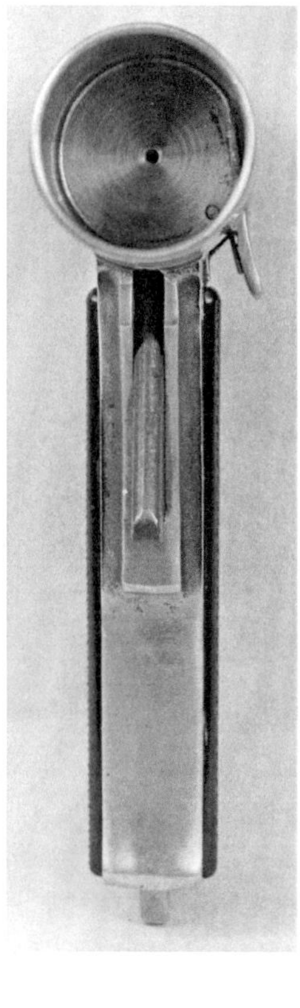

(Right) Design features of the M2. (National Archives)

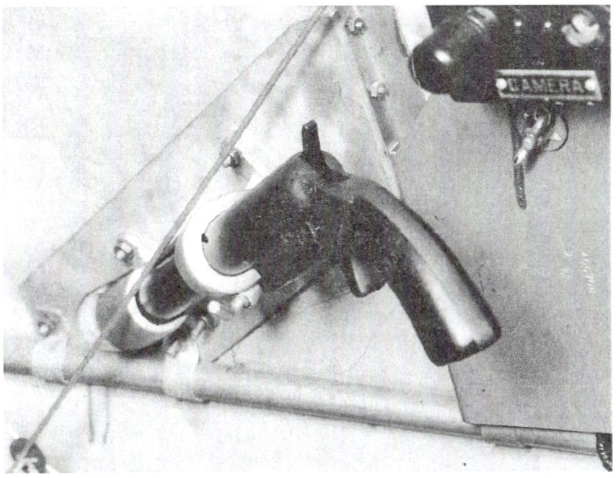

(Left) Prior to the adoption and integration of the M2 pistol, there was no single standard issue for the Air Corps, which utilized a variety of types of pyrotechnic projectors. A Mark-III pistol can be seen in a rack provided in a Douglas O-25A observation aircraft. (National Archives)

(Below) The signal casings for the Driggs-Faber signal pistol were designed in the form of a detachable barrel. (National Archives)

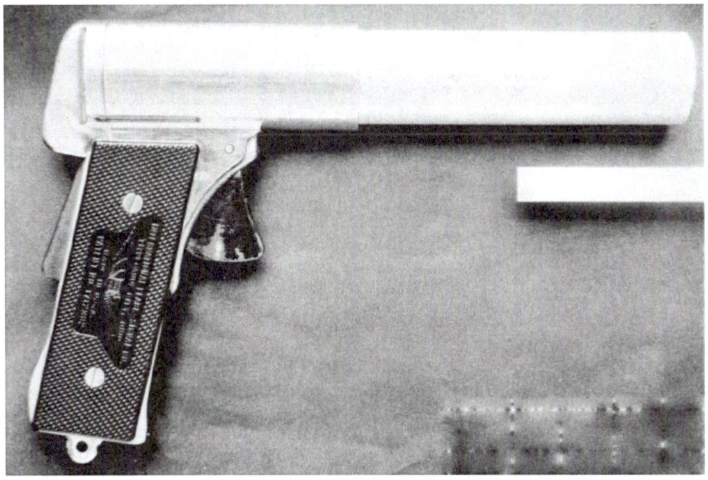

(Below) The M2 Pyrotechnic Pistol. On the left side of the pistol is the lever for extracting the flare casing, operated by the user's thumb. By eliminating the trigger guard and providing an oversized handle, it became more user friendly while wearing flying gloves. For safety precautions, a toggle is provided on the back of the grip, which much be depressed before firing, much like the .45 1911 automatic pistol. (Robin Bartel Collection)

The barrel of the M2 is more of a socket that secures the parachute flare in place, the barrel is the casing of the flare itself which extends well beyond the end of the pistol.

(Below) The patent number on the grip is for the breech-loading method, awarded to Louis Driggs and Henry Faber in 1929. Wartime production of the M2 reached 45,043 pistols delivered by the International Flare-Signal Company. (Robin Bartel Collection)

(Right and below) USAAF members demonstrate the M2 Pistol. The M2 was on allowance for air and ground units for identification and communication purposes. Lt. W.L. Atwell is firing an M2 Pistol at night, signaling a cargo plane approaching the drop zone for supplies at the air warning station on Mt. Atwell, India, August 1943. (Courtesy Charlie Flick)

(Below) The Signal, Aircraft, Red Star, Parachute M11 was principally reserved for distress purposes and in April 1936, an allowance of one M2 pistol and 6 M11 parachute flares was authorized for inclusion in life rafts, to be contained in a rubberized case Dwg. No. 37G2993. Illustrated is the standard compliment for USAAF pneumatic life rafts, with the M2 Pistol and M11 flares in the respective rubberized case, September 1936. The container was to be cemented to the floor of the raft. This example has four lift-a-dot tabs for attachment. (National Archives)

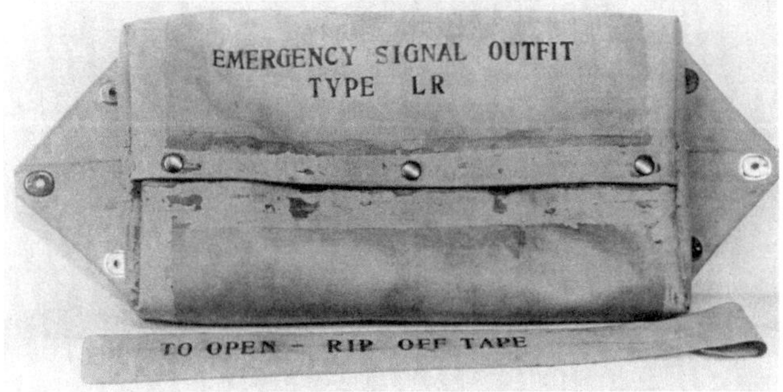

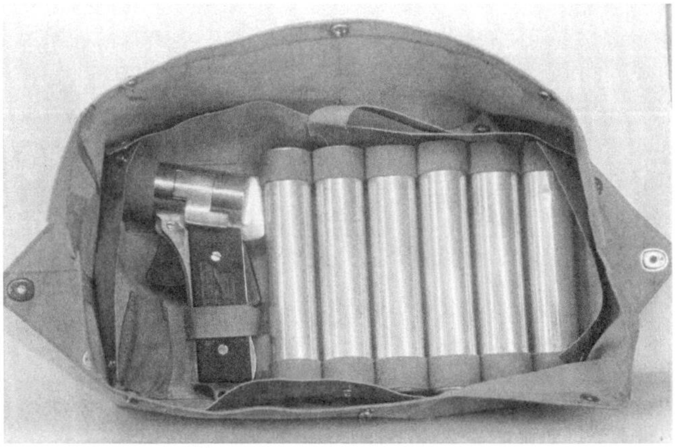

In 1940, the US Navy decided to evaluate all pyrotechnic projectors, both those that were standard and those no longer standard but still in supply. At that time, two types of pyrotechnic projectors were on standard status for the US Army: Pistol, Pyrotechnic, M2 (aircraft use); Projector, Signal, Ground, M3 (Ground use). The Very Signal Pistol, Mark-III (ten-gauge) was the US Navy standard.

During 1941, more attention was devoted to obtaining a better type of pyrotechnic projector for ground forces and aircraft, and in particular the British 1-1/2" signal pistol for aircraft use was being reviewed. By July 1941, the British Signal Pistol, No.4 Mark-I, and Signal Pistol, No.3 Mark-II, had been tested with consideration for adoption as substitute standard for use in airplanes. The Ordnance Department had neither the funds, nor requirements for such a gun, but was in the process of approving or disapproving a British requisition on Lend-Lease for 4,300 Signal Pistols, No.3 Mark-II, to be manufactured by the Auto Ordnance Corporation. These were intended for aircraft being sent from the United States to Britain on British contract.

At this time the Navy Department expressed interest in the British 1-1/2" signal pistol for their airplanes. The Ordnance Department was concerned that Navy procurement was almost certain, and that such action would overshadow Army procurement taking precedence on requirements and delivery schedules. It was felt advisable that the Navy and Army coordinate their actions with the Army for Lend-Lease and with the Navy for a new standard for aircraft. To further complicate the issue the USAAC became interested in procurement of a pistol for utilization of the 1-1/2" signal as well.

By the end of 1941, it was practically assured that the Navy Department would procure the 1-1/2" British type signal pistol, and the USAAC was at the point of standardizing the same, or a similar model, stating that they desired a pistol that could be fixed to the airplane. Within a year, the review of new types of pyrotechnic projectors went from no established requirements to a matter of urgency. It was determined that all phases of pistol design should be scrutinized, and that procurement must start from scratch. This created more complications in the decision-making process.

A meeting of Air Corps, Navy Department, Ordnance Department and British representatives was called on 12 November 1941, to select a common design for future manufacture. At the meeting, it was decided that instead of the Bureau of Ordnance, Bureau of Aeronautics, Army Air Corps, British Air Forces and Army Ordnance Office all working independently towards procurement of a 1-1/2" signal pistol, responsibility should be assigned to the Army Ordnance Department for the design and procurement of a pistol suited to the needs of all the services. It was further decided those needs would best be met by a pistol having the military characteristics of the British Signal Pistol, No.4 Mark-I.

Following the joint agreement, the standard for air forces would be a 1-1/2" signal pistol along the general lines of the British type. The Army Ordnance Department was to be responsible for determining production of a standardized model. Three companies submitted samples that met the requirements of the committee, and these were tested at Aberdeen Proving Ground in February 1942.

R.F. Sedgley Inc. submitted a sample that was designated as the M6, having the appearance of the ten-gauge Mark-V Pistol but of all steel construction. Kilgore modified the M2 with a sleeve for mounting, designated as the M7, but still utilized the rimless

(Left) An 8th AF officer of the 401st BG is holding a British, 37mm, Webley, No.4 Mark-I flare pistol preparing to signal returning B-17's from the airfield control tower, December 1943. In his left hand is the 1-1/2" shotgun style signal cartridge, available in single and double star configurations. This British pistol was compatible for use in a mount or fixed position in an aircraft, which was found to be a desirable feature by US forces. (National Archives)

cartridge. A sample submitted by the Eureka Vacuum Cleaner Co. was designated as the M8, and also included a mount. At the time of the joint agreement to adopt the new signals, the International Flare-Signal Co. had an outstanding order for 58,753 M2 Pistols. With the new direction, the Ordnance Department cancelled the contract in December 1941, but only after 45,000 had been completed. This obligation was met in June of 1942.

The M6 pistol had been expected to replace the M2 when sufficient M6 pistols and rimmed type cartridges became available to meet the requirements of the Army, Navy, Marines and Defense Aid. When this occurred, the M2 with rimless cartridges would be withdrawn from service and made available for non-military use, such as merchant ships and civilian aircraft. However, the tests conducted at Aberdeen in February illustrated significant flaws in the M6, with the M7 and M8 considered superior models. The contest now was between the Kilgore M7 and Eureka M8, but both required improvements. In March 1942, each aeronautical service established their projected requirements for pyrotechnic pistols, with the US Navy submitting requisitions for 40,000 M7s and Army Ordnance for 46,000 of the M8. In order to avoid delay in production, both the M7 and M8 were authorized for procurement until developments definitely indicated which was the better model.

Although the Modified M2(M7) was developed by Kilgore, the contract for the production of the M7 was awarded to the Johnson Tool & Engineering Co. for the remaining balance of pistols from the cancelled contract from December 1941. Johnson Tool had actually manufactured the M2 for Kilgore in the past. Additionally, the Eureka Vacuum Cleaner Co. was awarded a contract for 10,000 pistols. With these interim contracts established, both companies revised their models to the required characteristics as detailed by the Aircraft Technical Subcommittee. Johnson modified the pistol to be made of steel and adaptable for breech-loading British and American rimmed signals. Eureka improved their model, making minor refinements improving its function and reliability. In a meeting held 30 May 1942, between representatives of the Ordnance Department, BuAer and USAAF, samples of the Johnson M7 and Eureka M8 were inspected. It was the committee's decision that the M8 was a better pistol than the M7, both in workmanship and smoothness of operation. Production plans were further along than on the M7, and the M7 had no mount as of yet, while the M8 had a mount which was completely interchangeable with that of the British mount. It was shown that with this combination of the M8 pyrotechnic pistol and mount, complete Army-Navy standardization could be obtained and that Army-Navy-British equipment would be completely interchangeable. The committee recommended that all Army and Navy Air Forces requirements be that of the Pyrotechnic Pistol, M8. Based on the committee's recommendation, the Army Ordnance department designated it as the Pistol, Pyrotechnic, M8 under drawing number 42-44-1A.

On 8 June 1942, the new signal pistol was adopted under joint Army-Navy-British standardization as Pistol, Pyrotechnic, AN-M8 w/Mount M1. Upon standardization, the M2 became limited standard, and in September 1944 it was classified as obsolete.

Concerns regarding signal pistols used in aircraft had been that they were found impractical to fire through an open canopy or window while operating at high speeds, and also the control of trajectory to ensure the signal cleared the aircraft structure. Improvements in the new design included provision for being mounted and fired from a fixed position inside the aircraft. A further improvement included use of stamped parts – making it lighter in weight, at 2.13 pounds. Total wartime production of the AN-M8, from 1942 to 1945, would reach 342,611.

(Above) The AN-M8 was a single shot, double action, breech loading pistol, specifically designed for aircraft and used in conjunction with the Mount M1. USAAF aircraft manufacturers and life raft contractors were advised to switch to the AN-M8, replacing the M2 by November 1942, when available, under Ordnance stock number B-001-04-38813. The BuAer notified service personnel and aircraft contractors of the switch from the Mark-III (ten-gauge) to the AN-M8 as early as September 1942. However, quantities of the AN-M8 did not appear until early 1943. (Robin Bartel Collection)

(Left) The Eureka Vacuum Cleaner Company was the principle contractor for the AN-M8 pyrotechnic pistol until 1944. Stamped on the right side of the frame was a triangle that included the abbreviation of the company name, property mark and nomenclature. Most examples were dated but some have been found without a date.

(Below) Army Ordnance photographs of the M8 Pistol and Mount M1. Lugs were provided along the rim of the barrel which allowed the pistol to lock into the mount that was secured in the aircraft fuselage, the pistol could be loaded without removing it from the fixed position. (National Archives)

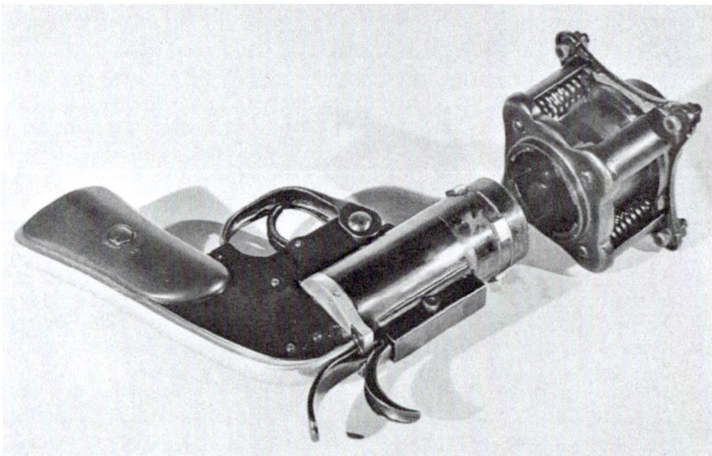

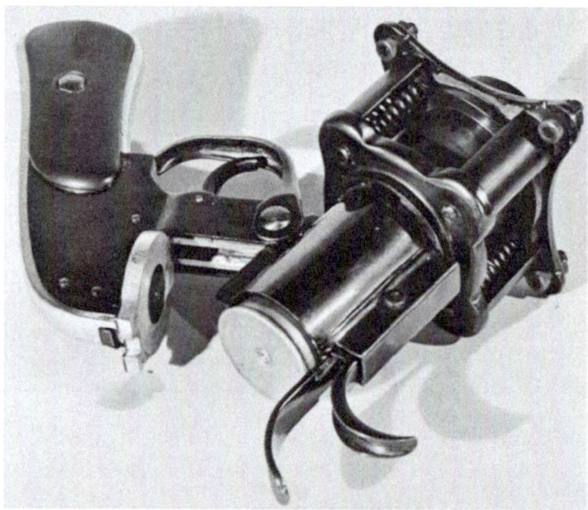

Right) Production at the Eureka plant was bogged down with labor troubles, slowdowns and strikes, over a period extending from December 1942 until late spring 1943. They were again delayed due to difficulty in obtaining steel tubing for the barrels, defaulting on delivery schedules through the remainder of 1943 and into 1944. Due to delays at the Eureka plant, it seemed advisable to attempt to establish another facility for the manufacture of the AN-M8 Pistol. The McInerney Spring & Wire Co. submitted samples to Aberdeen in November 1943, which were approved for manufacture, but production was slow – even to getting a satisfactory pistol that would meet inspection criteria. Ultimately, they had more serious failures meeting production schedules than Eureka. They produced the AN-M8 Pistol through 1944 up to April 1945. Illustrated is an example of the markings on an AN-M8 manufactured by the McInerney Spring & Wire Co.

Drawings were distributed to contractors and service activities for the proper placement of the mount, specific to the type of aircraft. It was important that the trajectory would sufficiently clear the fuselage. For the USAAF, the allowance of a signal pistol in fighters was eliminated, being replaced by the remote-control Discharger M5 which would remain the standard.

The BuAer found the discharger to be problematic, and in February 1943 they standardized the AN-M8 for all aircraft including all carrier types.

In May 1943, Naval aircraft service personnel were ordered to remove the discharger AN-M5 from VF class aircraft and replace it with the AN-M8, stock number 2-P-3499 w/ mount and 2-P-3499-250 w/o mount. In July 1945, this allowance was eliminated from all carrier type aircraft.

(Above left) The AN-M8 is locked into the mount in the cockpit of a TBF Avenger allowing the pistol to be fired without opening the canopy.

(Above right) Gunners in VB class aircraft were charged with the launching of flares, to include parachute types that were dropped through two ports under the aircraft near their position. Seen under the twin .30 Cal. machine guns manned by Robert Davilla ARM/2c of VB-5 is the port for the AN-M8 Pistol aboard the USS Yorktown, February 1944.

(Right) External view of the M1 Mount just below the right arm of pilot LCDR. Robert Isley of VT-16 aboard the USS Lexington, December 1943. The ports on both aircraft pictured appear to be covered with friction tape, which would serve as a waterproofing application.

(Below) Above the navigators table in this US Navy PV-2 Harpoon is an AN-M8 signal pistol in the M1 mount. Navigator Ens. Philips would be responsible for the discharging of the appropriate signals when necessary, April 1945. (National Archives)

(Above) Two USAAF personnel communicating and directing air traffic. Aside from radio communication, a signal lamp and pyrotechnic pistols are being utilized. An AN-M8 is being used by the individual on the right and an M2, loaded with a parachute flare, is on the deck to the left of the image. (National Archives)

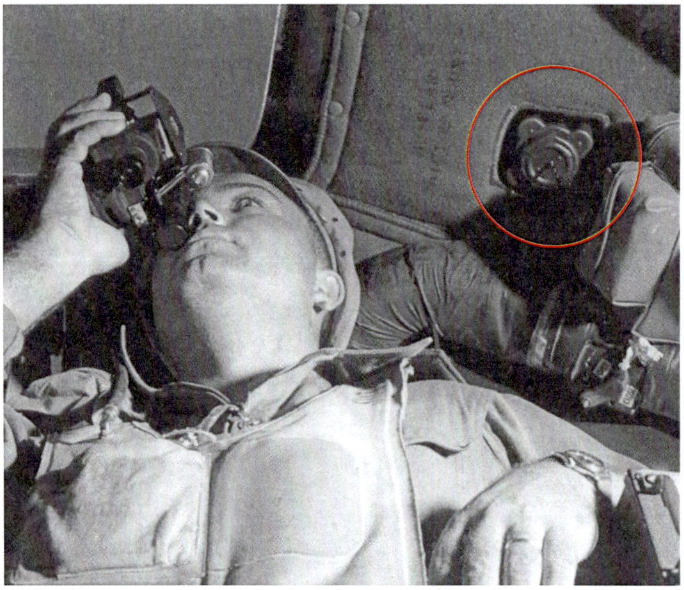

(Left) USAAF navigator aboard a B-24 Liberator of the 864th BS takes readings from the sun en route to Kwaialein, Marshall Islands October 1944. Above his left shoulder is the M1 mount, through the top of the airplanes fuselage. (National Archives)

(Below) Included in the June 1942 acceptance of joint Army-Navy-British standardization was the Discharger, Pyrotechnic, M5/AN-M5. This was a remote controlled fixed apparatus, also known as a Molin Type, to be installed in aircraft of 3 persons or less, it was adopted for both the USAAF and BuAer with the AN-M8 intended for bombardment or transport type aircraft. The device held six signal cartridges which were cocked and fired by the Remote Controller M2 manufactured by the Eureka Vacuum Cleaner Co. This design was based on a similar six-barrelled discharger, used by the British for airplanes for night identification to their landing crews. Deficiencies of the discharger included the inability to reload while in flight and to specifically select star arrangements of the cartridges. (National Archives)

(Below) The AN-M8 signal pistol as carried by US Navy PBY aircrew members during training exercises at NAS Jacksonville, c.1944. A trainee prepares to board the PBY. Draped around his neck is the AN-M8 Pistol.

(Right) An aircrew member undergoing life raft drill brings aboard all the emergency provisions available from the aircraft. The signal pistol serves a double function as a communication device on operational duty and as a distress signal in the event of an emergency. In July 1942, and as an expedient measure, the BuAer requisitioned 20,000 M2 Pistols from Army Ordnance to replace the Mark-III (ten-gauge) until the AN-M8 became available. The purpose was also driven by the recent joint Army-Navy standardization of rimless signal cartridges, which at the time were only compatible with the M2 (later for use in conjunction with the AN-M8). Upon the availability of the AN-M8, the Mark-III and Mark-V (ten-gauge) was relegated to training and ground purposes only. Because of the velocity and elevation gain of the ten-gauge star, the BuAer still found it had value for ground units. (National Archives)

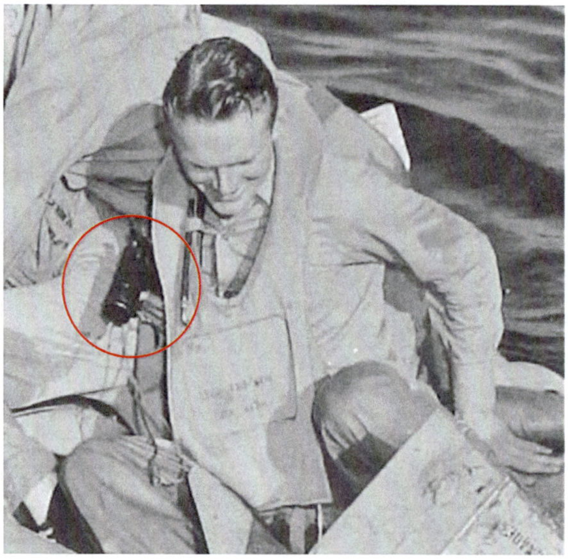

(Below) The AN-M8 gradually replaced the M2 for use in USAAF life raft equipment, but due to its size, only five M11 parachute flares were included. Pictured is the updated case, Dwg. No. 37G2993, with the AN-M8 Pistol. (National Archives)

(Left) An aircrew member prepares to fire the AN-M8 with the M11 red parachute signal under a simulated survival situation. (National Archives)

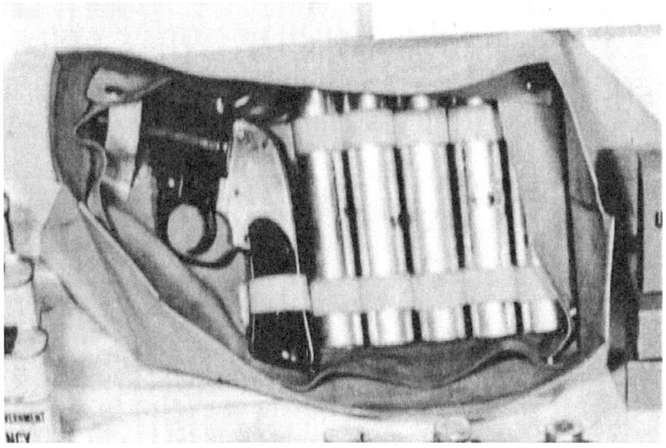

At the time of standardization of the Ground Signal Projector M1, the Very Pistol Mark-III (10 gauge) continued to be carried on the Ordnance Book of Standards as Limited Standard. The M1 projector was a much heavier weapon than the Mark-III and required heavier ammunition, which prompted the using services to request the issuance of the Mark-III for rife platoons, pending the design and standardization of a lighter alternative. In the summer of 1938 a simple projector for use solely for the projection of pyrotechnic signals was being designed. Pending the standardization approval for issuance of the Mark-III it was approved, transferring it to a standard status for rifle platoons.

The new type of projector for use by ground troops was approved for standard in August 1939 as the Projector, Signal, Ground M3, placing the Mark-III back to a status of limited standard along with the M1 Projector.

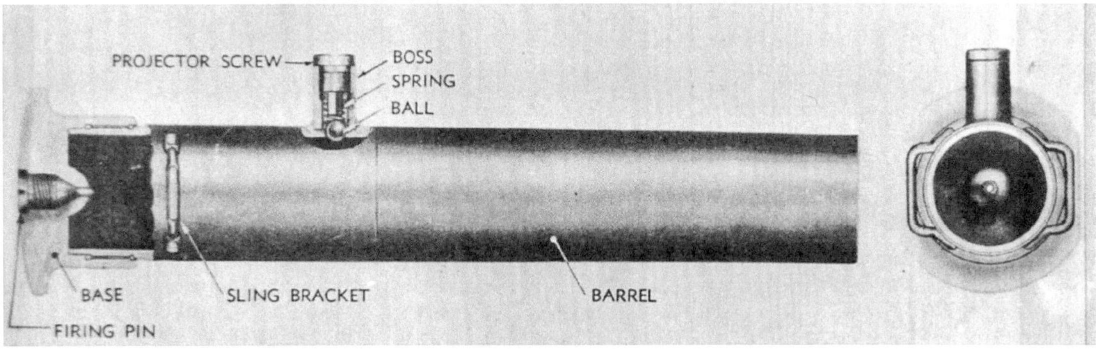

(Right) The M3 Ground Projector consisted of a barrel and base assembly 12 inches in length with a 1.656 diameter. To fire, the signal was muzzle loaded locked into place by a spring and ball. When the base was struck against the ground a fixed firing pin engaged the primer on the signal. The Krebs Manufacturing & Equipment Co. was the initial contractor for the manufacture of the M3.

(Right) In 1940 Krebs MFG. developed an improved modification of the M3, which was tested proving sufficiently superior and standardized as the Projector, Signal, Ground, M4 in July 1940. The improvements included a simplified base and signal locking mechanism, using the same function as the M3. Illustrated is the M4 Projector with the improved M17 through M22 parachute ground signals that included stabilizing fins. Additional contractors for the M4 included the Oliver Farm Equipment Co. and Barlow & Seelig Manufacturing Co. From 1940 a total of 22,635 M3 and M4 Ground Projectors were manufactured.

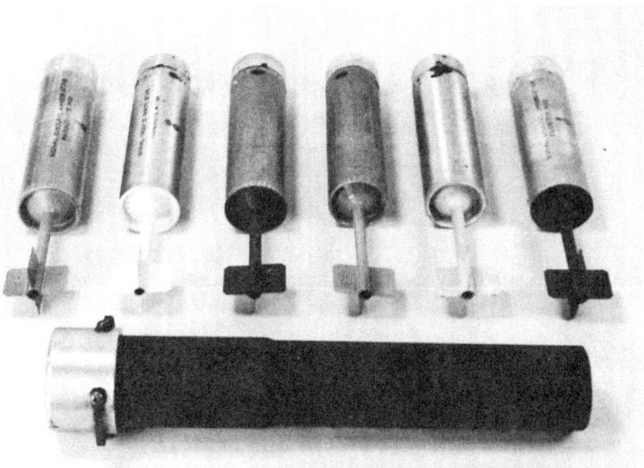

During 1941, continued studies of development of tactics, both in the US Army and of troops engaged in the ETO definitely indicated the necessity for taking all possible measures to provide adequate means of communication between combat echelons of small units and headquarters.

The M4 Projector, then newly standardized, was too unwieldy for the use of small units, and a very definite need existed for a small, easily handled, signal projector. With all its obvious faults the Mark-III Pistol had these military characteristics and in February 1941 it was re-designated as standard once again.

A considerable number of these pistols were still in inventory. Before preparing complete manufacturing data and drawings an investigation was made to determine the possibility of commercial procurement to insure interchangeability of replacement parts and for economy on new procurement. Remington Arms Co. were no longer producing of the ten-gauge Mark-III and had not retained their equipment to commence new fabrication. The Navy department had procured commercial Very pistols in recent years from Remington Arms Company and the Harrington & Richard Arms Company, and offered the use of their drawings and specifications, together with the list of bidders they had used.

Harrington & Richardson Arms Company submitted samples of a commercial pistol they designated as the Mark III-A one-inch Signal Pistol which was tested at Aberdeen and the Infantry Board, to determine if it would be acceptable as an alternative to the Mark-III Very Pistol. The biggest factor in favor of the Harrington model was the fact that they were already tooled up and could get into production on a ten-gauge pistol within a very short time. Based on the favorable overviews, the Mark III-A Pistol was recommended for standardization in October 1941, and approved as substitute standard in December of that year designated as the Pistol, Very, M5. A contract was awarded for the manufacture of 13,000 pistols to meet service requirements.

Production by Harrington was very slow and by 8 September 1942 only 5,260 had been delivered. Since standardization of another signal pistol to replace the M5 was imminent, they were ordered by the Ordnance Office to cease all work on the M5 as of noon 12 September 1942.

(Above and right) The 1" Mark III-A pistol tested as manufactured by Harrington, additionally, an existing example of the M5 adopted as substitute standard. (Robin Bartel Collection)

Upon instituting the M5 Pistol as a substitute standard plans were under way for adopting a 1-1/2" pistol for ground forces, so there would uniformity for both air and ground units, and making the ten-gauge Very pistol limited standard.

The Small Arms Division of the Army Ordnance Department offered the suggestion that a 1-1/2" pyrotechnic pistol for firing signals from tanks be developed, which would simplify procurement both of the pistols and signals. It was thought probable that a signal pistol could be developed for use in tanks that would be more simple that the M8 Pistol. The Infantry Board carried on extensive tests in an effort to obtain a satisfactory type of signal pistol or projector, one that would serve as a means for air-ground recognition and identification.

In June 1942, a comparative test between the M4 Projector and the M2 Pistol and also the between the ten-gauge types Mark-III and M5, established that the M2 was decidedly superior. Recommendation was made by that board that the M2 Pistol be substituted for the Mark-III and M5 Very Pistols. In doing so, it was overlooked that the M2, originally developed for aircraft use, had been replaced by another design making the

M2 limited standard. When the matter came before the Ordnance Committee for decision, they were opposed to the adoption of the M4 Projector for issue in lieu of the Very Pistol and in preference of the M2 Pistol, which left infantry without any suitable approved pyrotechnic device for use of front line platoons. As an interim the Small Arms Division suggested that the M4 Projector be recognized as standard until development could be completed on additional items already under test.

Out of the development work being pursued to fulfill the need of the Infantry Board for a better projector for ground-to-air signals, a hand signal projector designated as the T4 evolved and was adopted to firing the interim type signals (M26 thru M36). It was superior to the M2 Pistol and M4 Projector in weight, size, and ease of carrying. The rimless cartridges were designed to meet an immediate need for compatibility with the British signals, and until a pistol could be developed to handle the rimmed cartridge casings. Production of the rimless cartridges was intended only for limited procurement, to accommodate the new Army-Navy shotgun style rimmed type cartridges the T4 was significantly modified. The principle change was configuring the projector to a breech loading method.

The T4 was adopted as standard for ground troops 1 October 1942 designated as the Projector, Pyrotechnic, Hand, M5. Because of the confusion which might arise, due to the existence of two other signaling devices bearing this number, the Very Pistol, M5 and the Pyrotechnic Discharger M5, the nomenclature was changed to Projector, Pyrotechnic, Hand, M9 on 26 November 1942, referenced under DWG. No. 78-6-1. The M9 Projector was now the standard for ground forces placing all ten-gauge pistols and the M4 Projector as limited standard.

In April 1943, the Infantry Board conducted tests on a comparative study utilizing the M9 Projector and the AN type signals M37 thru M45, which they concluded was not satisfactory for infantry use. It was their recommendation to adopt the new experimental signals that were adaptable to the rifle grenade launcher with which many infantrymen were to be equipped. Replacing the M9 Projector for infantry use was the newly adopted M17A1 through M22A1 parachute ground signals, modified for projection from the service rifle. The requirement for infantry use of the M9 was eliminated by the summer of 1943 and was recommended to be dropped, but found new life with the USAAF.

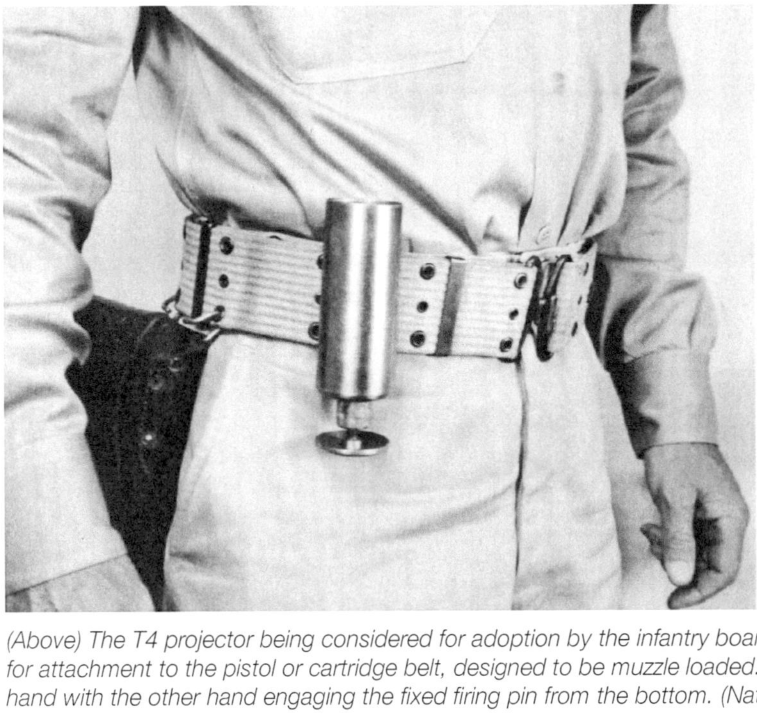

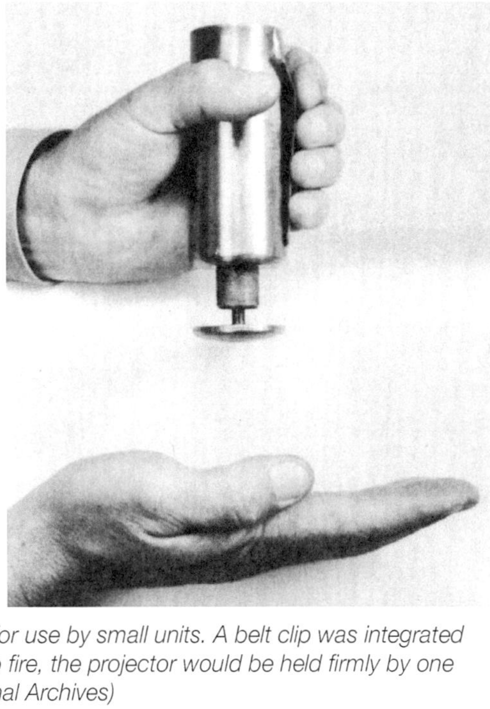

(Above) The T4 projector being considered for adoption by the infantry board for use by small units. A belt clip was integrated for attachment to the pistol or cartridge belt, designed to be muzzle loaded. To fire, the projector would be held firmly by one hand with the other hand engaging the fixed firing pin from the bottom. (National Archives)

Upon standardization of the M9 Projector, a contract was awarded to the Eureka Vacuum Cleaner Company with expected deliveries of at least 10,000 by April 1943. In August 1943, the Grand Rapids Brass Company took over production; Eureka was directed to transfer all raw materials and partially completed components to Grand Rapids. This transfer was initiated to relieve Eureka from the stepped-up production demand of the M9 for inclusion in USAAF life rafts, they were already significantly behind on the scheduled deliveries of the AN-M8 Pistol. By the close of 1943 Grand Rapids Brass Co. was directed to complete all outstanding orders by May 1944.

In June 1944, the All Metal Products Company took over production, remaining the sole manufacture until September 1945. In April 1945, a change order was issued for the balance of 8,200 M9 Projectors to be replaced with M9E4 Projectors, to fulfill a requirement of the USAAF because space limitations in fighter planes made the AN-M8 Pistol impractical for use.

The Air Technical Service Command requested that attempts be made to modify the M9 Projector for use in Mount M1, so that it could be installed in aircraft to replace the Discharger AN-M5. This was in response to reports of aircraft being lost, due to the pilots' inability to fire recognition signals. The modification was the application of the lugs as seen on the end of the barrel of the AN-M8 Pistol. 1,840 M9E4's were delivered before the contract was cancelled in May 1945, when hostilities in the ETO ended and their requirement was eliminated.

The USAAF in Britain found it necessary, at first, to use pyrotechnics of British manufacture. While requirements were small, the British Air Ministry was able to supply those requirements, however, by 1944, such requirements had reached a volume which made it impossible for the needs to be met from British sources.

In particular, there was a need for a 1" distress signal projector. The development of such a projector was undertaken and standardized as the Projector, Pyrotechnic, M10 on 10 August 1944. The M10 retained all the same characteristics of the M9, but chambered for a 1" signal. However, before it entered quantity production, the USAAF decided there was no longer a requirement for the type and it was declared obsolete.

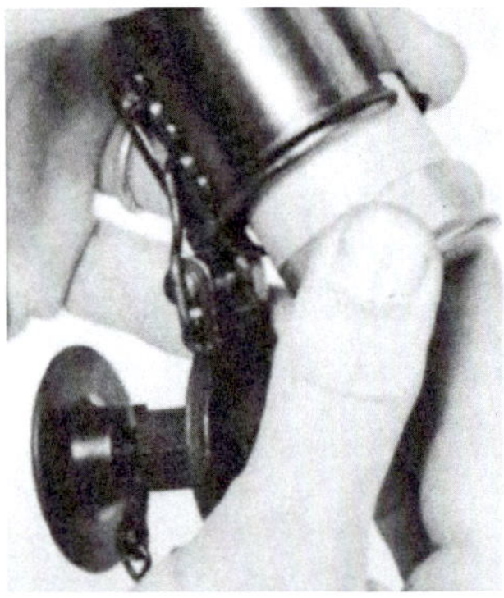

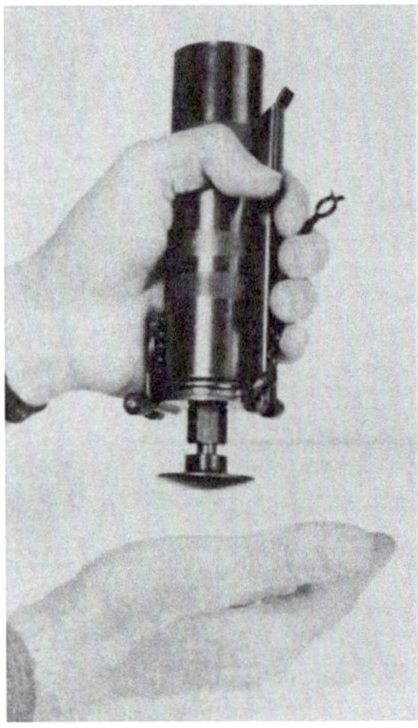

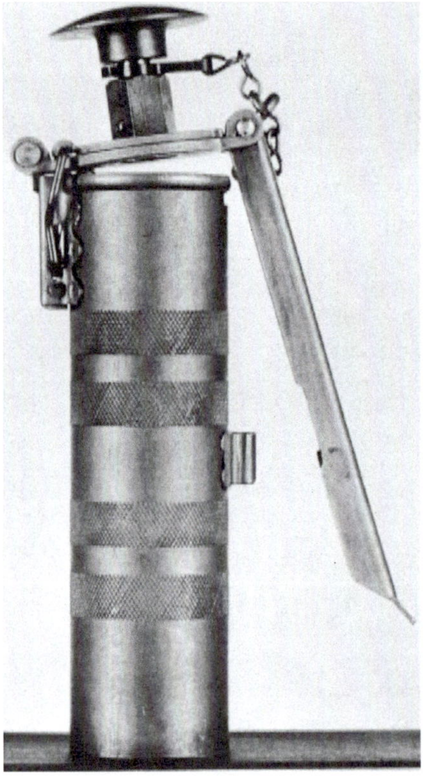

(Left and below) The M9 Projector was incorporated in USAAF life raft equipment with the emphasis towards weight reduction and space restriction remaining on standard status, with all future procurements allocated to meet USAAF requirements. What was developed was a simple device with minimal movable parts that could be held in one hand and fired a distress signal via a fixed firing pin plunger with a swift pop by the palm of the other. Once quantities became available, by July 1943 life raft contractors were notified to switch to the M9 superseding both the M2 and AN-M8 Pistols. The AN-M8 was technically supplied as an Interim until the M9 became available. By the end of 1943, life raft specifications were being amended to reflect the use of the M9 as standard for all multi-person distress signal kits. With the introduction of the M9, the weight of the signal kit was reduced by one pound and provided room for six M11 parachute signals. The M9 can be found with a blued or a Parkerized finish. Total production of the M9 was 287,422 from 1943 to 1945. (National Archives)

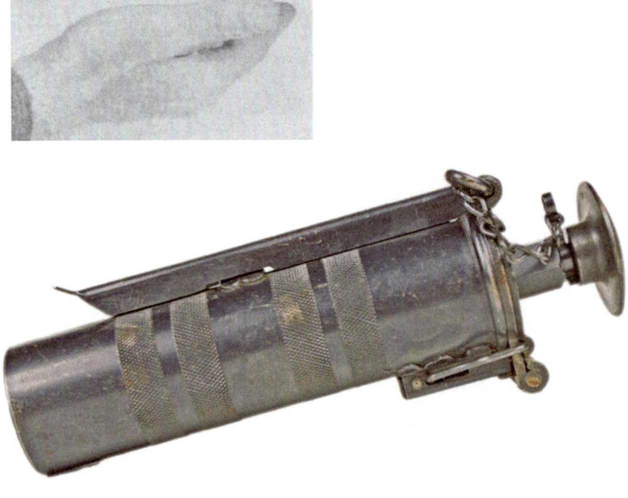

(Above) The Emergency Signal Kit provided in USAAF multi-person life rafts, Dwg. No. 37G2993. The units were either fabricated by the life raft contractor or by service activities, which makes available an array of minor variations. Dwg. No.37G2993 consists of two components, an inner and outer pouch, the outer is provided for protection and waterproofing and the inner for securing the projector and signals. Pictured are several physical examples that include the M9 and AN-M8 hand projectors with M11 parachute flares.

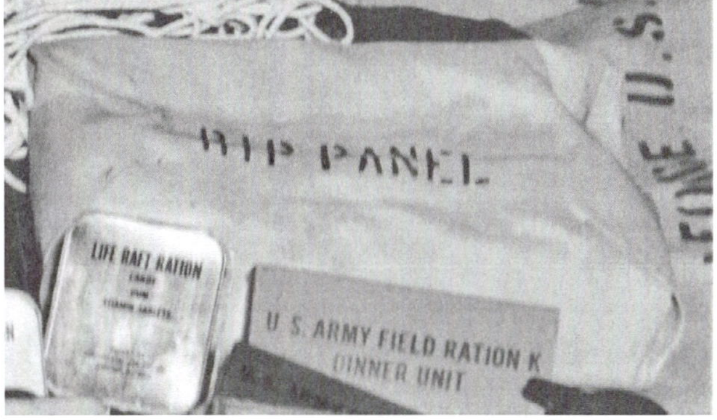

(Right) The signal kit being displayed with the accessories of USAAF life rafts, c. 1943 and 1944. Prior to the adoption of the central accessories container the signaling kit was detailed to be cemented to the floor of the raft. In the immediate post war years, the signal kit would be replaced by the AN-M75 two star red and MK-1 MOD-1 hand held smoke distress signals, these would then be superseded by the MK-13 MOD-0 combination smoke and star distress signal in the immediate post war years

Above) Units operating in the ETO often included RAF emergency equipment supplementing standard US life rafts. As always, with emergency equipment for aircraft, space restrictions were a concern, and it was found that the M2 pistol with M11 signals took up a fair amount of space. To better utilize the space, the 8th AF Service Command adopted a new signal kit incorporating RAF distress signals. As part of local Technical Maintenance Instruction 04-15-1, dated 5 January 1944. The signal kit was revised to include the British 1" signal pistol No.2 MK-V and nine tins of cartridges. The standard compliment of M11 signals only allowed six shots, the compliment of nine tins of 1" cartridges allowed for 35 shots, taking up an equal amount of space. This would become the standard compliment for USAAF multi-person life rafts in service in the ETO. Pictured is the revised signal kit including RAF distress signals, c. 1944.

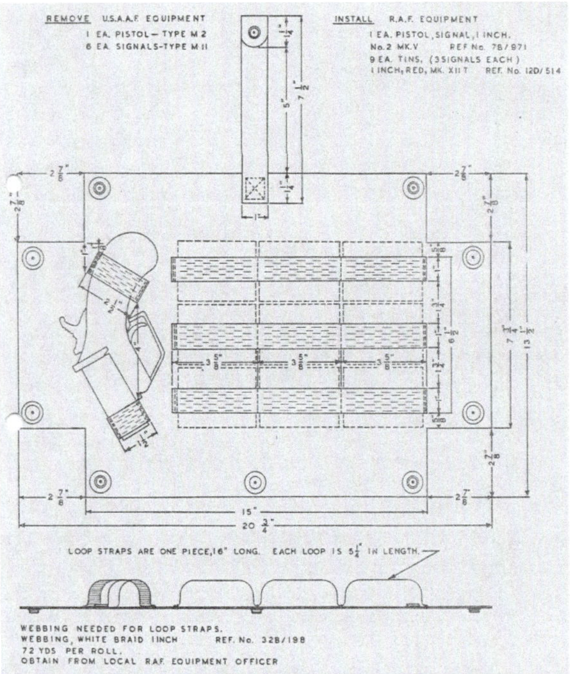

(Above) Fabrication drawing accompanying TMI 04-15-1. (National Archives)

(Right) A surviving example of the signal kit containing RAF distress signals, c.1944. (Kris Usherwood collection)

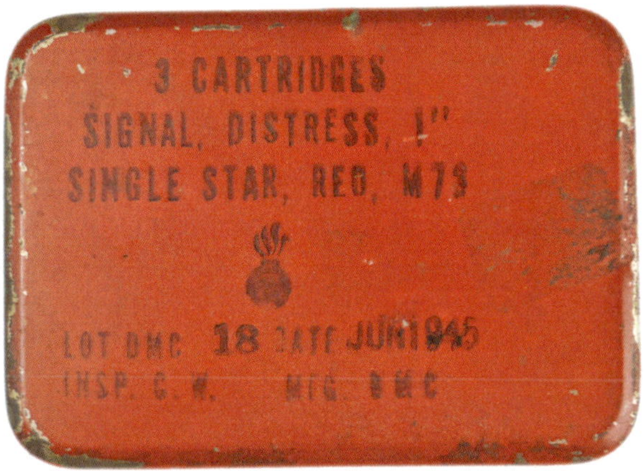

(Left and above) Due to the extensive use of the British 1" signal pistol by operational units in the ETO, the Army Ordnance Department developed a 1" signal, designated Signal, Distress, 1", Single Star, Red M73. These signals were adopted in 1945, intended to be used for M10 hand projector but compatible with British pistols.

(Above) The M11 parachute signal is highlighted due to its being a distress signal, with the others in the series being used for recognition and communication. With the adoption of the M2 Pistol, the new signal cartridges were slightly increased in diameter to 37mm (1.56") from the 35mm (1-3/8"), with a length of 7.69", and manufactured with aluminum and steel casings. The M10, M14, M15 and M16 share the same characteristics, differing only in color and nomenclature. Illustrated are examples of aluminum (top) and steel (bottom) casings of the Signal, Aircraft, Red Star, Parachute M11.

(Left) The M9 parachute flare, later revised to the M9A1 for use with the AN-M8 pistol, was primarily reserved for observation or reconnaissance aircraft. When fired, the white flare burned at 60,000 candle power for one minute illuminating an area for dropping messages. A message bag can be seen on the right of the picture. (Courtesy Charlie Flick)

(Above) Flares for the USAAC evolved to parachute types of a single color, expanding from the original three colors of red, white and green from WWI. Shown is the series of joint Army-Navy aircraft signals utilized with the M2 Pistol, adopted as an interim until the M8 Pistol and the new shotgun style signal cartridges became available. Within the development process, the color white was eliminated and replaced by yellow, making red, green and yellow stars the standard. The bulk of cartridges utilized a two-star configuration, as noted by the colored bands on the casings. The casings were manufactured in both aluminum and steel. Pictured from left to right are the M28, M29, M30, M31, M32 and M33. Not illustrated are the three single star cartridges M34 (Red), M35 (Yellow) and M36 (Green). Being a joint Army-Navy standardized article, the designation is preceded by "AN-". These signals were assembled much like the parachute types, with a color-coded end cap and raised markings for identification in the dark. The AN-M8, AN-M5 and M9 projectors were designed to universally fire these cartridges and the shotgun shell type. (Robin Bartel Collection)

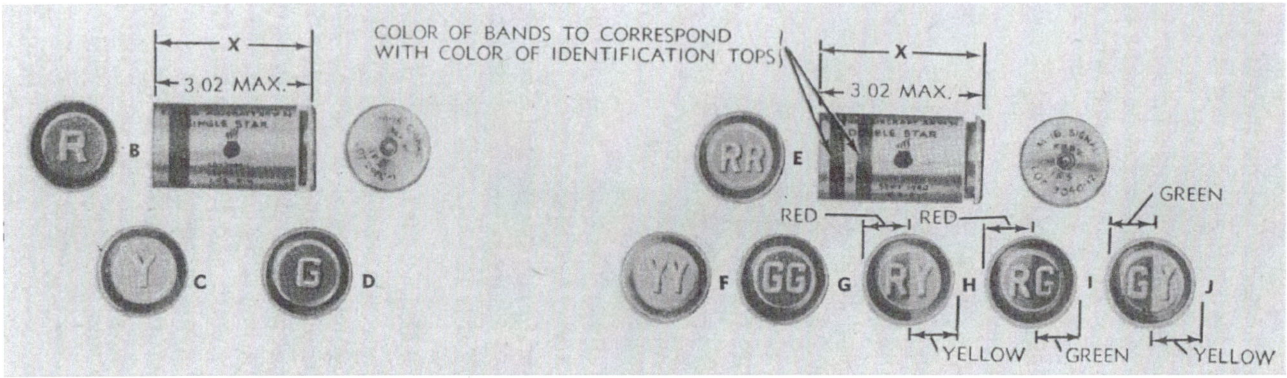

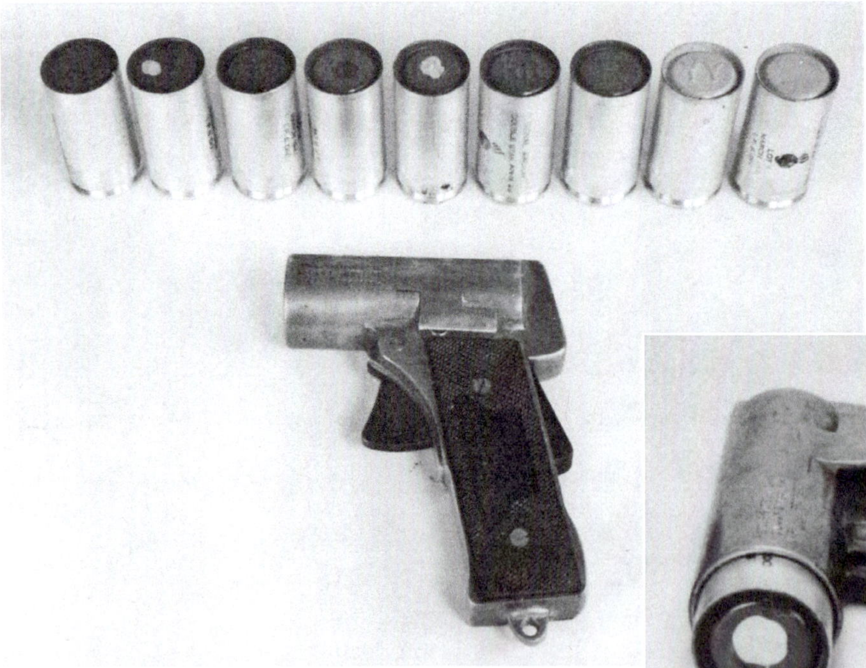

(Below) With the adoption of the British type 1-1/2" signal, a new series of 37mm shotgun shell type cartridges was adopted in June 1942, under joint Army-Navy-British standardization. These were not compatible with the muzzle loaded M2 Pistol, but could be fired from the AN-M5 Discharger, AN-M8 Pistol and M9 Projector, which were breech-loaded. Pictured left to right are the AN-M37, AN-M38, AN-M39, AN-M40 and AN-M42. Not pictured is the AN-M41 Red-Green star cartridge. These were made of standard rocket paper inserted into a brass base. Like the AN-M28 through 36 series, they featured colored bands on the case and on the paper closing wad. They were not provided with low light identification. (Robin Bartel Collection)

(Below) Three single star cartridges were additionally adopted, designated as the AN-M43 (Red), AN-M44 (Yellow) an AN-M45 (Green). Yellow and green are shown. (Robin Bartel Collection)

(Left) Demonstrating the new AN-M37 through 45 series signal cartridges being loaded in an AN-M8 Pistol.

(Below) A comparative view of the two series of signal cartridges. (National Archives)

(Above) The US Navy additionally used rounds designated as Signal Cartridges, Two Star which included a tracer. These cartridges are known as the Mark-3 and Mark-4, with the only difference being internal. The two wide bands are the star and the thin top band is the color of the tracer. Not pictured is the Yellow Tracer, Red-Yellow Star. (Robin Bartel Collection)

(Far left) In May 1943, the US Army Ordnance Department standardized an additional series of signal cartridges for aircraft, designated as the M53 through M58 that included a tracer with double star, like that of the BuOrd Mark-3 and 4, sharing the same star and tracer configuration. These were Army-Navy-British standardized in July 1943. Pictured is the AN-M53 Yellow Tracer, Red-Yellow Star. The series includes the AN-M54 Green Tracer, Red-Red Star, AN-M55 Green Tracer, Green-Red Star, AN-M56 Red Tracer, Green-Green Star, AN-M57 Red Tracer, Red-Red Star and the AN-M58 Red Tracer, Green-Red Star.

(Left) The revised casing of aluminum for the series re-designated as the AN-M53A1 made effective in 1945.

Signals aboard aircraft were stored in fixed racks. In 1940, a recommendation to develop portable cases for pyrotechnic signals was submitted. The principle reason for the recommendation was that, in the event of an emergency – such as a fire in the aircraft or safety precautions in a ditching – the signals could be quickly thrown overboard. This concept was approved initiating the adoption of pouches Types A-1 through A-4, designated under the classification of Container, Signal Flare, 16 July 1941.

The Type A-1 was designed to carry nine M11 type parachute flares referenced under Dwg. No. 41G9233; the Type A-2 was designed to carry 20-M11 type parachute flares referenced under Dwg. No. 41J9269; the Type A-3 was designed to carry 15-M11 type parachute flares or five M9 flares referenced under Dwg. No. 41J8919; and the Type A-4 was designed to carry four M11 type parachute flares referenced under Dwg. No. 42G2215. The Type A-4 was intended for pursuit type aircraft.

With the introduction of the M28 through M33 signals, new pouches were developed for their stowage. Drawings for the four pouches were completed and adopted as standard in the spring of 1942. They were also designed to accommodate the new M37 through M45 signal, designated as Types A-5, A-6, A-7 and A-8.

(Left and far left) Signal containers Type A-1 and Type A-4. Internal pockets separated the signals. Those in the Type A-4 were side-by-side and those in the Type A-1 were staggered. One side of the container included four studs designed for the attachment of the signal pistol holder Type A-1, Dwg. No. 41G8920.

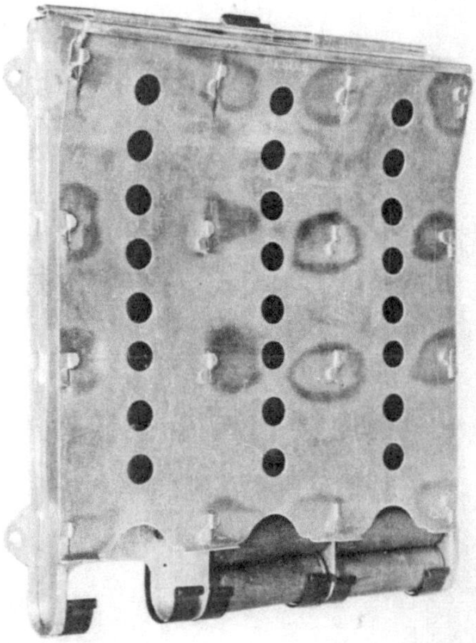

(Above left) Fixed rack for the stowage of 10 gauge signals utilized in pre WWII USAAC aircraft.

(Above right) T.R. Barnette AMM, USN is at his station in the blister of a PBY operating over the Aleutian Islands, March 1943. Over his left shoulder are examples of fixed racks utilized in Naval aircraft for the 10 gauge signal, additionally the Mark-III signal pistol can be observed secured to the bulkhead. (National Archives)

(Right) An example of a fixed storage rack for signals attached to the fuselage of an airship of Airship Squadron ZP-31, containing the AN-M28 to 36 series. Similar racks would be used for the parachute types. (National Archives)

Examples of fixed racks inside USAAC reconnaissance and bombardment aircraft:

(Below left) Rack used to secure large signals such as the Type M9.

(Below right) Box that would contain the smaller parachute flares designated for the Type M10 and M11. Containers A-1 through A-8 would replace the fixed racks or boxed methods. (National Archives)

(Left, upper and lower) Type A-5 pouch

(Above right, upper and lower) Type A-8 pouch. All pouches A-5 through A-8 shared identical construction details, differing only in their dimensions. Two male studs on each side were used to secure the pouch in place and also designed to secure the holder for the M2. The Type A-5, Dwg. No. 42G11966, held 6 signals for pursuit aircraft and was adopted as standard 21 March 1942; the Type A-6, Dwg. No.42G11967, held 12 signals for light and dive bombers, observer, cargo and amphibious aircraft; the Type A-7 Dwg. No. 42H11968, held 18 signals for heavy and medium bombers; and the Type A-8, Dwg. No.42H11969, held 20 signals for reconnaissance aircraft. Containers A-6 through A-8 were adopted as standard 5 May 1942. With the joint Army-Navy adoption of the AN-M8 Pistol, the BuAer would also utilize these containers.

(Above left and center) In conjunction with the flare containers, a pouch for the stowage of the standard M2 pistol was developed and designated as the Type A-1, referenced under Dwg. No. 41G8920. It was provided with four lift-a-dot female fasteners for securing to the cartridge pouches or fuselage of the aircraft.

(Above right) Four studs were provided on the back of the flare containers A-1 through A-4 for the attachment of the pistol pouch Type A-1.

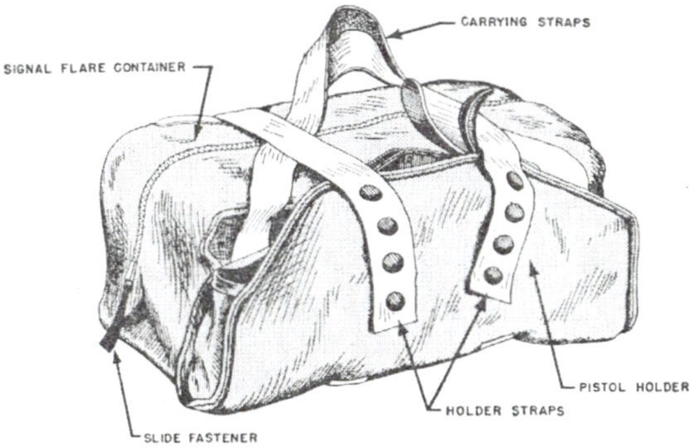

(Top) When the AN-M8 Pistol was not in use, it was to be stored in a holder designated as the Holder Assembly-Pyrotechnic Pistol, Type A-2, Dwg. No. 43G7044. The holder fastened to the A-5, A-6, A-7 or A-8 by means of a canvas strap that would loop around the pouches. The BuAer referenced the holder under ASO stock number R83-H-650100.

(Above) Illustration of the cartridge bag and A2 holder from Technical Manual TM 1-416, Miscellaneous Aircraft Equipment, printed March 1944.

(Above right) A cartridge bag can be seen in the navigators' position aboard a B-29 somewhere in the Marianas.

(Above and right) Often associated with the AN-M8 Pistol is a molded leather holster. It should be noted that no reference to this item could be found in any WWII dated USAAF or USN documents, technical data booklets or stock listings. All reference to stowing the pistol details the Holder A-2 as illustrated above. Some examples of these leather holsters are stamped MADE IN CANADA. No data has been found to determine if this was a US Government issue item.

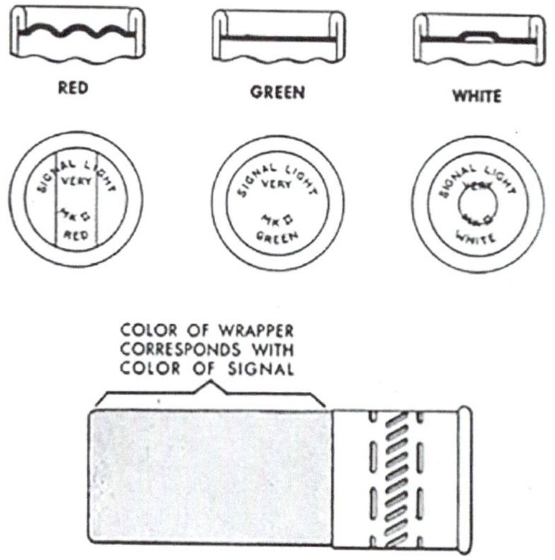

eft and above) US Navy Very Signal Lights. Ten-gauge shotgun ıells that include a star of either red, green or white. The wad ır each shell will correspond to the color of the star, As can be ɜen by the chart, the red included three ridges, green smooth and 'hite with a single bump for identification in blackout conditions. he paper casing was also intended to be manufactured with the orresponding color but that is not always consistent. The Very signal remained unchanged from the WWI era through WWII.

(Below) The standard packaging method was for 10 signals in a wax coated Kraft carton, later changed to a sealed tin for waterproofing. (Robin Bartel Collection)

In September 1941, the BuAer initiated a project to provide a personal signaling device intended to be carried in pockets in life jackets, for both aviators and shipboard personnel. 300 small hand-held projectors were developed, in cooperation with the Winner Manufacturing Company, of a plunger type firing device, that they had labeled as the Blair-Flare.

These 300 hand-projectors underwent field testing at NAS San Diego and in San Francisco by the 1st Marine Air Wing in the months of April and July 1942. Upon conclusion of the tests, the hand-projector proved to be highly effective with the recommendation that the ten-gauge Very signals be provided with some form of waterproofing, and that a change be made holding the safety clip to the body of the projector. As this projector underwent field testing, the BuOrd Naval Gun Factory contemplated adoption for a new distress signal kit, in April 1942.

Shortly after, the NGF adopted the new hand-projector as the Mark-III under Dwg. No.240767 Ordnance Specification No. 2399.

(Right) The adopted MK-III Hand Projector, which measured 5-3/8" long, with a diameter, at its widest point, of 1-1/4". To load, the breach was unscrewed from the plunger and reassembled with a Very signal. To fire; the plunger was pulled back retracting a spring. Upon release, the plunger with fixed firing pin, returned back to its original position, striking the primer of the signal. In June 1942, the BuOrd procured a quantity 22,612 from the Winner MFG. Co. under Navy Department contract 140s-5147, for inclusion in a new signal kit for the Bureau of Ships. This is the only known contract for the Mark-III Hand Projector.

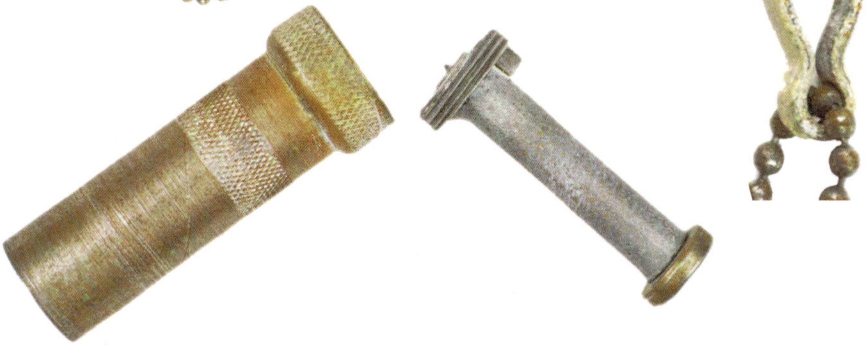

(Left and below) The new BuOrd signal kit was designated Abandon Ship Outfit, containing one hand projector and 12 Very's signals (4 red, 4 green and 4 white). Prior to the procurement of the projector, the BuAer suggested that the cartridges be wrapped in some form of Pliofilm for waterproofing. The method adopted was to encase them in a two-piece cellulose acetate container, taped at the seam, then wax coated. The packaging method for the Abandon Ship Outfit added further steps towards waterproofing by containing the cartridges and projector in a cellulose plastic canister with a sealed screw-top cap, which in turn was housed in an outer key opening tin. (Robin Bartel Collection)

In August 1942, NAS San Diego drafted and submitted a suggested modification to prevent the loss of the safety clip, by attaching the chain to the breach. In addition, the BuAer established that a pyrotechnic projector be incorporated in life raft equipment, steering away from the idea of providing the signal on an individual basis. In the succeeding months, the BuOrd made further revisions based from the drawing submitted by NAS San Diego.

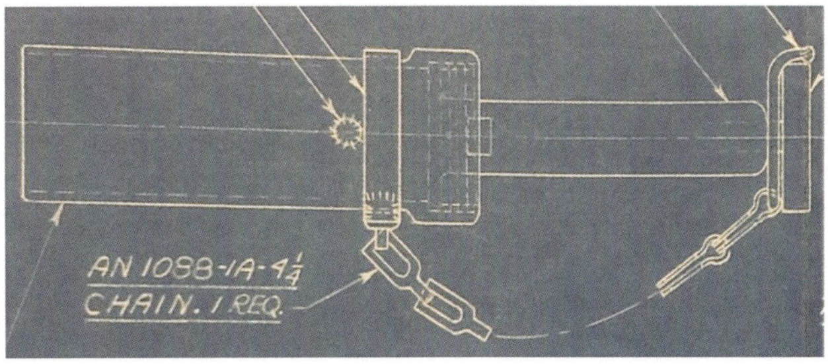

(Right) Dwg. No. 2-3667 with suggested revision to the hand projector NAF Dwg. 240767 Mark-III. (National Archives)

(Left and above) In October 1942, the BuAer requested that the BuOrd initiate procurement of the hand projector as revised under BuOrd Sketch No. 108074. This revision provided a chain connecting the breach to the plunger, an integrated safety mechanism and lug adapter to secure the breach to the plunger. This projector assembly was detailed under Dwg. No. 388751-54 Ordnance Specification No. 1416 and designated as the Hand Projector Mark-IV. The Eureka Vacuum Cleaner Co. Inc. was the first to submit a sample of the Mark-IV in November 1942, followed by a contract in December for 78,000 projectors under BuOrd contract No. 656. Additional contracts were awarded to the Eureka Vacuum Cleaner Co. Inc. in July 1943, BuOrd contract 4039, July 1944, BuOrd-6711, and in January 1945, BuOrd-8135. The Bell Gossett Co. would receive awards in February 1943, BuOrd-758, and July 1943, BuOrd-4102. Production of the Mark-IV Hand Projector would reach an estimated 550,000 from 1942 through 1945 at a cost of $1.80 each.

(Above) At the same time, the BuAer specifically requested that 100,000 projectors and 600,000 red Very's signals be procured for a distress signal kit. For the signaling kits, the projectors and Very cartridges, in plastic capsules, would be requisitioned from Naval Ammunition Depots with the ASO responsible for negotiating contracts on packaging. On 22 February 1943, the BuAer awarded a packaging contract to Cuneo Press Inc. for 100,000 signal kits under ASO contract 288s-11320. One Hand Projector Mark-IV and six red Very signals were packaged in a sealed wax coated carton, to be supplied as standard equipment for life rafts, available under ASO stock number R83-K-710309 referenced as Projector Kit. Additional requisitions were made from NAD to supply contractors of back pad kits, shipwreck kits, pararaft PK-1s and air droppable assemblies. Cuneo Press received another packaging contract 288s-30122 in the spring of 1945 to supply contracts for newly adopted air droppable kits and stock purposes.

(Right) By July 1943 projector kits were becoming available at central supply points for requisition, slowly integrating into life raft equipment for the Pacific fleet, for which it was intended. By the end of 1943, the need for individual emergency equipment was emphasized, and as the war progressed, the focus shifted towards signaling devices for rescue. Many local commands recommended that projector kits be drawn and kept on the person, especially for those units conducting night operations. Pictured is a member of VMSB-245, c. April 1945, enjoying a cup of ice cream. On his lap is a Mark-IV Projector secured to the waist strap of his life vest. VMSB-245 conducted night operations and within standard night flying procedure it was recommended that night distress signals be kept loaded and in an easily obtainable location. (National Archives)

(Right) A distress signal kit with its concept purportedly originating within CASU-27. As typically prescribed, the projector and shells were to be carried in the flight suit pocket, but for convenience a pouch was fabricated. Being locally fabricated, many variations are found. The kit pictured here contains the standard projector kit of one hand projector and six red Very signals. The pouch is secured closed making for a compact package that can be placed in a flight suit pocket.

(Above) A member of VF-34 has such a kit sewn to his flight suit on his left thigh. (National Archives)

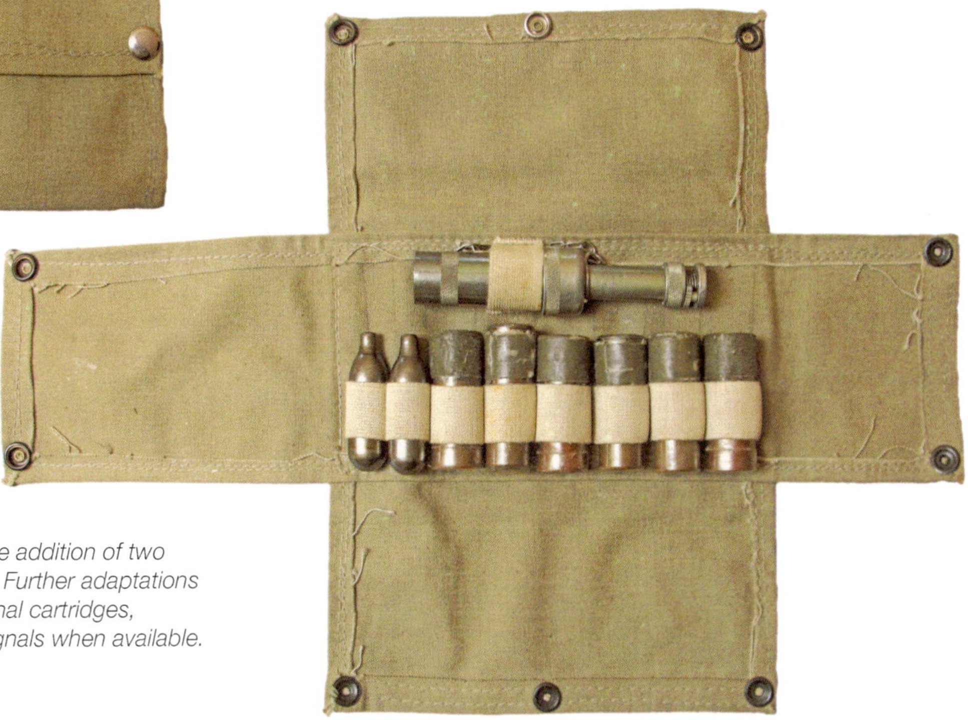

Above and right) This unit has the addition of two loops for life vest CO2 cylinders. Further adaptations would be the removal of two signal cartridges, replaced by two T-49 distress signals when available. (Paul Faulkner collection)

(Above and above right) Another method implemented was the use of a pouch that could be tied to the life preserver. This packaging was fairly efficient being just slightly larger than a standard life jacket dye marker.

(Far right) Two Avenger crew members ARM3/c T.E. Taylor and Ens. Homer P. Kallsen of VC-12 each wear a distress signal kit, USS Core April 1945.

(Right) A VC-19 aircrew member is utilizing a distress signal kit attached to his life preserver aboard the USS Guadalcanal, August 1944. (National Archives)

(Right) A member of VB-83 has a Mark-IV Projector secured to his sea marker packet aboard the USS Essex, May 1945. He also carries four sea markers in addition to a whistle and life preserver light. (National Archives)

(Above) VMF(N)-553 Capt. Robert Baird USMC is congratulated by General Louis Woods for holding the distinction of being the only Marine night fighter ace. Secured to the waist belt of Baird's flight suit is a Mark-IV Hand Projector. (National Archives)

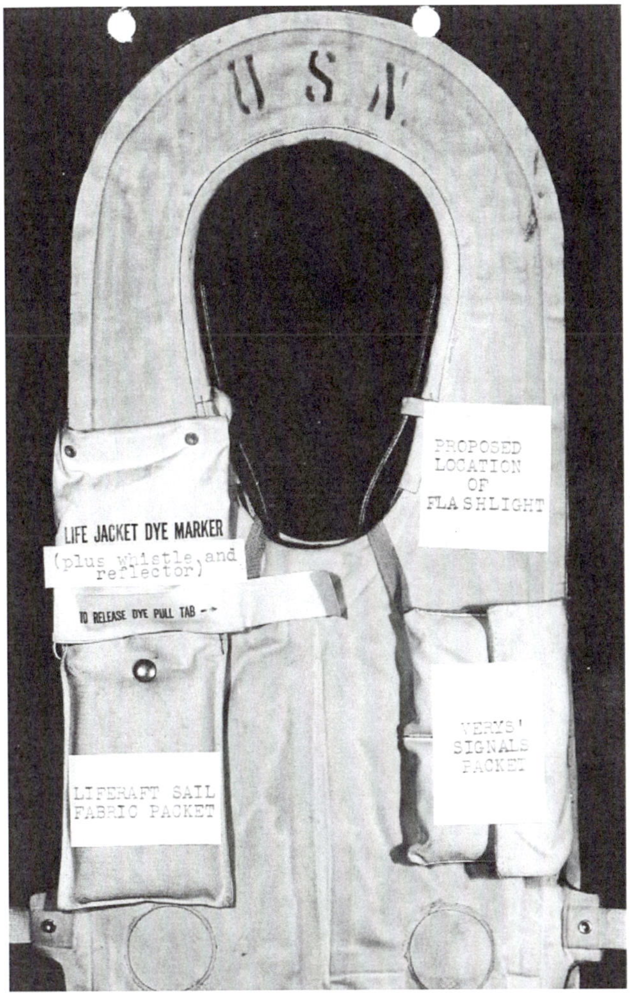

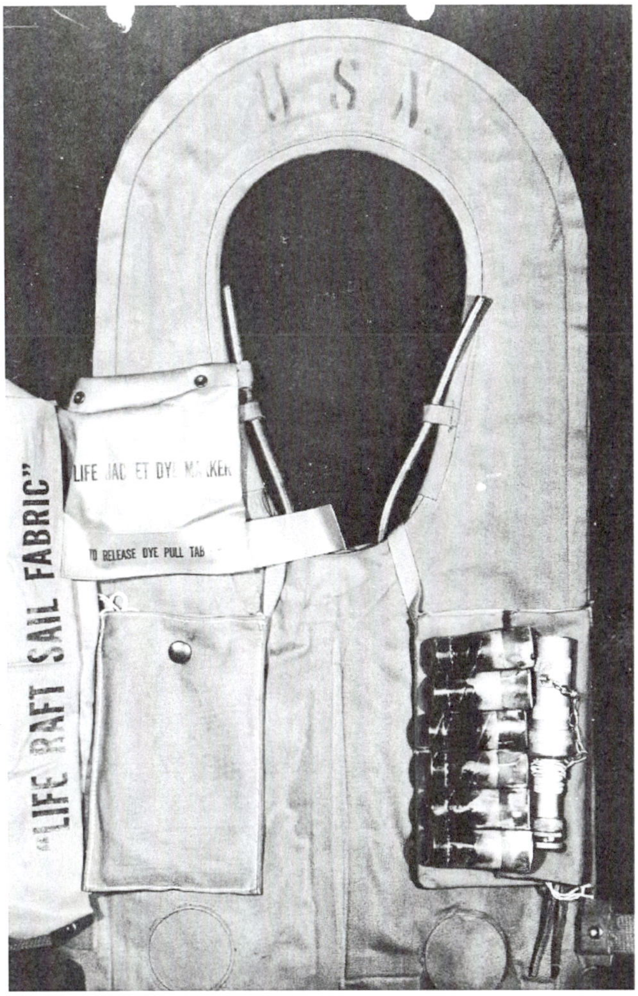

(Above) In September 1943, the ordnance section of NAS San Diego submitted a method for attaching signaling devices to life jackets, to better increase the chances of rescue of aviators adrift in only a life jacket. This configuration was field-tested for six weeks with positive results, and it was stated that squadrons operating in the area had adopted this principle concept. The flashlight, sea marker and projector were newly introduced articles at this time for the BuAer. (National Archives)

(Below) In August 1942, Carrier Replacement Group 10, which additionally field tested the Blair-Flare, suggested that the hand projector and up to five signals be contained in a web pouch on the pistol belt. No further consideration was given to this concept by the BuAer at that time, but as the emphasis on personal equipment progressed, the concept was revisited. Shown is a locally fabricated distress signal kit worn by TBF/TBM Avenger radio operator Ross H. Morgan who served with VC-8, VC-15 and VC-58 aboard the USS Tulagi, USS Croatan and USS Core. Designed to be worn on the belt, the kit contained a penlight, projector kit and MK1 Mod-0 smoke signal.

(Left) VF-41 pilots aboard the USS Ranger undergoing a morning briefing on air strikes to be conducted over North Africa, November 1942. Two pilots have modified M1936 pistol belts with loops containing ten-gauge Very's signals. A MK-III Projector can be seen in the last loop on the pilot to the right. (National Archives)

(Below) The two principle signaling devices utilized by the BuAer: the AN-M8 pistol and Mark-IV hand projector. The AN-M8 was utilized in aircraft for communication and recognition purposes. The hand projector was reserved for use as an emergency signaling device utilized in emergency kits, life rafts, and issued to individual flight personnel. (National Archives)

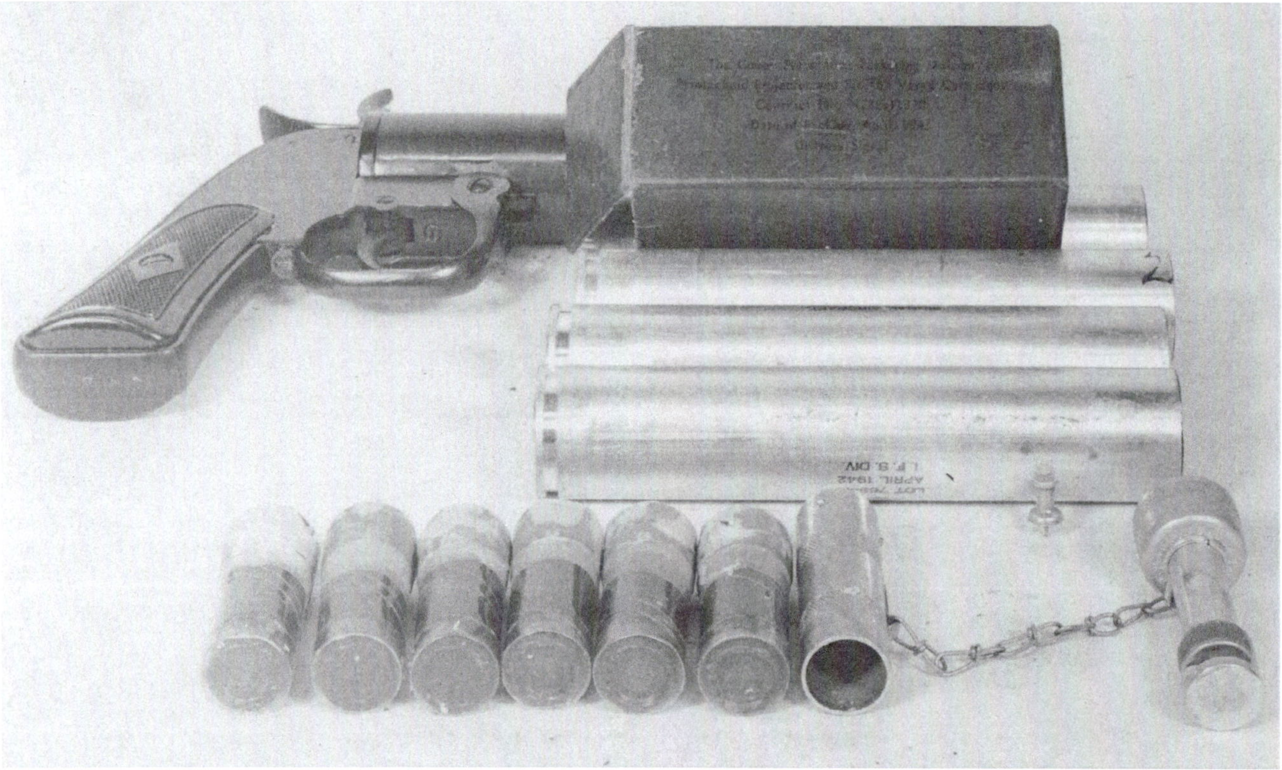

At the time of adoption of the hand projector for the BuAer, the USAAF became interested in a small distress signal utilized by the RAF. The introduction came by way of the adoption and use of the RAF Type K one-man dinghy. Part of its standard compliment was 6 small hand held signals 5" in length and 1-1/4" in diameter, designated Signal, Distress, Two Star, Red, Mark-I. The self-contained signal would eject two red stars; once ignited, the first star would launch after a delay of two seconds, followed four seconds later by the second star, reaching an altitude of 50 feet.

As early as January 1943, the 8th AF Command recommended that these signals be included in the USAAF one-man life raft, to be furnished for signaling at night, at a ratio of six like that of the British counterpart. Recommendations were forwarded to the Materiel Command in the continental United States for future consideration.

The Type K Dinghy was the principle one-man raft in use by American units operating in the ETO, and these were already equipped with the signals. In 1943, American one-man rafts reaching the ETO were either replaced by, or modified to, include the articles of the Type K Dinghy. Following the modification of the American one-man raft in June 1943 for attachment to the parachute harness, and the increased requirement for bombardment aircraft, the inclusion of this type of distress signal was emphasized. 60 samples of the

British signal Mark-I were received by the Materiel Command in August 1943, for evaluation by the Air Ordnance Officer to assist in development of a suitable copy.

The Materiel Command evaluated the British signals through the rest of 1943, deciding to incorporate them into the Type C-2 pararaft then in development, making room by eliminating the sea marker can. The raft unit would be supplemented by the packet provided as standard for all life vests.

Following the authorization for manufacture of the Type C-2 life raft, the Materiel Command established a requirement for a quantity of four two-star distress signals per pararaft in April 1944.

Initial deliveries of the Type C-2 life raft were earmarked for the ETO, and upon arrival they were to be supplied with the British Mark-I from local depots or RAF Stores. This was an interim measure to meet service requirements until signals of this type were manufactured in the United States.

The signal being developed under Ordnance Dwg. No. 78-0-82 was designated Signal Distress, 2 Star Red, T-49. T-49 was an Army Ordnance designation with the "T" denoting it as a test item.

(Left) The background to the British two-star distress signal is not known, however its original designation was Mark I, available under RAF Stores Reference 12D/551. Subsequent revisions were made to the signal internally, with each given a new mark number. The outer container remained unchanged. The Mark II was available under Stores Ref. 12D/670. The signal pictured is that of the Mark III with a manufacture date of June 1945, Stores Ref. unidentified. The RAF standard was to supply six signals, packaged in various containers, along with emergency equipment. On the right is an example of a round tin type.

Production of the T-49 commenced in August 1944, with awards to two contractors, the Kilgore MFG. Co. (International Flare-Signal Division) and the Unexcelled MFG. Co. Deliveries were expected in November-December 1944. Contracts were awarded through US Army Ordnance, available under stock number S005-90-28015.

By late 1943, carrier based night fighter squadrons were being introduced to the combat theater in the Pacific, which in turn established new requirements, specifically signaling devices for night or low visibility. As night operations continued through 1944, the problem of locating survivors became a growing concern and it was apparent that a personal signaling device was needed. Recommended procedure was for the individual to have the hand projector pre–loaded, on his person and at his immediate disposal in the event of a ditching or bail out. Also to have several rounds of tracer in his side arm. The BuAer emphasized individual equipment, as illustrated, with the Mark I Mod-0 smoke signal.

The BuAer became extremely interested in the T-49 to address immediate service requirements regarding the issue of pyrotechnics for individuals, and by December 1944 they had negotiated a requisition of 25,000 signals to be allocated to carrier based units. The signals were to be issued two-per person, for all personnel engaged in night flying.

The T-49 was reaching Naval supply points in January 1945 with all 25,000 being in service by April of that year. The bulk, 22,000, were expedited to Hawaii to make them available to carrier personnel as soon as possible. The remaining 3,000 were shipped to Norfolk.

The new signal was found to be highly satisfactory, especially over the hand projector and Very signals in terms of weight and space limitations. In March 1945, the BuAer established a requirement for an additional 200,000 more units, to be allocated from US Army Ordnance contracts.

By this date, a combined night and day distress signal was under development, which would alleviate the bulk of having four separate apparatuses while offering the benefits of both the hand-held smoke and star distress signals.

Within the BuAer, the T-49 was issued directly to the individual to be carried as personal equipment. The T-49 was eventually standardized by the Joint Aircraft Committee, 25 May 1945, as the Signal, Distress, Two Star, Red, AN-M75. The T-49/M75 service life was short-lived, being superseded by the MK-13 Mod-0 combination signal in 1946. Total production of the signal was 1,179,000 with 129, 000 of those produced in 1944.

(Right) P-47 pilot Maj. W.C. Beckham of the 351st FS has two red star distress signals attached to the waist strap of his Type B-3 life preserver. England c.1944. (National Archives)

(Below) Two views of a signal manufactured by the Unexcelled MFG. Co. measuring 5"in length and 1-1/8" in diameter. The ignition sequence was the same as the British counterpart but differed slightly in time delay. Upon ignition, the first star launched after five seconds, followed by the second star in seven seconds, reaching an altitude of 125'.

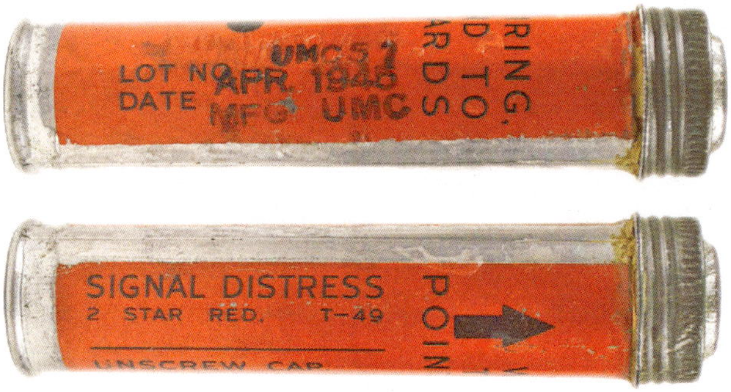

(Far right) Two views of the signal manufactured by the International Flare Signal Division of the Kilgore MFG. Co. The initial 25,000 units delivered to the BuAer between January and March 1945 were allocated through this contractor. No delivery records have been located for the additional 200,000 allocated to the BuAer, which could have been from either contractor.

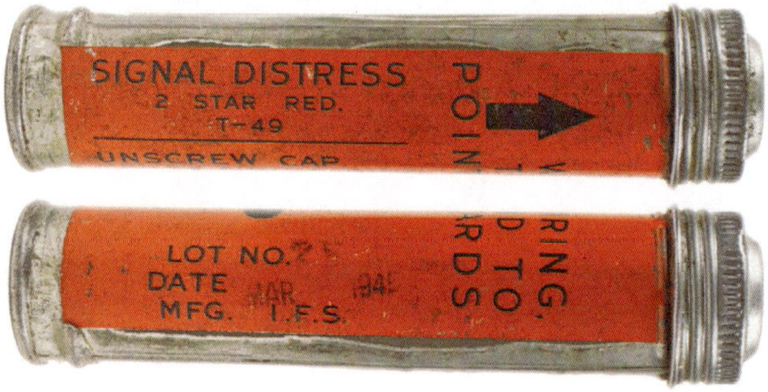

(Right) This signal has had the designation T-49 blacked out and over stamped with M75.

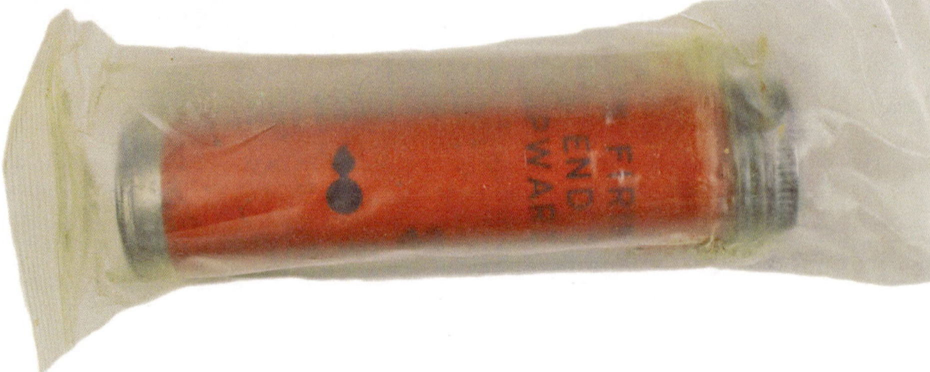

(Right) As per specifications, each signal was sealed in a Pliofilm bag or a treated paper wrapper).

(Left and above) Pockets for the T-49 have been applied to these USAAF contract life preservers as outlined in Technical Instruction TI-252 of March 1945 and Technical Order 13-1-31 of July 1945. Both orders describe the attachment of two pockets for either the hand-held smoke Mark-I Mod-1 or the T-49 distress signals. The pockets detailed in those orders were to be universal for either signal. Pockets shown here were made only to accommodate the T-49. BuAer Technical Note 15-45 was utilized prior to their release. The pockets on the vest pictured above are placed as prescribed in T.O. 13-1-31. Only one pocket has been applied to the vest pictured to the left, but note the addition of a tab for the attachment of the life preserver light.

(Above) Mark-I Lights were standard Bureau of Ships signaling devices utilized on life boats and rafts. Available in two colors, red and blue, they were stored in containers containing six or twelve lights. Red lights were utilized by the BuAer (since red is the international color for distress) and often used to supplement pre-war life raft equipment as a night signaling device, but useful in daylight due to the brilliant burn. They were essentially a road flare with a wooden handle, ignited much like striking a match. Attempts were made to include these lights in life rafts prior to the introduction of the hand projector, but they often failed to function after exposure to moisture, and could emit sparks that burned holes in the rubber bladders. They were used only for lack of a better option.

In April 1941, during the development process for the Type B-2 emergency parachute kit, it was requested that a fusee be developed for inclusion. What the Materiel Command adopted was a standard commercial 15-minute road flare, 13" long with a 1" diameter, however during field testing it was found to make the kit too rigid and uncomfortable.

Following standardization of the B-2, a new fusee was adopted on 4 July 1942, Dwg. No. 43B3554 titled Flare, Parachute Emergency Kit, Signal. This new five-minute flare measured 6-5/8" long x 1" diameter, and was available under stock number 8300-215250, with shortened nomenclature: Flare, Signal. One flare 43B3554 was included in the Type B-2 jungle kit, two flares 43B3554 in the Type B-4 parachute kit and the Type C-1 sustenance vest.

(Above) The 15 minute fusee flare utilized in USAAF parachute emergency kits. This example is a commercial type by the Chemurgic Corp. The metal spike allows the flare to be impaled in the ground to hold it vertical.

(Right) The five minute fusee has a commercial appearance but displays the month and year of manufacture.

DISTRESS SIGNALS

Part IV. Signaling Mirrors

Prior to the United States' entry into WWII, there was no real collective or exchange of information between services, with each seeing to their own requirements. Upon entering WWII, the implementation of the signaling (or heliograph) mirror was proprietary to the nautical services, namely the BuAer (US Navy) and the Merchant Marine.

The BuAer was the first aeronautical service to include a reflective device as standard for life raft emergency equipment, in November 1939. The mirror included was contractor-furnished equipment that was to be fabricated as detailed in NAF Dwg. No. 214299, approved 15 September 1939. This mirror was described in NavAer Specification M-3N, which simply stated, "The reflector shall be 4" in diameter made from .010 gauge brass and cadmium plated," with no other details for a sighting or a lanyard hole.

Subsequently, the mirror was revised in February 1940 to be of a thicker .030 gauge brass, chromium plated, with the addition of an 1/8"sighting hole along the outer edge, and contained in a chamois bag.

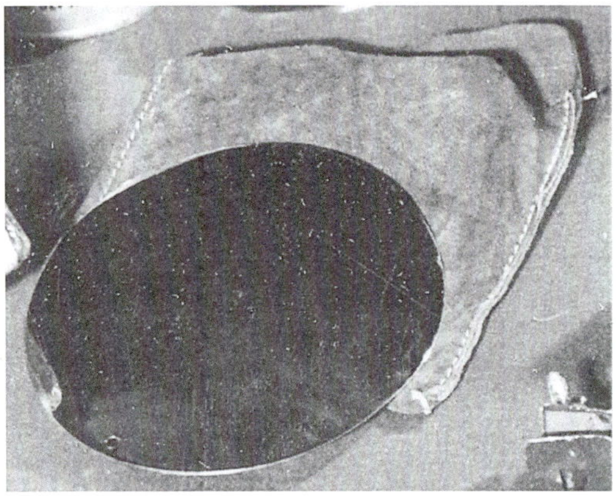

(Left and below) First generations of the round chromium mirror only included a 1/8" sighting hole located at the edge of the mirror. A variety of chamois or leather bags were provided of a basic rectangular construction. This reflector would remain standard until being officially superseded by the emergency signaling mirror NavAer Specification M-580. (National Archives)

(Left and below) Based on service requirements for personal signaling devices, the NAF began fabrication of a metal reflector for aeronautical stock and replenishment purposes in 1942. The mirror and bag were produced under NAF Dwg. No. 214299 and typically stamped under the flap of the chamois bag with Assembly Dwg. No. 214299-2. The chamois bag shown is as detailed in Dwg. No. 214299, conforming to the round shape of the mirror. This reflector was available under ASO stock number R83-R-20500 titled as Reflector, Chrome Plated. By late 1942 it was found that a center sighting hole was more effective in targeting potential rescuers, and incorporated into the NAF reflector with a diameter of 3/16", with the 1/8" hole along the edge repurposed for a lanyard. In February 1943 it was prescribed through the Naval Aviation Newsletter that the mirrors provided in life rafts from contractors were to be updated to the center sighting method. In March 1942, the USAAF Engineering Division initiated action to provide signaling mirrors for all life rafts, and by the end of that year, NAF reflector stock number R83-R-20500 was being provided as supplies permitted.

(Below) The signaling mirror shown has a cotton cover, consistent with the type provided in the Original back pad kit.

(Left) USMC Pilot 1st Lt. Charles Bourget of VMSB-235 wears a round signaling mirror secured to his life preserver. By late 1943, it became a typical practice to remove the signaling mirror and whistle from the back-pad kit having them in a more readily available location.

(Right) Two Marine Corps pilots wear round reflectors secured to their life preservers, c.1945. (National Archives)

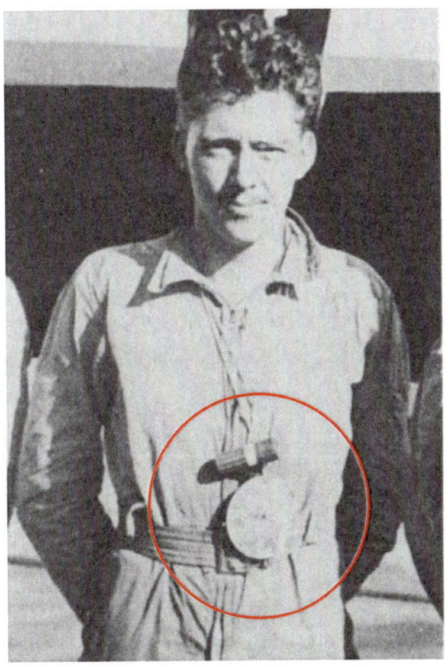

(Left) Ensign W.J. Miller of VT-10 has a round reflector and match safe M-554 draped around his neck. Miller and his two crew members were forced to ditch in their TBF Avenger after conducting supporting strikes in the Hollandia area, New Guinea, on 21 April 1944. They were rescued by a Kingfisher float plane from the heavy cruiser USS Boston.

(Right and below) An example of a metal mirror approved by the USCG for use on Merchant Marine vessels. Regulations stipulated that two mirrors were to be provided for each life boat for ships over 3,000 gross tons, and an instruction sheet was also to be provided. The mirror illustrated here is of polished stainless steel measuring 5" x 6". The packaging method from this company wrapped the mirror in tarred paper then sealed it in a wax coated carton.

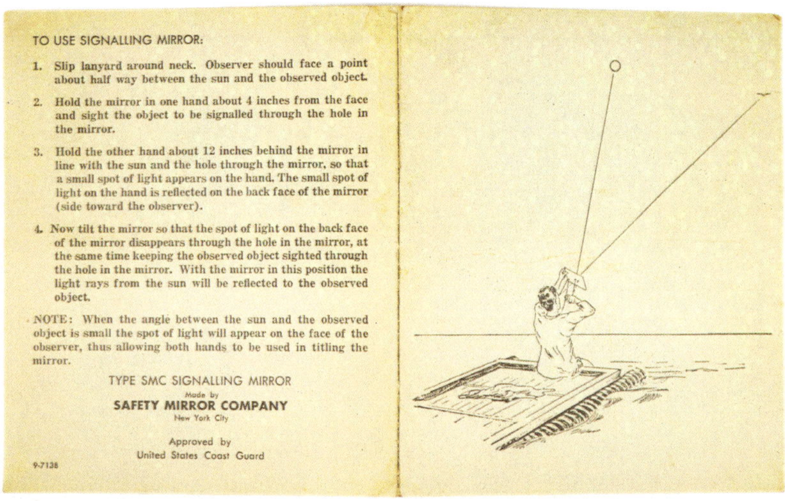

TO USE SIGNALLING MIRROR:

1. Slip lanyard around neck. Observer should face a point about half way between the sun and the observed object.
2. Hold the mirror in one hand about 4 inches from the face and sight the object to be signalled through the hole in the mirror.
3. Hold the other hand about 12 inches behind the mirror in line with the sun and the hole through the mirror, so that a small spot of light appears on the hand. The small spot of light on the hand is reflected on the back face of the mirror (side toward the observer).
4. Now tilt the mirror so that the spot of light on the back face of the mirror disappears through the hole in the mirror, at the same time keeping the observed object sighted through the hole in the mirror. With the mirror in this position the light rays from the sun will be reflected to the observed object.

NOTE: When the angle between the sun and the observed object is small the spot of light will appear on the face of the observer, thus allowing both hands to be used in titling the mirror.

TYPE SMC SIGNALLING MIRROR

Made by

SAFETY MIRROR COMPANY

New York City

Approved by

United States Coast Guard

9-7138

(Left and below left) USCG approved metal mirror. Chrome plated and measuring 5" x 7". Two mirrors were packaged in foil lined Kraft paper and provided with a two-sided instruction sheet. Due to War Production Board restrictions, brass was often not utilized in metal mirrors for the Merchant Marine with the typical types being chromium plated or polished steel.

(Below) Also provided to the Merchant Marine, this mirror featured a dimpled type sighting hole mirror. Rectangular shaped mirrors are found in many locally assembled emergency kits, typically adapted from the abundant availability of shaving mirrors or toiletry sets.

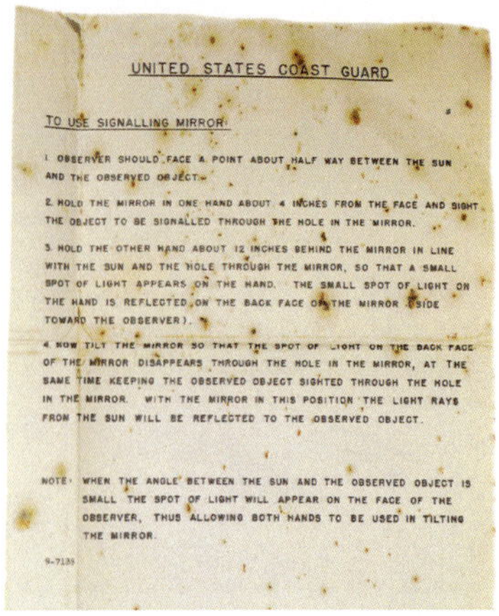

UNITED STATES COAST GUARD

TO USE SIGNALLING MIRROR:

1. OBSERVER SHOULD FACE A POINT ABOUT HALF WAY BETWEEN THE SUN AND THE OBSERVED OBJECT.
2. HOLD THE MIRROR IN ONE HAND ABOUT 4 INCHES FROM THE FACE AND SIGHT THE OBJECT TO BE SIGNALLED THROUGH THE HOLE IN THE MIRROR.
3. HOLD THE OTHER HAND ABOUT 12 INCHES BEHIND THE MIRROR IN LINE WITH THE SUN AND THE HOLE THROUGH THE MIRROR, SO THAT A SMALL SPOT OF LIGHT APPEARS ON THE HAND. THE SMALL SPOT OF LIGHT ON THE HAND IS REFLECTED ON THE BACK FACE OF THE MIRROR (SIDE TOWARD THE OBSERVER).
4. NOW TILT THE MIRROR SO THAT THE SPOT OF LIGHT ON THE BACK FACE OF THE MIRROR DISAPPEARS THROUGH THE HOLE IN THE MIRROR, AT THE SAME TIME KEEPING THE OBSERVED OBJECT SIGHTED THROUGH THE HOLE IN THE MIRROR. WITH THE MIRROR IN THIS POSITION THE LIGHT RAYS FROM THE SUN WILL BE REFLECTED TO THE OBSERVED OBJECT.

NOTE: WHEN THE ANGLE BETWEEN THE SUN AND THE OBSERVED OBJECT IS SMALL THE SPOT OF LIGHT WILL APPEAR ON THE FACE OF THE OBSERVER, THUS ALLOWING BOTH HANDS TO BE USED IN TILTING THE MIRROR.

9-7139

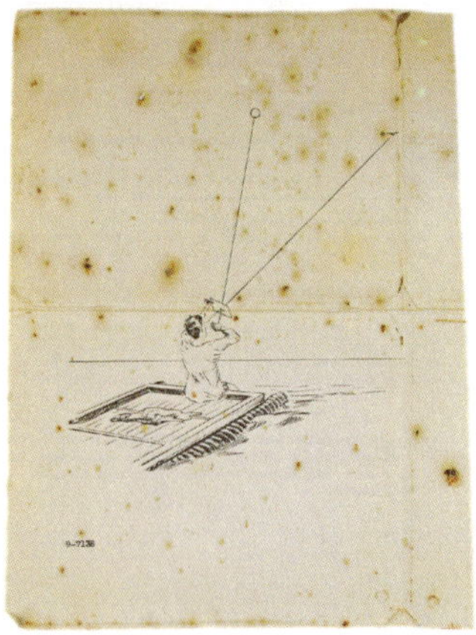

(Left) A metal mirror provided in a waterproof scrim wrapper measuring 5" x 3-1/2". This mirror accompanied the manual, "Survival Hints," distributed by the intelligence sections operating in the Southwest Pacific Area from late 1943 through the duration of conflict.

(Below) While the Americans were principally utilizing plated metal mirrors, the British RAF began with circular glass mirrors, 4" in diameter, mounted in either a brass or Bakelite frame with a leather strap for securing to the life vest, in late 1940. These were an emergency response during the Battle of Britain, when so many fighter pilots finished up in the English Channel. Shown (left) is the Bakelite framed model and (right), the brass version. These were later carried by American flyers who were members of the RAF Eagle Squadrons, but there is no record of their use by the AAF or USN. Both versions were labeled with the same RAF Stores Reference Number of 6E/271. Mick Prodger collection

(Left) RAF pilots in the Western Desert. The central figure wears a Bakelite framed circular glass signal mirror attached to his "Mae West" life vest.

(Below) By the end of 1942, the British Air Ministry had adopted a polished stainless steel type, available in a 2" and 4" size, with the latter being preferred, but when space restrictions prevailed the 2" was used. American mirrors incorporated a rear sight method, whereas the British used a foresight method, which assisted alignment with a target by way of a key held in the forward position. Shown is the 4" mirror in use.

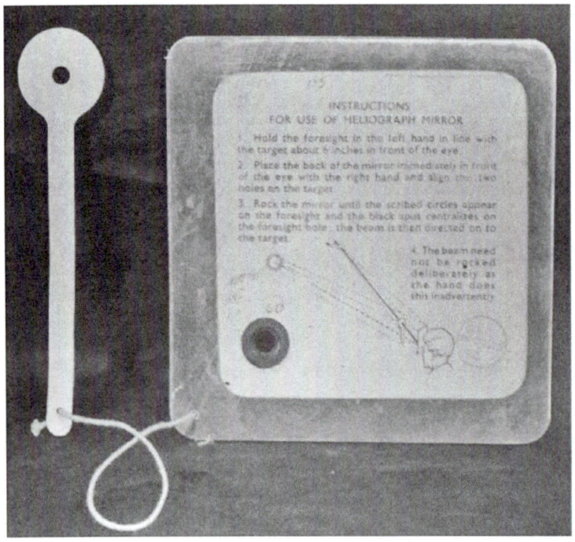

(Below) The 2" model was standard for the RAF Type K Dinghy and the 1941 Pattern life preserver, with the 4" for multi-person emergency equipment. Both sizes used mirror and key tethered together, with printed instructions on the back of the mirror, all contained in a cotton pouch. These mirrors would have been used by USAAF personnel operating in the ETO.

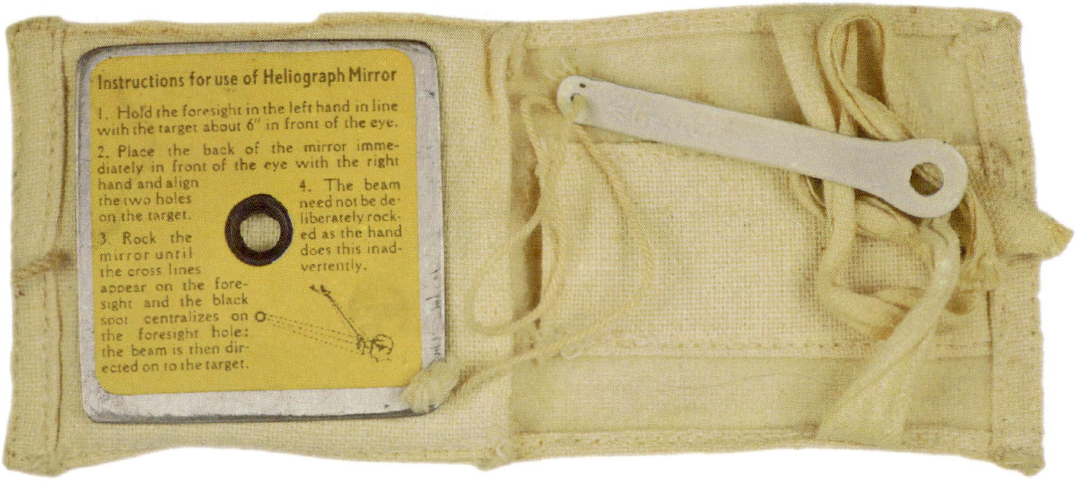

(Left) In January 1943 the Hawaiian Air Depot at Hickam Field, submitted a proposed modification for the Type B-3 life preserver for the installation of a signaling mirror. This modification was in response to demands from combat units for a personal signaling device. The modification was referenced under their Dwg. No. XHAD43D224, which detailed that the threads be removed from the top half of the leather reinforcement patch to create a pocket. The pocket was to be secured close by a glove snap to prevent the loss of the mirror and stenciled with MIRROR INSIDE. The mirror provided was locally fabricated, with a lanyard and a looped parachute shock cord to secure around the wrist when in use. The submitted drawing to the Materiel Command proposed that it be adopted for all life vests and to be incorporated into production. (Paul Faulkner Collection)

(Above) USAAF Major C.E. Taylor wears an AD43D224 Type B-3 modified life preserver on Makin Island c.1943. (National Archives)

(Left) Commander of the 20th Reconnaissance squadron, Maj T.A. Ahola wears a AD43D224 Type B-3 life preserver. (National Archives)

(Left and above) The mirrors provided for the modification were polished sheet metal, treated with a protective coating of lacquer and green paint on the reverse. The mirror measured 4" x 2-1/2" with a 1/2" sighting hole. Some had the Hawaiian Air Depot Dwg. AD43D224 stamped on the back

(Right) Two fabrication styles have been observed: one being a completely flat surface, and the other having a ridge along the outer edge formed by a stamping method. Both were trimmed with friction tape to prevent chaffing from the rough finished edge.

(Above) USMC Sgt. Elmer Daniels has the 43D224 modification on his B-3 life preserver. Daniels was a rear gunner on dive bombers and later served with VMTB-143 on Guadalcanal again serving as a gunner in TBM Avengers. (Courtesy Adam Lewis)

(Right) Hickam Field submitted Change-A to the Materiel Command in May 1943, which increased the pocket size by adding a 1-1/4" leather strip and placing the snap to the corner, as the mirror was found difficult to remove on the original pattern. (Paul Faulkner Collection)

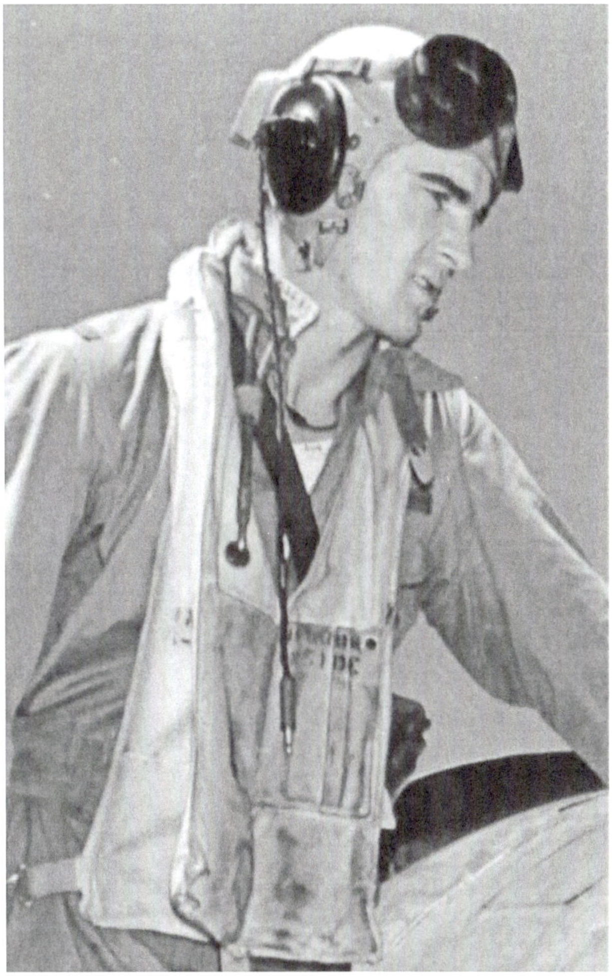

(Above) Fellow aircrew members of VT-16, Lt. jg William McCarthy and Paul Bonilla AOM2/c, are wearing Change-A B-3 life preservers aboard the USS Lexington, November 1943. (National Archives)

(Right) USAAF P-47 pilot Lt. Stanley J. Lustic wears a Change-A Type B-3 life preserver on Saipan, July 1944. (National Archives)

(Right) An additional modification on Type B-3 life preservers for a signaling mirror include a cotton patch sewn to the outer shell.

(Far right) Lt. Cdr. J.H. Reidy of VBF-83 is wearing such a modified life preserver aboard the USS Essex, August 1944. (National Archives)

(Left) As the B-4 life preserver became available in the combat theater, it too was modified as per Dwg. AD42D224, but due to the construction method the pockets were sewn to a rubber patch then cemented to the vest. There were two variations of pockets: one with the mirror tethered by a length of cord to a grommet located on the pocket as pictured. The other style utilized the original mirror, as designed, with its shock cord.

(Right) A variety of materials were used in the construction of the pockets, namely colors. Shown is the type that housed the shock-corded mirror. All modifications would have the stamp MIRROR INSIDE stamped on the pocket.

(Above left) USAAF P-61 crew member Sgt. George Duncan is wearing a modified Type B-4 life preserver with the shock corded mirror in Iwo Jima, March 1945. (National Archives)

(Above right) USAAF P-51 pilot Lt. John D. Grant is wearing a modified Type B-4 life preserver of the lanyard type mirror on Iwo Jima, July 1945. (National Archives)

(Left) Use of modified life preservers circulated service-wide in the PTO and can be seen worn by both US Navy and USAAF flying personnel. A multitude of examples exist. Pictured is a B-3 with a patch pocket worn and named to J.E. Downing GM2/c, USN. It was not an unusual occurrence for bridge officers and deck personnel to wear inflatable life preservers.

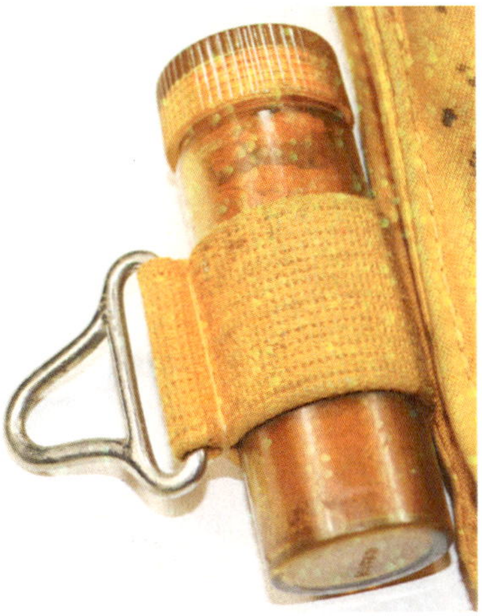

(Above) Another modified AD43D224 B-3 life preserver. The mirror included with this vest (top center) utilizes the shock cord but has not been treated on the back nor does it have the friction tape applied to the edge. This vest had been additionally modified to secure a waterproof match safe, 42A6930, filled with fluorescein powdered dye. From 1942 through 1943 only cans of sea marker were readily available for the USAAF. Sea marker packets were not introduced until late 1943. As an interim measure for personal use, modifications like these were often devised. Other uses included filling with chlorine for shark chaser.

(Left and far left) Surviving examples of modified B-4 vests.

(Top left) A group of volunteer 7th AF members examine their newly issued, modified Type B-4 life preservers, prior to conducting field tests on life saving equipment on Oahu Hawaii, August 1944. (National Archives)

(Above) USAAF aircrew members Lt. William J. Verbeck, Lt. Robert E. Bemis, Sgt. Ralph V. King, and Sgt. Albert A. Pettifer, wearing modified Type B-3 and B-4 life preservers. (National Archives)

(Left) B-24 Liberator aircrew members of the 42nd BS, 7th AF, wearing modified Type B-4 life preservers. Okinawa, July 1945. (National Archives)

So far, the only types of mirrors provided were of a metal type, which were found to be subject to corrosion and too flexible. When bent they created distortion, reducing their reflective properties.

By late 1942, Maj. William Donovan of the OSS submitted a proposal to form a coordinated committee on emergency rescue equipment, based on the recommendations of Dr. Henry Field and Maj. Harold Coolidge (who worked on initiating the development of a shark repellent). Field and Coolidge conducted much of the leg work, communicating with many agencies, specifically Larry Young of the National Bureau of Standards. Young presented a rear sighting glass mirror which improved aiming capabilities. Being glass it was corrosion resistant. The glass mirror was developed under a coordinated effort between Young and the General Electric company between October 1942 and March 1943.

The development of the tempered glass mirror was being overseen by a committee comprising a multitude of government agencies, spearheaded by the US Navy Department under the Office of the Coordinator of Research and Development.

In March 1943, the National Bureau of Standards drafted tentative specifications requiring three different sizes, 2.5" x 4", 4" x 5", and 8" x 1 0". It was believed at the time that the amount of surface area related to the strength of the reflective properties. This theory was proved false by the British, when they tested the American glass mirror in conjunction with their 2" and 4" polished stainless steel models in February 1943. The mirror tested was the 4" x 5" type, which proved to have a greater brilliancy over their polished steel mirrors, but the sighting range was equal to that of the 2" heliograph.

By the time of the drafted specification, all agencies that incorporated signaling mirrors were interested in the new glass type, due to many problems encountered with inferior styles being provided by contractors essentially cutting corners.

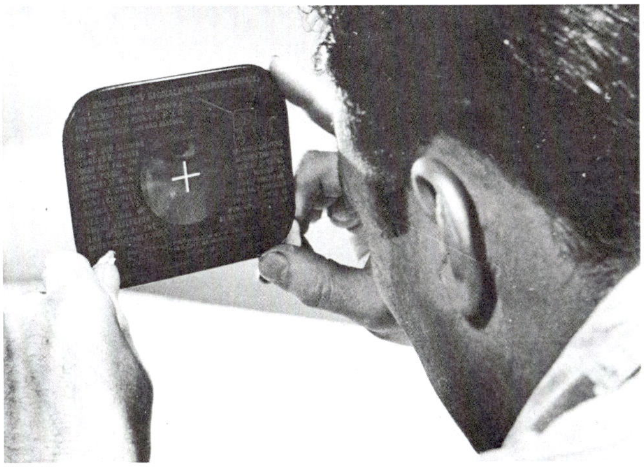

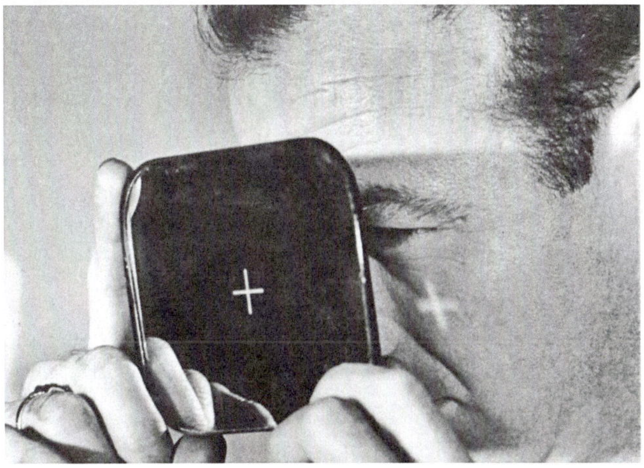

(Above) Young's mirror consisted of evaporated aluminum on tempered glass, with a cross-shaped viewing hole in the center. The user would know the mirror was reflecting at the target when the sun shining through the cross could be seen on their person,via the mirror on the back side.

(Right) The results of the field tests concluded that the glass mirror was far superior to the metal types, with sightings up to 11 miles both by aircraft and surface vessels. (National Archives)

The glass tempered mirror was universally adopted by all nautical and aeronautical agencies requiring this type of emergency equipment. In June 1943, the newly established Liaison Committee on Emergency Rescue Equipment recommended the glass tempered mirror for adoption, but the mirror was already in production with several agencies having already drafted their own specifications and in preparation for standardization.

The BuAer standardized the glass tempered mirror on 28 June 1943 as NavAer Specification M-580. Three sizes of mirror were specified, as outlined in the tentative draft by the NBS, but only the Type-II, 4" x 5" was procured. This mirror would also be unofficially referenced as the Mark-I in post war literature, and by the General Electric Company as the ESM/1 (Emergency Signaling Mirror). The larger 8" x 10" (Type-III) mirror was to be procured by the Bureau of Ships for life boats but, with the Type-II proving as efficient, they too only procured the 4" x 5" size.

One recommendation was to have printed instructions on the back of the mirror versus a separate sheet. Recommendations were made to provide instructions for the BuAer round chromed mirror but none were printed. Some local units printed material to include with life raft equipment.

By August 1943, the BuShips had 200,000 mirrors on procurement and 20,000 by the BuAer. This would just be an initial procurement for those two services, with production increasing to 80,000 mirrors a month by the end of the year, delivered to all procuring agencies. Production of the ESM/1 mirror would exceed 2 million by VJ Day.

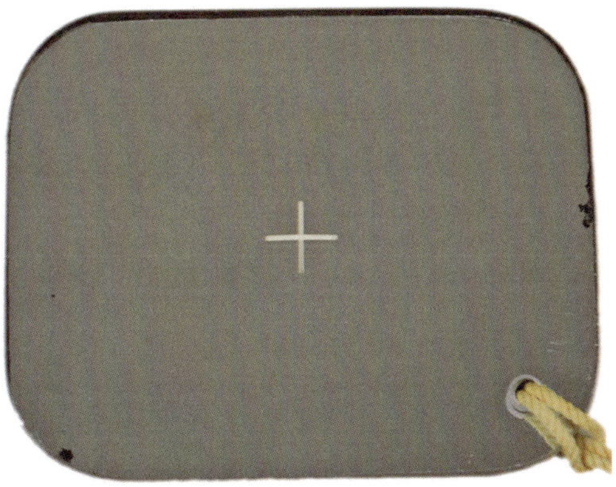

(Left and top of following page) The three known variations of the M-580 or ESM/1 mirror as procured by the BuAer and BuShips. The initial production (left) included the words PATENT APPLIED FOR. The M-580 was available under ASO stock number R83-M-525525 as Mirror, Signaling.

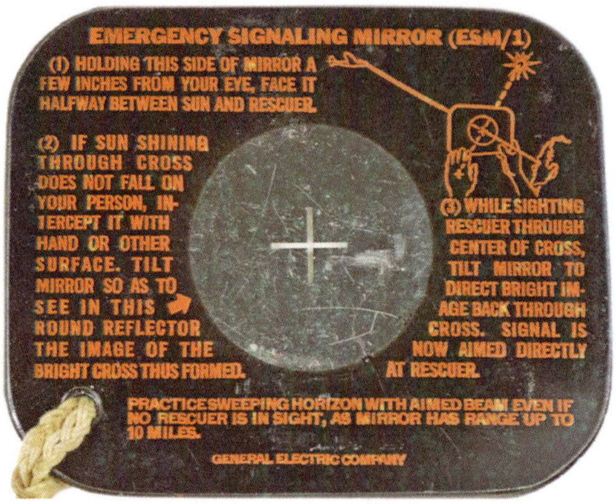

(Far left) The words PATENT APPLIED FOR were eliminated from the M-580 after Larry Young was assigned Patent No. 2395605 on 7 December 1943.

(Left) The last production mirror dropped the manufacturer's address. The ESM/1 was officially patented on 26 February 1946 – after it was no longer in production.

(Above) Packaging was a factor after the problems encountered with the metal type mirrors and with the US Navy regarding waterproofing. The mirrors delivered to the Navy would have a cardboard master carton with the mirror sealed in foil lined Kraft paper. The mirror was packaged with a single, flat piece of cardboard for padding.

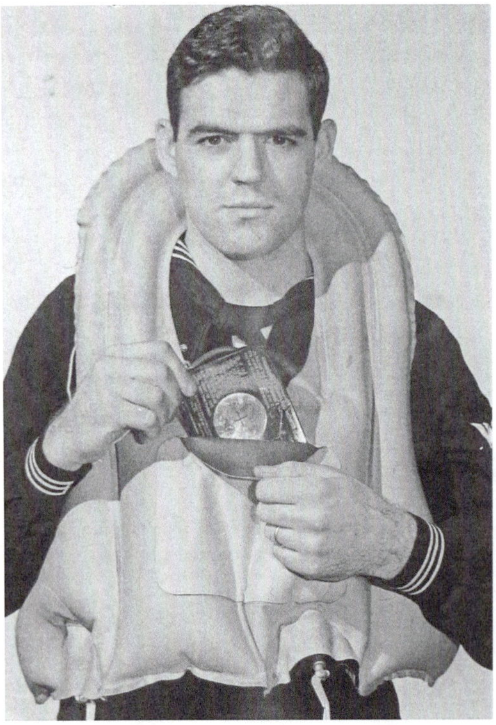

(Left and far left) In line with providing personal signaling devices, a proposed modification to include the tempered signal mirror as a standard for life preservers was presented in November 1943, from an undisclosed Naval Air Station. The simple application of a rubberized fabric pocket was to be attached to the life preserver to assure all pilots or aircrew had the means to signal potential rescuers. (National Archives)

(Above) A Naval airman utilizes an M-580 signaling mirror for field tests in the spring of 1943. (National Archives)

(Below) A USAAF serviceman demonstrates the use of the signaling mirror aboard one-man life raft, c.1944. As of August 1943 the USAAF had 275,000 ESM/1 signaling mirrors on procurement with the bulk of procurements allocated to them from 1944 through 1945. (National Archives)

(3 pictures below) The USAAF standardized the tempered glass mirror, 15 June 1943, under their Specification No. 94-40653 and Dwg. No. 43B21574 available under stock number 4500-580650. Illustrated is the lineage of the USAAF procured ESM/1. Like that of the Navy's M-580, the initial production would include PATENT APPLIED FOR, later eliminated. In addition, Army Air Force mirrors included their specification number. As can be seen, there is some variation in the printing of the manufacturer's name in the second and third mirrors. It should be noted that Pittsburgh Plate Glass Company was also involved in the manufacturing process, and the mirror was also exported to Canadian forces.

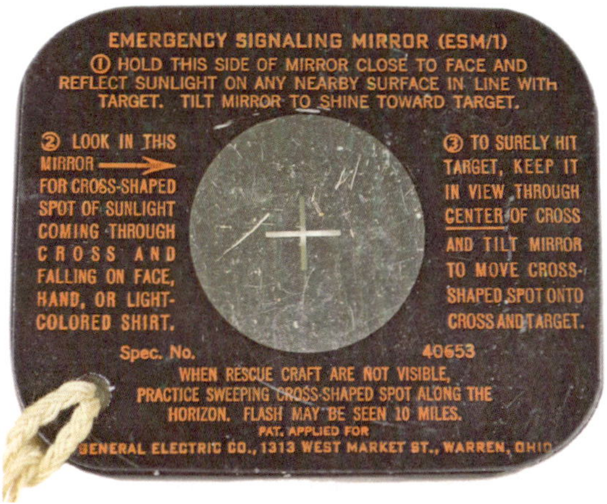

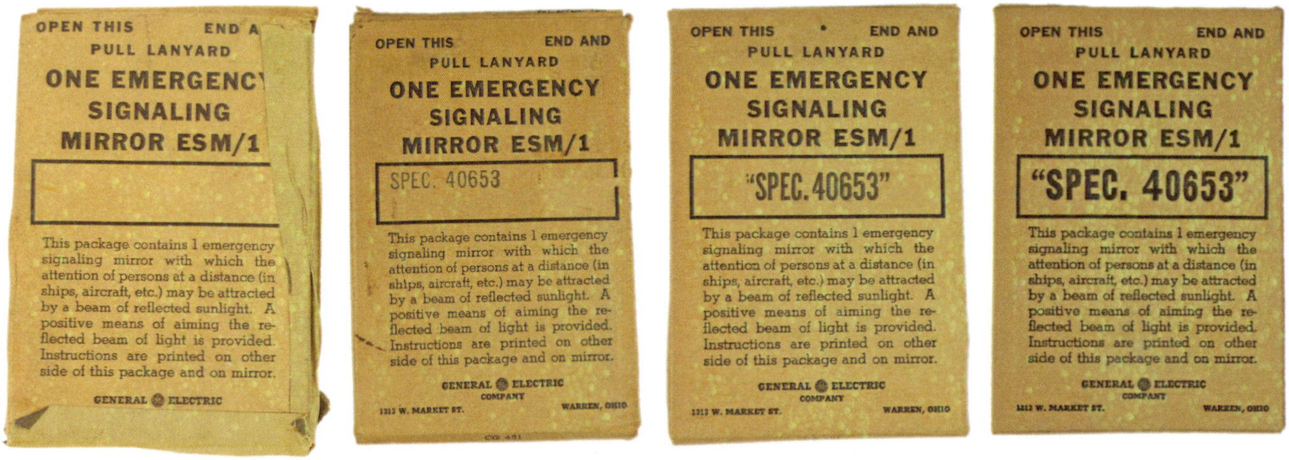

(Above) The development of packaging for the USAAF signaling mirror. The carton on the first deliveries for both the M-580 and 40653 would include a blank bar on the front for adding nomenclature later. For the USAAF the specification number was first hand stamped, then retrofitted on existing cartons, with the eventual listing of the specification number as part of the overall fabrication process.

(Left) Mirrors provided to the USAAF did not include the waterproofing application of the foiled lined Kraft paper. Instead, each mirror was packaged with a length of cardboard folded over the mirror for padding.

(Above) The USAAF amended the packaging method as part of Amendment-A 1 December 1944. They eliminated the carton and cardboard padding in lieu of a simple manila envelope. Pictured are two variations of the manila envelopes as part of Specification No. 94-40653-A. By this date the BuAer had adopted the M-580A. The ESM/1 signaling mirror was only sold to the US Government or affiliated agencies with production ceasing at war's end.

(Left) USMC Major Jefferson Dorrah of VMF-323 has an M-580 signaling mirror secured to his life preserver, Okinawa April 1945. (Courtesy Gregory Pons)

(Right) Variations of the 4' lanyard for the ESM/1 signaling mirror.

(Below) USAAF Specification No. 94-40653 covered two sizes of mirrors referencing them as Size-20 for the 4" x 5", and a Size-15 for a 3" x 5". This Size-15 mirror was known by the manufacturer's nomenclature as the ESM/2. The only known procurement and utilization was for the USAAF C-1 Emergency Sustenance Vest, and due to space requirements for the vest, the mirror was packaged in a manila envelope. The ESM/1 was the preferred sized mirror which was procured in large quantities, however further adoption in the immediate post war years would be the size of the ESM/2.

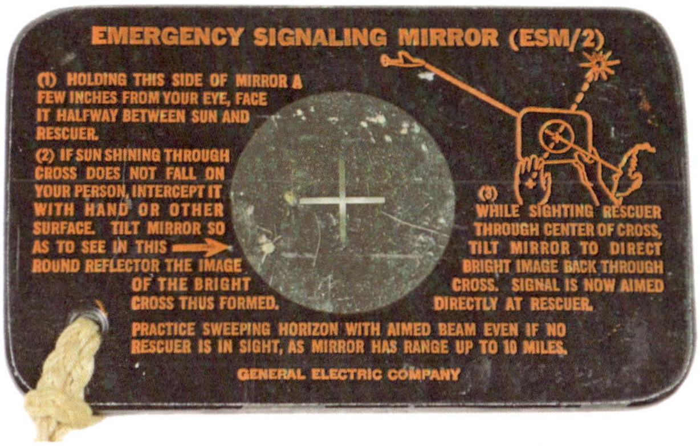

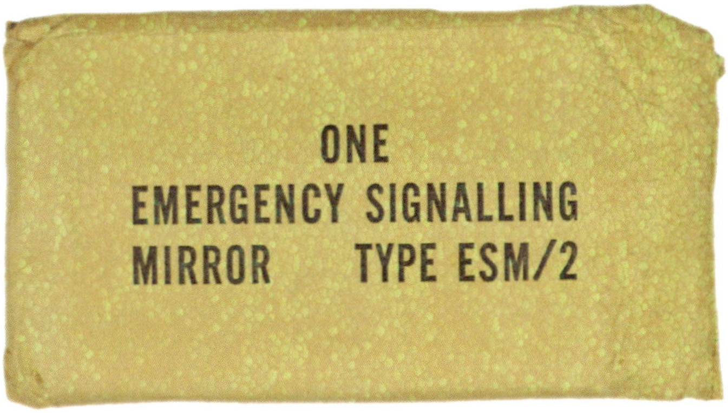

Two drawbacks were encountered with the ESM/1 signaling mirror. It was difficult to find line of sight on a moving target with an unstable base, and it essentially required ideal stationary conditions for efficiency. The ESM/1 could be viewed as an expedited standardization, decided by a joint committee and some bureaucracy, but was far superior than the metal type mirrors then in use.

In early 1943, in the middle of the development process of the tempered glass mirror, a gentleman by the name of C.H. Learned of the National Inventors Council, submitted his Reflex-Button aiming method. The rear sighted mirror required the user to aim at the target and observe the reflected light on a surface, having to concentrate on two points at once – which can be a difficult task, pitching in a life raft or life preserver.

The Learned mirror utilized a red, retroreflective button set at a 30-degree angle. When the sun shone through the window, the button reflected the light, creating a red dot on the window. The user merely held his eye behind the mirror slightly manipulating it until the red dot appeared in the window and lined up with the target. The eye no longer had to focus on two different perspectives. The retro-reflective method was better suited for less than ideal conditions, but was swept under the rug with the committee already supportive of the tempered glass type.

By late 1943, the BuAer revisited the Learned method and worked in conjunction with the Signal Service Corp. to refine the mirror, eventually standardizing it as NavAer Specification M-580A on 5 July 1944, available under ASO stock number R83-M-525520. The M-580A comprised a silver-backed glass mirror, placed in a cushioned metal frame with eight orange Stimsonite reflectors integrated on the back for night signaling purposes. The M-580A would replace the M-580 when supplies permitted.

(Right) The M-580A comprised two stamped sheet metal covers: one half included directions for use, and the other half instructions not to tamper with the reflex button. The mirror measures 3-1/4 x 4-1/2 inches and was unofficially known as the Mark-II in post war literature.

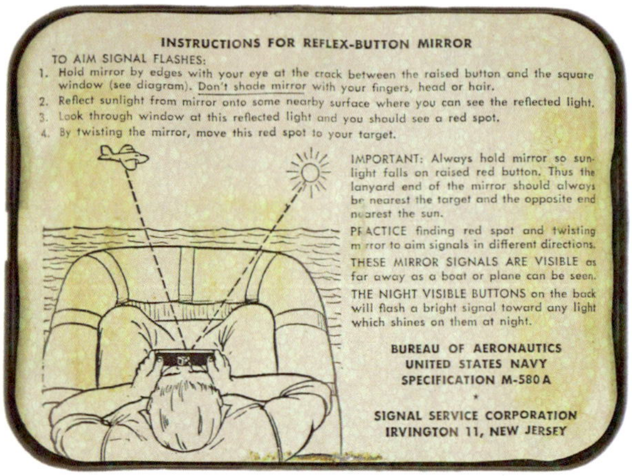

(Above) Instructions on the M-580A. Due to continued emphasis on proper use, later production units eliminated the tampering label replacing it with the instructions.

(Below) The two halves were secured to the mirror by a strip of friction tape, making for a complete assembly which was then kept waterproof in a foil lined wrapper that additionally included a set of instructions.

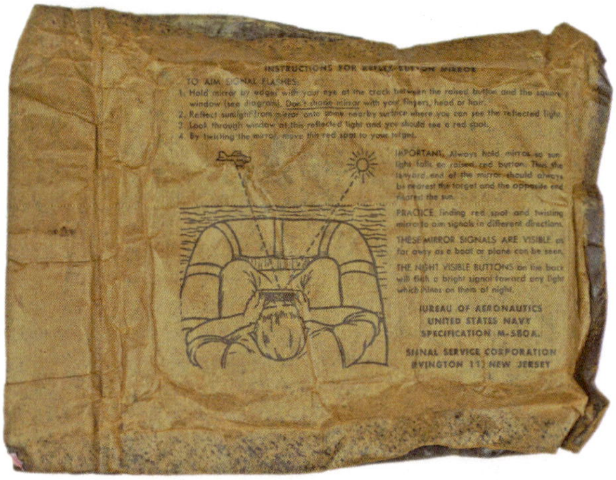

The final generation of signaling mirrors developed during WWII utilized a new material called Scotchlite. Scotchlite consisted of sheets with small spherical glass beads that would appear illuminated when reflecting light. In December 1944, tentative specifications were drafted by a coordinated effort between the National Bureau of Standards and the Minnesota Mining & Manufacturing Co. to include Scotchlite for use in signaling mirrors. A series of Scotchlite washers around the sighting hole improved the retro-reflective (or Learned) method, making it easier to use – functioning on the same principle as the M-580A.

The Minnesota Mining & Manufacturing Co., General Electric Co., and the Cruver MFG. Co. submitted samples for testing by spring, 1945. Actual quantity fabrication of the mirrors was delayed due to the use of new materials, refinement methods of Scotchlite, and plastic materials required in the manufacture.

A great deal of research and development was conducted on signaling mirrors during the war years, but they played a very limited role in real time rescue operations, and their significance faded with the introduction of hand held smoke signals, flares and the universal use of the sea marker.

(Above) The M-580A being field tested at NAS Lakehurst in October of 1944. The Learned mirror was much more user-friendly in life rafts and proved highly effective when used by individuals afloat in life preservers. (National Archives)

(Below) The USAAF standardized the Scotchlite type signaling mirror which superseded 94-40653, on 4 March 1946, as the Type B-1, Specification No. 94-41063. The Type B-1 closely resembles the ESM/2, updated with the Scotchlite retroreflective method.

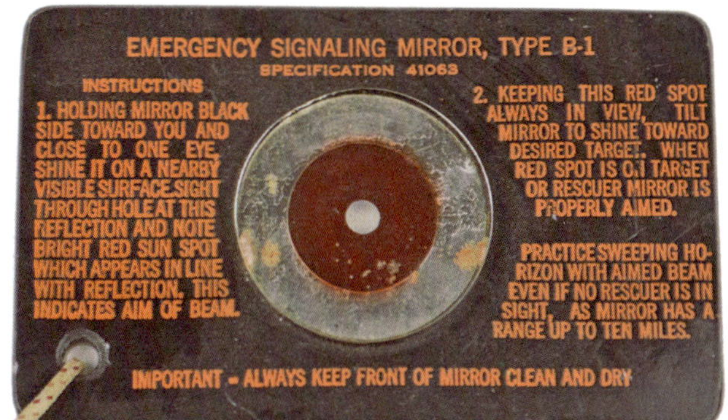

(Below) The BuAer continued the research and development of the Scotchlite type mirror through 1946, eventually standardizing their model Type Mark-III, Aeronautical Specification 23M5, 20 March 1947, superseding the M-580A.

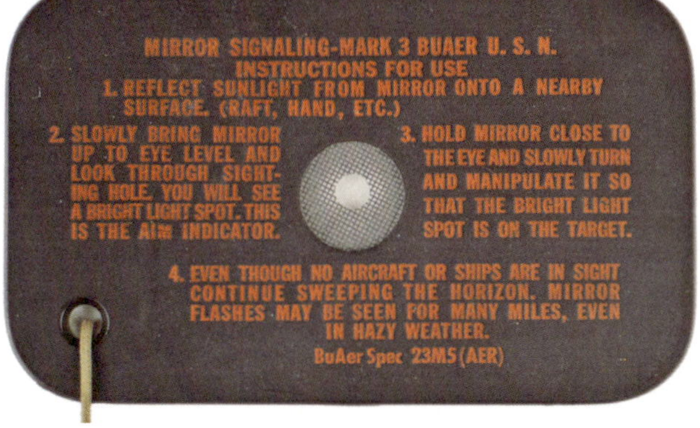

In early 1943, the BuAer investigated Scotchlite for use as a personal emergency locating device for night rescue, with a series of tests conducted at NAS Norfolk. One aspect they were looking at was the idea of buttons that could be attached to life preservers in various sizes, and used with life rafts. Following tests in June, it was decided not to pursue buttons for life preservers but adopt Scotchlite for use in life rafts.

To aid in the location of survivors, Scotchlite was applied to one side of a life raft oar, providing aircraft with extended visibility of up to 1,100'.

The BuAer drafted and released Technical Order 127-43 on 10 December 1943, to modify life raft oars in service with the retro-reflective material (Scotchlite), directing contractors to apply the material at the factory.

(Below) Various concepts that underwent evaluation at NAS Norfolk in the spring of 1943. (National Archives)

(Right) Modification kits were procured from the Minnesota Mining & Manufacturing Co. to modify and apply the retro reflective material to life raft oars available under ASO stock number R83-K-710380.

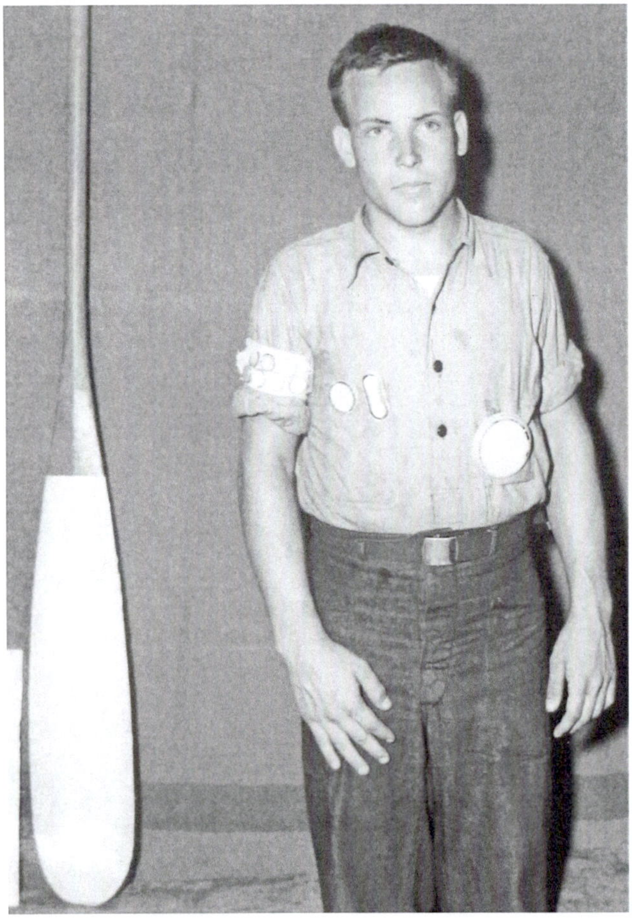

(Below) The modification kit with three sheets of Scotchlite to apply to a set of oars in multi-person life rafts. These photographs accompanied Technical Order 127-43 for the prescribed application.

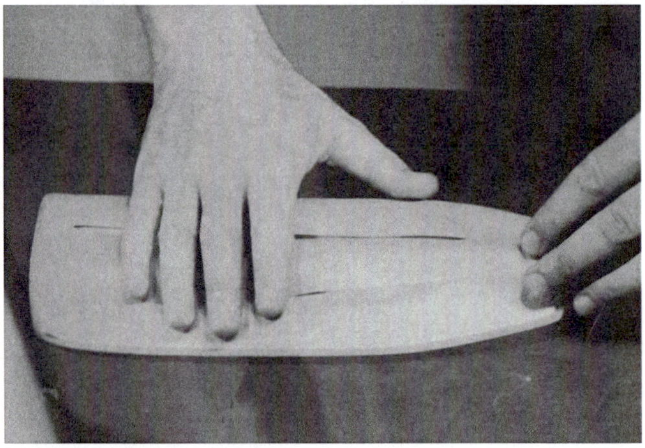

(Below) Examples of multi-person life raft oars with Scotchlite. After the issuance of Technical Order 127-43, Scotchlite was applied at the factory as illustrated. The paddle on the left is BuAer and on the right is a USAAF example. The USAAF adopted the concept in early 1944.

(Right) At the time of the adoption, the USAAF Type C-2 pararaft Scotchlite was incorporated to the mast paddles as shown.

(Right) As part of Amendment-A, in 1945, and with the change in design of paddles, the application was retained.

(Right) Scotchlite was applied to the new BuAer telescoping paddle, M-705, in 1945.

(Right) In the immediate post war years, Scotchlite was replaced by a new improved material called Reflexite. Scotchlite had a rough surface, much like sand paper, whereas Reflexite was smooth. Reflexite is still utilized today on safety equipment in the construction industry.

DISTRESS SIGNALS

Part V.
Flashlights

There were two principle electric flashlights utilized by the BuAer and the USAAF, categorized as a one cell and a two cell, with an array of types or patterns. The USAAF was the first aeronautical service to add flashlights to their compliment of emergency equipment, with the BuAer soon to follow. Each service integrated lights per their service requirements.

In June 1942, the USAAF added the floating lamp, Type A-7, as an accessory for their multi-person life raft. It was officially adopted as standard, 17 July 1942.

With the outbreak of total war and ships crossing the Atlantic on an almost daily basis, being harassed by German U-boats, many seamen were losing their lives. In part, it was a failure to locate survivors at night when the German submarines would often strike. A personal locating device became an immediate demand. The Merchant Marine had an established regulation requiring the use of self-igniting calcium lamps on life preservers and lifesaving suits, but these presented a hazard in wartime conditions. The British Merchant Navy was also having success in the rescue of survivors at night using personal locating lights. Based on the results of the British, coupled with demand, the US Navy began taking steps to i standardize a personal night time locating device for seagoing personnel, with the USCG also establishing a requirement for the Merchant Marine.

In the spring of 1942, a series of life preserver lights were presented by a variety of manufacturers, followed by approval from the US Navy and the USCG.

One common feature of the one cell flashlights was the red lens. At the time of its adoption, it was established that red lights would distinguish floating personnel from floating equipment which utilized a white light. In September 1942, these lights would be covered under Navy Department BuShips Interim Specification 17L16 (INT) assigned stock number 17-L-11793.

(Left and above) The Type A-7 floating lamp Specification No. 94-32410 was added to the compliment of USAAF multi-person life rafts as part of revised Technical Order 03-55A of June 1942, detailing the accessories that were to be standard. This identification lamp required two D batteries which acted as a ballast, keeping the light pointed up while floating. The Type A-7 was available under stock No. 7700-518591-5 and remained in service throughout the duration of WWII, although it was officially superseded by the hand-energized flashlight, Type A-9, in 1944.

(Right and above right) The Delta Electric Co. provided a metal bodied Type A-2050 one-cell life preserver light, approved in May 1942. The light as adopted could either be clipped on or secured by a cotton lanyard with a large safety pin, depending on the type of life preserver being worn. The light was activated by turning the lens cap.

The introduction of life preserver lights interested the BuAer, as it appealed to their service requirements for personal locating devices, and in January 1943 they drafted Naval Aeronautical Specification M-567, Light, Signaling; Life Raft and Life Jacket. This can be viewed as a general specification, as it outlines only requirements with the case or body being either metal or plastic, but they do specify the lens be clear and that the light include a keying button for signaling purposes.

One of the initial projects conducted by the Liaison Committee for Emergency Rescue Equipment, was to evaluate the color of life preserver lights and their effectiveness. In June 1943, they released their report. Recommendation No.1 stated that white light had the longest range of visibility, and that at distance, colored filters appeared white. It was only at close range that colors were distinguishable. There was no clear advantage to the use of colored filters.

By September 1943, the BuShips was evaluating and redesigning life preserver lights and, based on this report, adopted the clear lens. The BuAer had already seen the advantages of a clear lens and the BuShips was in the process of amendment, but the USCG would not grant a waiver to switching from red lenses to white. The opinion of the USCG was that the red lensed life preserver light was the international standard of merchant marine services and should not be changed or it would create confusion. They opted to stay with the established indoctrination.

(Left and above) Taunton Pearl Works introduced a metal bodied, single cell flashlight Type MSL 1263, approved in May 1942. All the same features applied as on the Delta model, except it utilized a chain lanyard.

(Below left and below) Colvin-Slocum Boats Inc. supplied a plastic bodied single cell flashlight as their Model-A, approved in June 1942. In specification requirements, either metal or plastic bodied lights were acceptable as long as they met corrosion and durability testing. A swivel type clip was provided and the light was activated by the same action of rotating the lens.

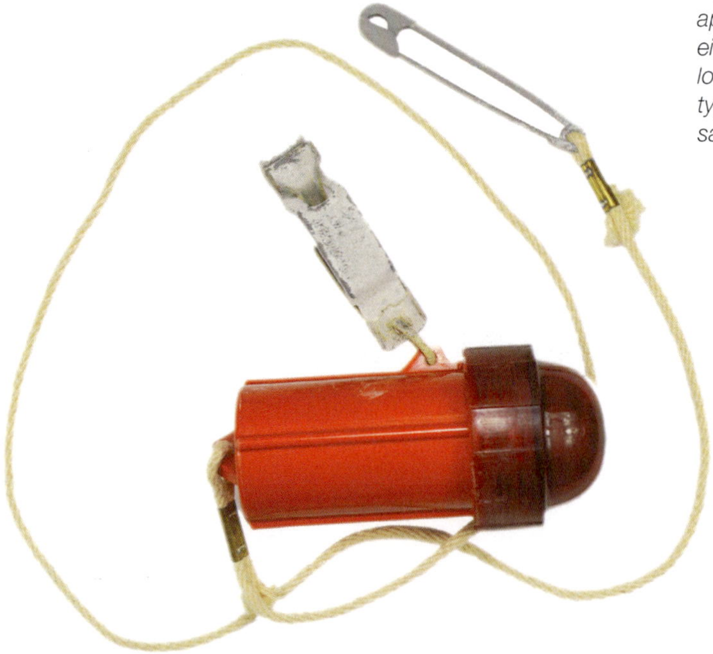

(Left and far left) William M. Lennan Inc. supplied a plastic bodied, one-cell light, listed as "Lennan Light Life Saver" Model No.2, approved at an unknown date. This light shares all the same features as the previous models illustrated.

(Left and two above) Adopted for experimental field testing was this design presented by Taunton Pearl Works, upgrading the existing model by adding a keying button, and re-designating it as MSL-1263A. In February 1943, 10,000 of these lights, covered under Specification M-567, were procured by the BuAer for field testing. 20,000 more were ordered the following month to supply the then new back pad kit. The light proved favorable in field testing and it was decided to adopt the M-567 as standard. It was however decided to remove the light from the back pad kit (now designated M-592) and place it on the list of Pilot's Personal Equipment. Additional procurements may have been awarded in 1943 but if so, they are not known. At the close of 1943, the BuAer adopted the life preserver light as standard and wanted to make a large procurement, but were declined by Taunton Pearl Works which did not have the capacity for production as they were consumed with other, unrelated contracts.

(Right and far right) The completely overhauled and streamlined life preserver light of the BuShips would include a metal body, clear lens, safety pin, and a switch with keying capabilities. Around December 1943, the BuShips began procurement of these new lights, still covered under their amended Specification 17L16 (INT). Declining to bid for the M-567, the BuAer was forced to look for an alternative and opted to adopt this model. In February 1944, 7,000 lights were requested for allocation to BuAer supply points from the BuShips contract that was awarded to the Bridgeport Metal Goods Co. In April 1944, the BuAer awarded a contract for 100,000 lights to the same company. This contract was through the Bureau of Supplies and Accounts, and would be allocated through standard Naval Supply Depot channels as stock number 17-L-11793. The lights would never be maintained in the ASO system. The two lights shown were manufactured by Bridgeport Metal Goods Co. illustrating two color variations. On the side of the body they would be stamped USN with manufacturer name and, in this case, BMG.

(Left) These lights were officially standardized 1 September 1944, dropping the Interim (INT) designation. At the time of standardization, the 17L16 had just undergone an additional amendment, slightly decreasing the height and slightly increasing the diameter of the case to accommodate the expansion, or swelling, of the batteries in extreme climate changes. In the original pattern, the batteries were nearly impossible to remove when expansion occurred. By this time, the Fulton MFG. Corp. was being awarded contracts. On the left is a BMG, and on the right, a Fulton manufactured light. Both have the USN property mark and manufacturer stamp placed on the bottom. One significant benefit these lights had over previous types was being waterproof, meeting a major criterion.

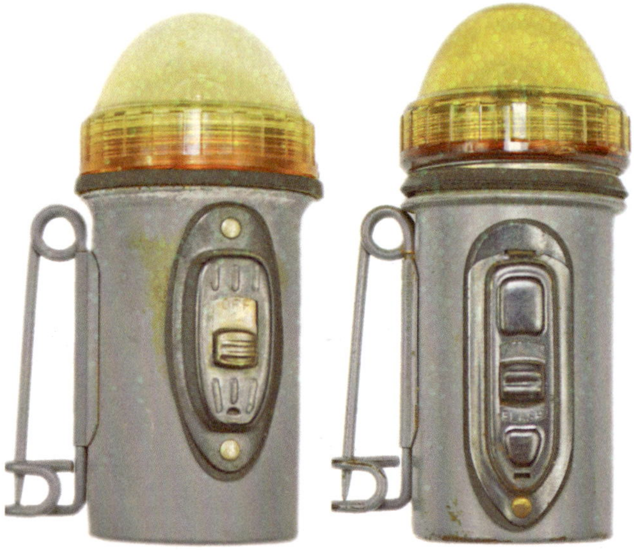

(Right) By late 1944, limited quantities of the 17L16 were available but received the universal complaint that the attachment pin was too flimsy, and the light tended to rip off even under simple tasks such as getting in and out of a cockpit. The light was again revised in March 1945 to increase the strength of the attachment pin, followed by a large procurement by the BuAer and the BuShips. Interestingly it was assigned a new stock number, 17-L-11792, retaining the same nomenclature, Light, Electric, Attachable, Small. These lights would not become available until summer, 1945.

(Above and left) As the war progressed, newly found applications of plastic materials were being discovered. The Delta Electric Co. applied them to the manufacture of life preserver lights. Their model A-2050 was upgraded to a torpedo-shaped body and re-designated under A-2050P, c.1944. Since it had a red lens, it was most likely marketed to meet USCG regulations for the Merchant Marine.

(Above and below) Colvin-Slocum Boats Inc. revised their Model-A light to a clear lens. The minor detail of placing their branding on the side occurred at an unknown date.

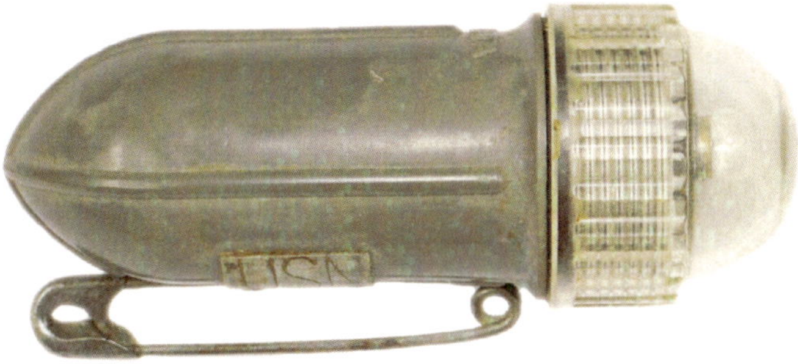

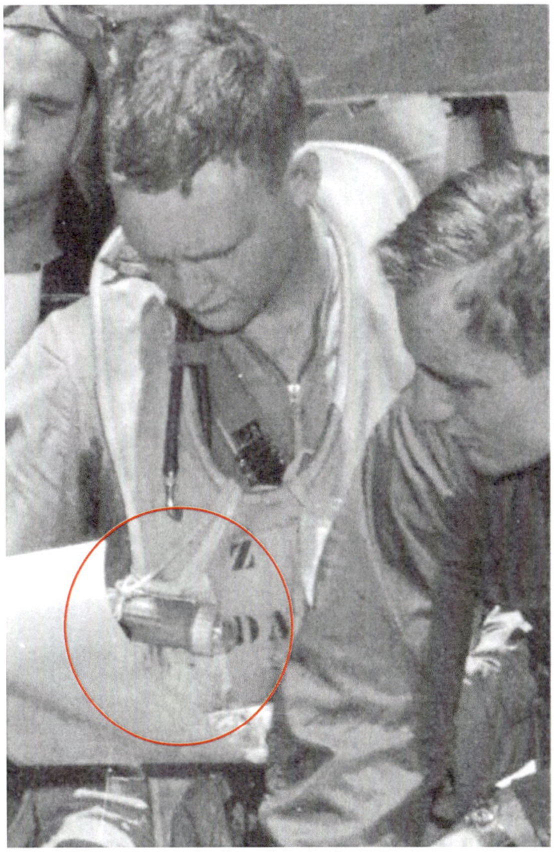

(Above) Delta used the same torpedo body style on their model A-2070 life preserver light, manufactured to meet BuShips Specification 17L16.

(Left) The Metal strap securing the pin to the body was stamped with USN or USCG according to its end-user.

(Right) TBF aircrew member on Okinawa, summer 1945. Hung from his sea marker is an A-2070 life preserver light.

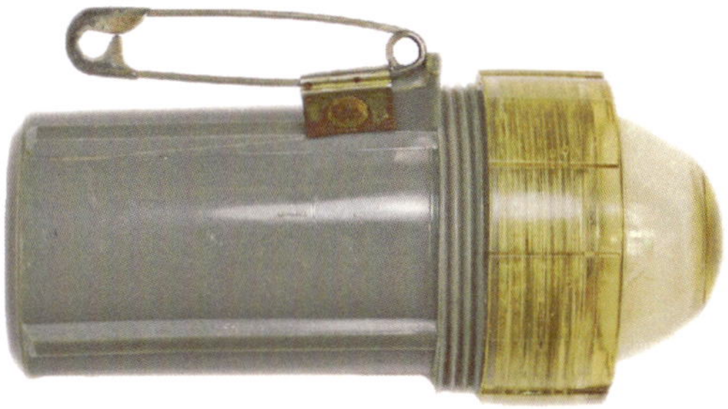

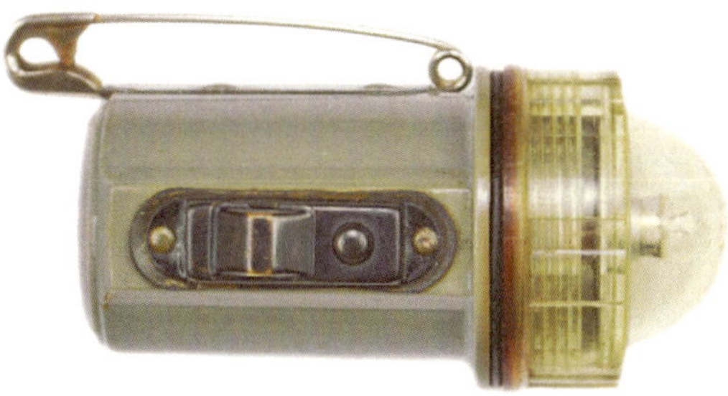

Additional lights manufactured by William M. Lennan Inc. No 1 and No.3. Unfortunately, no additional information is available except that each is a "Life Saver" model. No. 2 is illustrated at the top of page 347.

(Left top) No.1 in grey plastic

(Left center) No.1 in red plastic

(Left bottom) No.3 in grey plastic

To confuse the issue further, each light is marked with a different branding of the company name. It is not clear if these were procured by the US Government or merely marketed commercially and staged to appear as if they were military. The original box for the grey No.1 light is stamped Spec. 17-L-16 (INT).

(Left) This light was manufactured by the Henry A.S. Van Daalen Co., labeled as the "Dutch Admiral." These were listed as having been USCG approved for life preserver lights in September 1942 under the "Dutch Admiral" model. It is not clear if this particular model is the type approved in 1942 or not.

(Right) The USAAF adopted the single cell life preserver light outlined in the Statement of Military Requirements No. H-3-6, dated 9 January 1945, in which they requisitioned limited quantities from the BuShips under their Specification 17L16. The light was specifically adopted for use with the new Type B-5 life preserver. The use of the Navy 17L16 was to be temporary until a suitable model could be developed, but would ultimately become standardized as the third production 17L16, with large safety pin. The light supplied to the USAAF would be sea blue and stamped on the bottom with US and the manufacturer's name. This new light would not become available until 1947 as Specification MIL-L-5733A, Distress Marker Light for use with Type B-3, B-4 and B-5 life preservers. Pictured with the light is a pouch that was fabricated for use with the Type B-5 life preserver. The issuance of the light would be officially addressed in Technical Order 13-1-30, dated 2 May 1946, making it a mandatory item for all life preservers. This order references the light as Specification 17L16.

(Above) Lt. Cmdr. Charles Brewer, C.O. of VC-13, on the deck of the USS Core, August 1943. He wears an M-567 life preserver light. Brewer would later be listed as missing in action over Saipan while C.O. of VF-15 in June 1944. (National Archives)

(Above) VT-2 aircrew member J.F. Weber, AMM2/c aboard the USS Hornet, June 1944. He has an M-567 light clipped to his sea marker packet (National Archives)

(Left) VF-40 pilot Ens. Raymond Lebel, recently returned aboard the USS Suwannee after an engagement with Japanese aircraft, in which he accumulated three kills, May 1945. Attached to his sea marker is a 17L-16 (INT) life preserver light. (National Archives)

(Above) Well publicized photo of twin brothers, Andrew and Arnold Vanderwall, who both served with VT-7, illustrating both 1st and 2nd production 17L16 life preserver lights. On the left is the larger diameter 17L16 and on the right is a first production 17L16 (INT). USS Hancock, February 1945. (National Archives)

(Above) The Delta life preserver light can be seen worn by Merchant Marine seamen fixed to their anti-exposure suits, October 1942. (National Archives)

(Right) A 17L16 life preserver light is fixed to a talker's Kapok life jacket as part of the standard requirement. In addition, a whistle was to be provided. USS Ticonderoga, November 1944. (National Archives)

(Right) Lt. jg Levi Monteau of VF-40, aboard the USS Suwannee, with a Life Saver No.2 slung around his neck, April 1945. (National Archives)

(Far right) USMC Lt. Col. Donald Yost of VMF-351 utilizes a red lensed Life Saver No.2 life preserver light aboard the USS Cape Gloucester, June 1945.

(Below) 20th AF B-29 aircrew member's F/O Armand Bucchi, and 1st Lt. August St. George, wearing US Navy 17L16(INT) life preserver lights, summer 1945. Extensive use of life preserver lights for the USAAF would not occur until the spring/summer of 1945. (National Archives)

(Above) USAAF B-29 aircrew member 1st Lt. Clarence Presswood has a 17L16 2nd production life preserver light attached to the retaining strap of the oral inflation tubes, summer 1945. (National Archives)

(Far left) Aircrew member of the 81 mission B-24 named Bolivar has a clear lensed Colvin-Slocum Boats life preserver light attached to his life vest, October 1944. (National Archives)

(Left) Pilot Lt. W.H. House of VT-14 wears a clear lensed Colvin-Slocum Boats light aboard the USS Wasp, August 1944. (National Archives)

(Left) Though not a direct application to USAAF use, a Life Saver No.2 is fixed to a Kapok life preserver worn by 1st Sgt. John Dirauskas, 7th AF headquarters squadron, aboard the troopship USAT Sea Flasher en route to Saipan, December 1944. (National Archives)

(Above) Life preserver lights in the display room of the Liaison Committee for Emergency Rescue Equipment in Washington DC, December 1943. (National Archives)

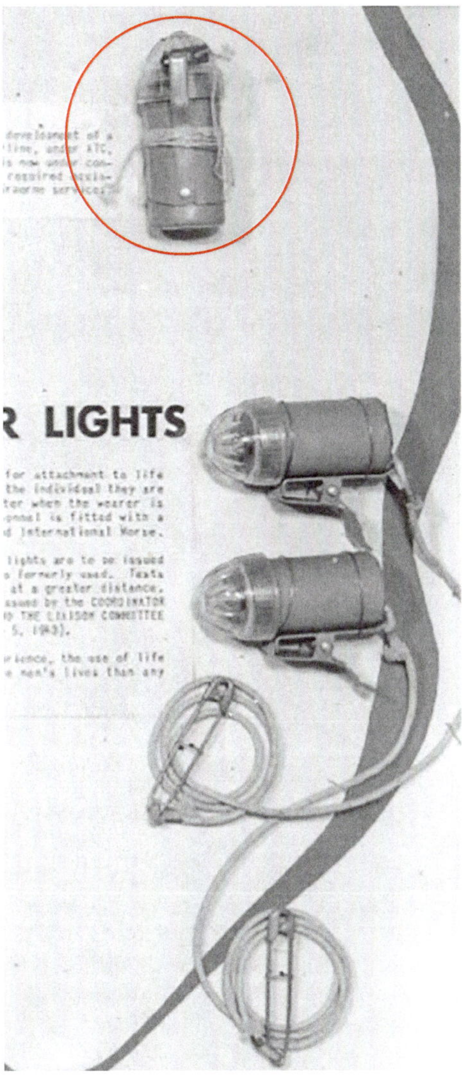

(Below left) More of the life preserver lights from the display by the Liaison Committee for Emergency Rescue Equipment in Washington DC, December 1943. Each of these lights had been demonstrated previously, with the exception of the example at the top of the picture, which was an experimental life raft recognition light developed by the USAAF Transport Command. This light automatically activated when it came into contact with water. (National Archives)

The single cell flashlight was considered for "emergency use only," while the use of two-cell types was for general use.

Going back to the 1930s, with the early development of emergency, or sustenance, kits, the USAAF had incorporated flashlights, with the Signal Corps type TL-122 being standard issue. This light would remain the standard in the USAAF for general use. The USAAF did develop a series of specialty flashlights designated under an "A" series. The Types A-1 and A-2 were commercial type flashlights from the 1920s and early 1930s. In 1941, three specialty lights were adopted, beginning with a headlamp model for aircraft maintenance, designated as the Type A-3, quickly followed by a three-cell flashlight for taxiing airplanes, designated as the Type A-4, and a two-cell light for aircraft inspection that incorporated a small, flexible arm, designated as the Type A-5. The Type A-6 penlight, (see page 360), was intended for minimal tasks but could be used in emergency situations. The Type A-7 floating identification lamp for life rafts has been covered (see page 345). The Type A-8 flashlight was a compact three-cell flashlight for aircraft maintenance, with the lens set at a 60-degree angle. When clipped to a waist belt or standing on the floor, the beam would point towards the work area. The Type A-9, hand-energized flashlight (see page 364), was developed and adopted for emergency rescue equipment.

The BuShips utilized two and three-cell flashlights, typically with a tubular metal body, under the designation of TB-8. Detailed information on the background of these lights is limited, but they predate WWII. These were the standard general purpose lights used by naval personnel throughout the conflict.

(Right) The primary general purpose flashlight for the USAAF during WWII was the TL-122 angle head flashlight, which met service requirements. Pictured is the standard Signal Corps TL-122A.

(Above) TL-122A on the pistol belt of bombardier 1st Lt. Rider, India March 1943. Most crew members required flashlights to conduct their duties in low light situations. The squadron PEO would be responsible for the requisition and distribution of flashlights obtained from the group Signal Officer. The US Army Signal Corps was charged with the procurement and issue of flashlights for the USAAF.

(Right) The TL-122B or C flashlight, introduced in 1943, was fabricated with a celluloid body versus the metal type of the TL-122A.

(Below) B-24 aircrew member of the 394th BS has recently returned from a bombing mission after dark. Attached to the waist belt of his flight suit is a TL-122B or C, Netherlands East Indies c.1944. (National Archives)

(Above) Three variations of the Type-D battery, also referenced as the BA-30 for the Signal Corps and Type-C for the US Navy.

(Left) Two examples of the US Navy TB-8 two-cell flashlight. Above is an example manufactured by the Fulton MFG. Corp. and below is a Bridgeport Metal Goods MFG. Co. sample. Other manufacturers included Bright Star Battery Co., Blake MFG. Corp., and the National Carbon Co. Inc. under the brand name Eveready. There are also variations within the manufacturer, such as the types of switches, making for a large array of samples available. All would be stamped with the property mark USN. These were available under NSD stock number 17-F-13452, tilted Flashlight, Electric, Hand.

(Left) In 1942, service requirements within the BuAer dictated that a two-cell flashlight be incorporated for use on all naval aircraft. The first logical choice was the TB-8 type, being readily available as standard stock. These lights are often seen worn as personal equipment, typically with pilots. Pictured is a VB squadron pilot who has recently returned from air strikes on Tarawa, September 1943. Draped around his neck is a BMG TB-8 flashlight. (National Archives)

(Below left) Pilot Lt. jg William Burckhalter of VF-16, aboard the USS Lexington, has a TB-8 suspended from his oral inflation tube. December 1943. (National Archives)

(Below) PBY aircrewman utilizing a TB-8 type flashlight secured to his pistol belt, April 1944, on Samarai Island, New Guinea. By this date, flashlights were being incorporated into units flying night operations. (National Archives)

(Above) Pilot Lt. jg Chas Richardson of VC-12 aboard the USS Core has a Bright Star model 1617 flashlight draped from his life preserver, April 1945. (National Archives)

(Top) A large variety of types of flashlights were available to Naval personnel and only a few examples can be illustrated. This example is referenced as Flashlight-Focusing, two-Cell, Tubular, Spark Proof, available from the ASO under stock number R17-F-13485. This flashlight has a celluloid body. The ASO Class 17 supply catalog lists two manufactures, Bright Star model 1617, and Eveready model 1359.

(Below) The USN property mark as stamped on an example of the Bright Start light. (National Archives)

(Above) A crewmember of VT-2 Bernard Stoub AMM1/c has a TL-122A secured to the tie tape of his sea marker packet aboard the USS Hornet, June 1944. (National Archives)

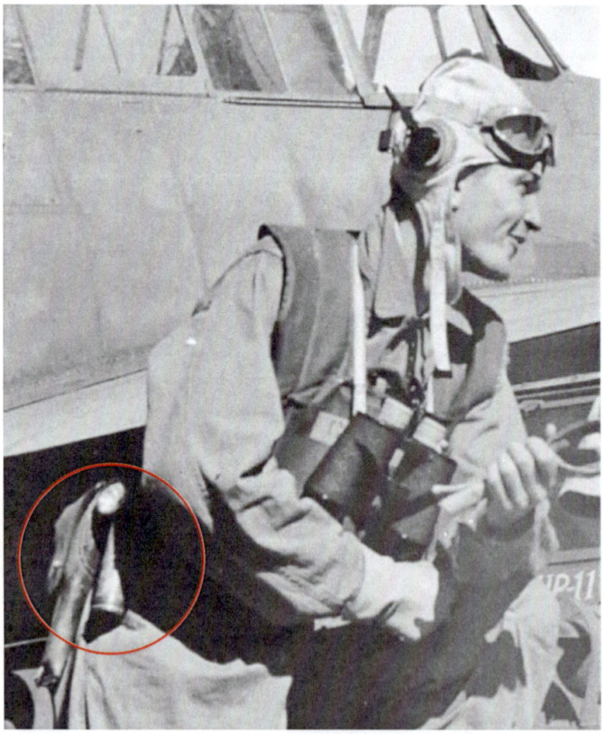

(Above) To meet demands, units secured and utilized all types of flashlights, both commercial and from other services. Pilot Lt. jg. George Gay of VT-11 is using a US Army Signal Corps TL-122A seen clipped to his pistol belt, Guadalcanal May 1943. (National Archives)

Two serious problems were encountered with the TB-8 type flashlights. Firstly, they were not durable (being made with a thin crushable body) and secondly they were not waterproof. In November 1942, the US Navy Department proposed an all-purpose flashlight be developed to meet sturdiness and waterproofing requirements being detailed in Specification 17F7 (INT). In February 1943, the US Navy adopted a design presented by the Delta Electric Co. as their model A-2060.

The A-2060 flashlight was an improvement on the TB-8 models, but in service use they were found to have waterproofing issues leading to corrosion internally. An improved model was introduced by the close of 1943 as Delta model A-2066, remedying the waterproofing issues.

By the end of 1944, it was decided to provide one waterproof two-cell flashlight 17-F-13550 per aircraft closely connected to night combat flying, to be carried and used by combat crew. The penlight had been provided for this purpose, but was found to be highly unsatisfactory. Due to supply deficiencies, only the pilot was to be issued the light initially. As supply permitted, all crew members were to be equipped.

Penlights were seen as a useful tool within both aeronautical organizations. The BuAer standardized their type as Naval Aeronautical Specification M-533, titled Flashlights, Electric, Penlight, 9 March 1942. This was a general specification outlining construction requirements that the body and internal parts to be either brass, copper, or phosphor bronze. The penlight was compact and could be used for many applications, whether in the cockpit or conducting duties aboard an aircraft, but it could also be used for emergency signaling purposes.

(Above) The Delta model A-2060 was fabricated with a corrosion resistant durable celluloid body and being of a single mold it reduced water seepage into the light. A large durable clip eliminated the use of a flimsy ring, as on the TB-8. The light was available under Naval Supply Depot Stock number 17-F-13452 titled as Flashlight, Electric, Hand, Plastic, two-cell w/lamp. This light technically superseded the metal tubular model, however the metal types were still procured and utilized in great numbers by the Bureau of Medicine & Surgery and BuShips. In April 1943, the BuAer field tested the A-2060 flashlight, considering it for inclusion in emergency rescue equipment. With the results proving satisfactory they incorporated it into their newly developed air droppable Shipwreck Kit, Specification M-594. (National Archives)

(Above) Batteries utilized by the US Navy were known as Type-C, and by the US Army Signal Corps as BA-30 (known commercially as Type-D).

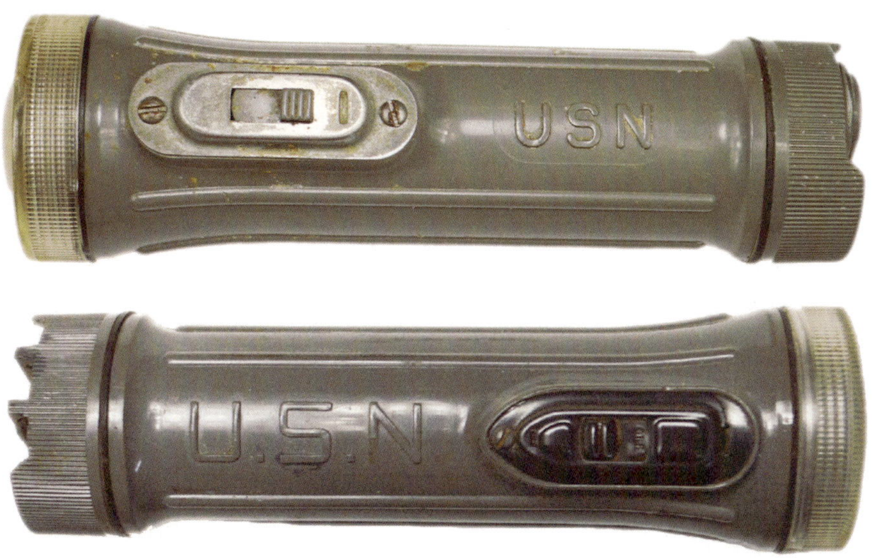

(Left) Production of the upgraded model A-2066 commenced in February 1944 and it was re-designated as Flashlight, Electric, Hand, Watertight, two-cell w/ lamp, under NSD stock number 17-F-13550, officially standardized as Specification 17L7 in April 1944. In April the BuAer requisitioned 100,000 A-2066 flashlights for inclusion as standard equipment for life rafts. This model would eliminate the clip, reverting back to a ring located on the end cap. Shown are the two known manufacturers. Delta Electric Co. (above) and the Fulton MFG. Corp. (below). Variation of the raised USN property mark can be seen. By October 1944, 1,021,000 A-2066 flashlights were in stock with an additional 1,343,000 on order.

(Above left) Going over last minute details in the ready room of the USS Ticonderoga, pilot Ens. George McIntosh of VT-87 has a A-2066 flashlight draped around his neck utilizing a M1917 pistol lanyard, July 1945. Note that he also wears a single cell light which was viewed as only having an emergency function. (National Archives)

(Above right) Standing on the wing of his TBF pilot John McMonigle of VT-40 has an A-2066 flashlight hanging from his life preserver aboard the USS Suwannee, April 1945. (National Archives)

(Right) With specification M-533 being "general," any light that met the basic requirements would have been procured and distributed for fleet use. When penlights were observed they were in a easily accessible location, like that seen on the 23P6 life preserver of pilot Ens. Richard Murray of VB-10 aboard the USS Enterprise, February 1943. (National Archives)

(Far right) Pilot Lt. jg. Henry Dearing of VF-12 aboard the USS Saratoga, November 1943. He wears a penlight attached to his life preserver. (National Archives)

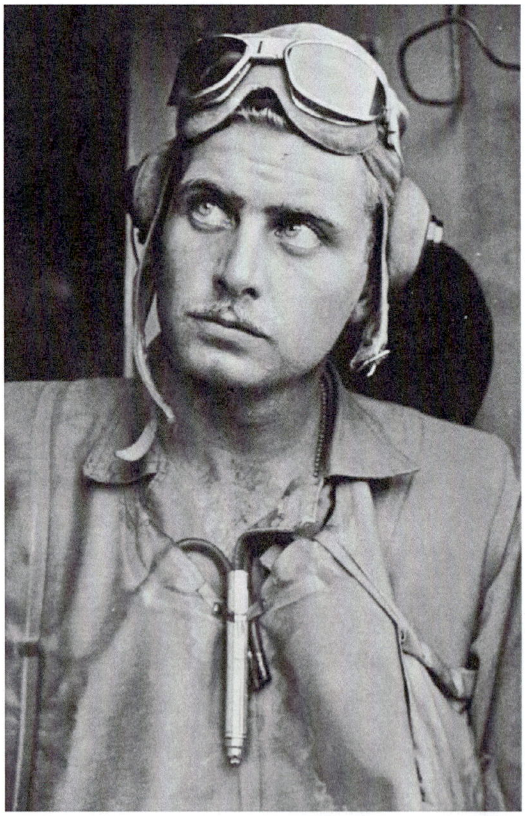

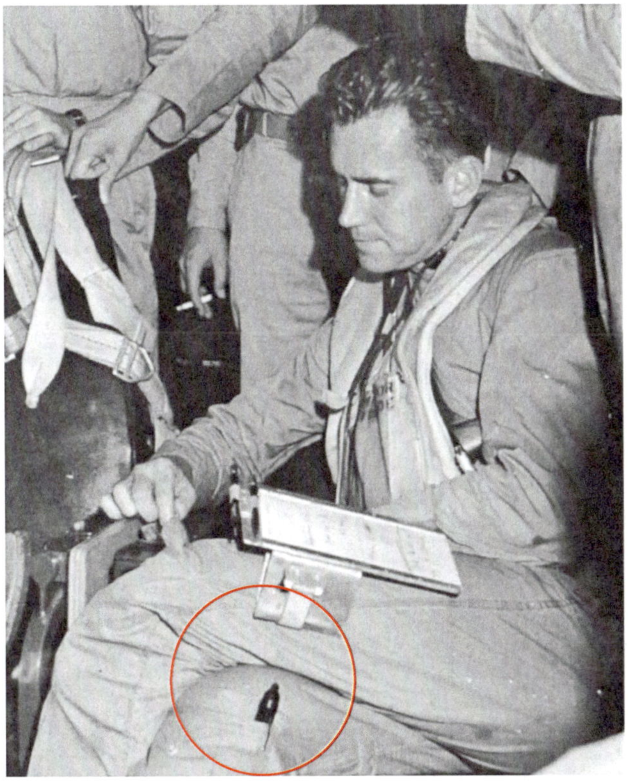

(Left) Penlights would have been a standard article distributed with plotting boards, which had an inner compartment with charts and notes. Again, some chose to place them in a more convenient location. A pilot in the ready room of the USS Yorktown has a penlight secured to his flight suit shin pocket, September 1943. (National Archives)

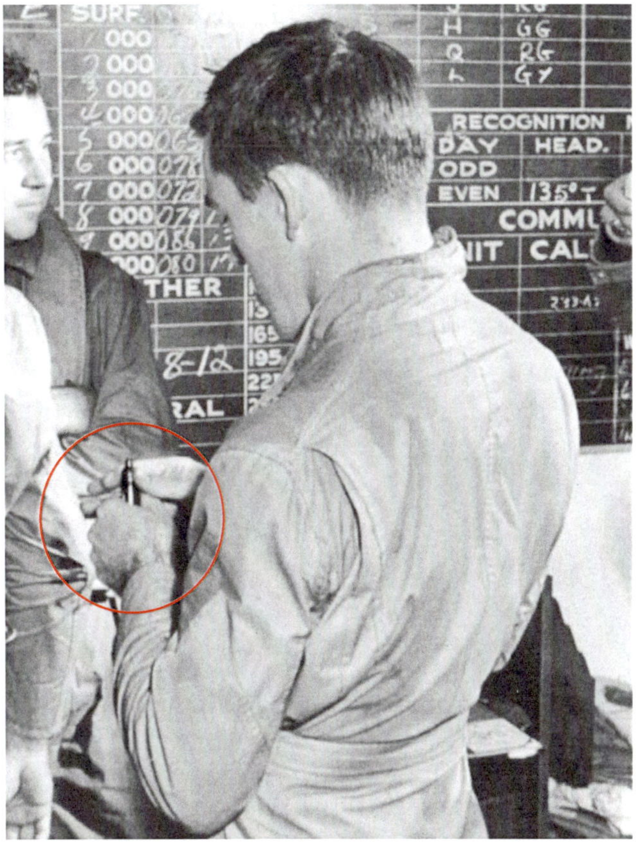

(Above) A pilot of VF-5 is checking his penlight in preparation for operations of the day, in the ready room of the USS Yorktown, October 1943. When studying contemporary images, a variety of types and models can be observed. (National Archives)

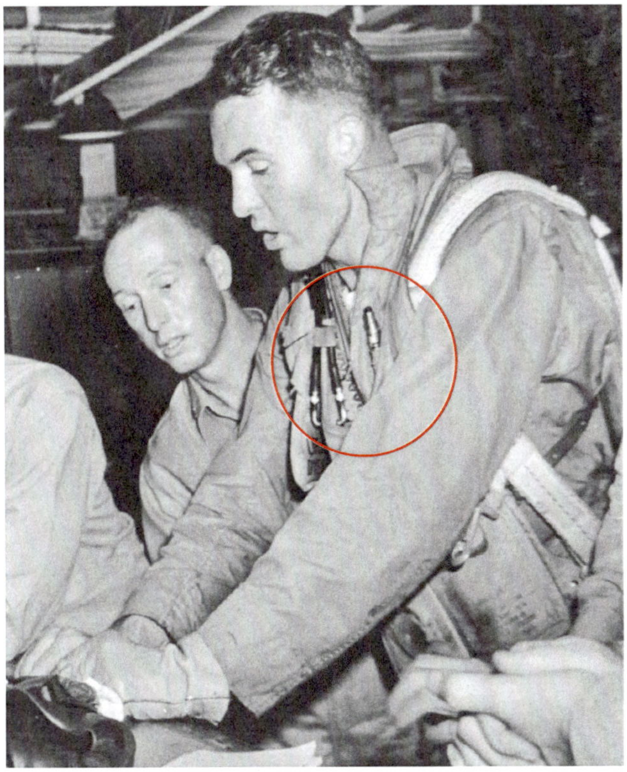

(Above) VF-15 pilot L. Prince Gordon illustrates a partial view of another model in a readily accessible location in his flight suit, USS Essex December 1944. (National Archives)

(Right) A VB pilot has a penlight secured to the lanyard being used to sling a TB-8 flashlight from his life preserver. This view illustrates a distinct pattern on the light itself. USS Yorktown, October 1943. The penlight was a typical compliment to a Naval aviator's flight gear compliment. (National Archives)

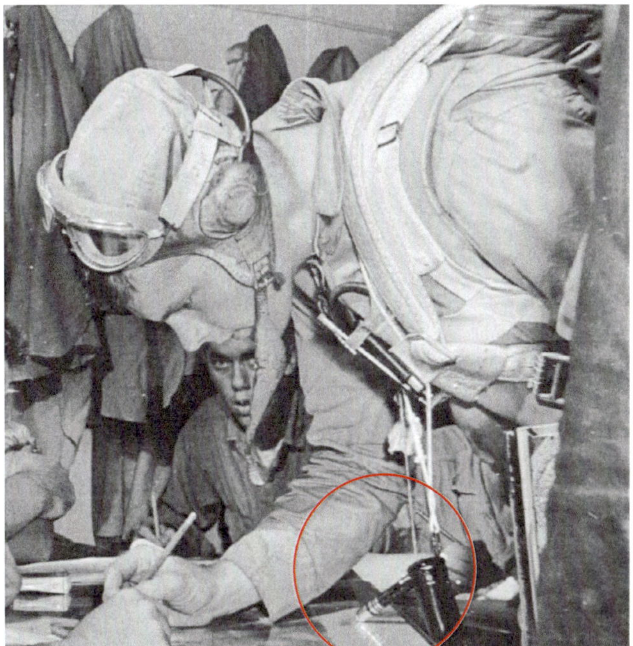

The USAAF standardized its penlight on 11 September 1942, as the Type A-6 Specification No. 94-32378. It was a metal bodied light manufactured under the same guidelines as the BuAer M-533. In January 1943, the USAAF was evaluating a possible change for the light to be of a celluloid body and the addition of a red filtered lens.

Working in coordination with the BuAer it was found that the red filter should be adopted and considered for joint Army-Navy standardization. The new plastic penlight with red filtered lens was adopted and placed into production 15 February 1943, as Amendment-A. Consideration for joint Army-Navy standardization continued into 1944.

Two additional design features were being investigated, on the Type A-6A penlight the red filter was a separate cap which could be lost and, second, the BuAer requested that a lanyard be provided so the user could drape the light from around the neck. The revision officially incorporated was an integral red filter that could be slid into position by a small toggle located on the side of the nose cone. It was standardized as the Type A-6B 8 March 1944.

Joint Army-Navy standardization was finally approved 14 March 1945, re-designating it under Army-Navy Specification AN-L-30 Part No. AN8023-1, however production lights continued to reflect the USAAF Specification 94-32378-B. As part of the Army-Navy specification, the light was to have provision for attachment of a lanyard, which was accomplished by adding a flat round tab on the back end of the light.

Drawing AN8023 was officially cancelled in June of 1946 being re-formatted under MILSPECS.

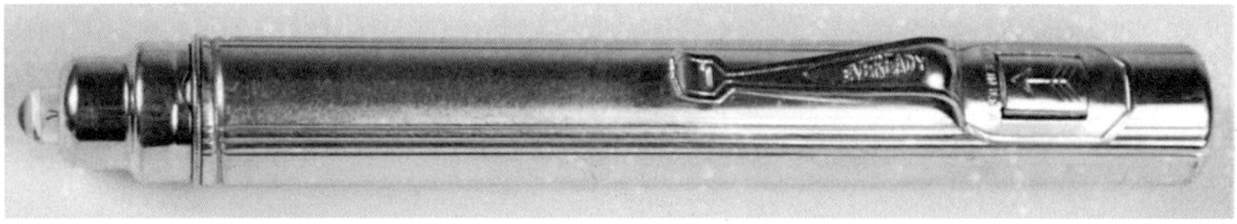

(Left) An existing (top) and laboratory example (middle and lower) of the Eveready model No.210 flashlight procured by the USAAF as the Type A-6. This was initially issued with the aircraft to illuminate maps and instruments, and contained in a bracket in the cockpit. It was later determined to issue it to all air crew. In the catalog for Class 17 Naval aeronautical materials this model of penlight was listed for NavAer issuance under stock number R17-F-13475. (USAF)

(Below) Type A-6A penlight standardized in February 1943. It was a major revision over the Type A-6 being made of a celluloid body and having an optional feature of a red filtered lens. When the filter was not in use it was stored at the back end of the light. (USAF)

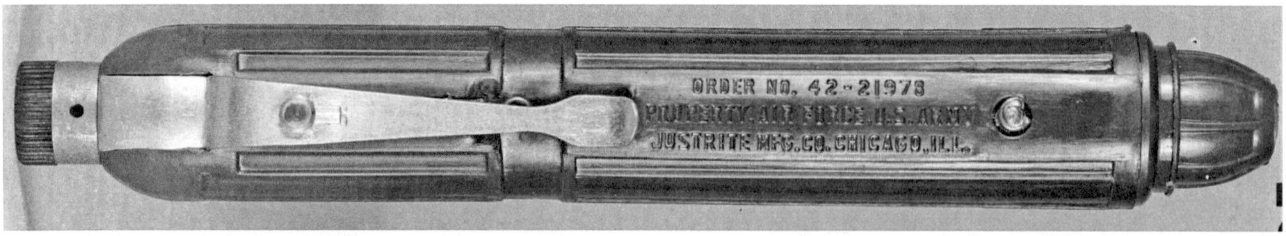

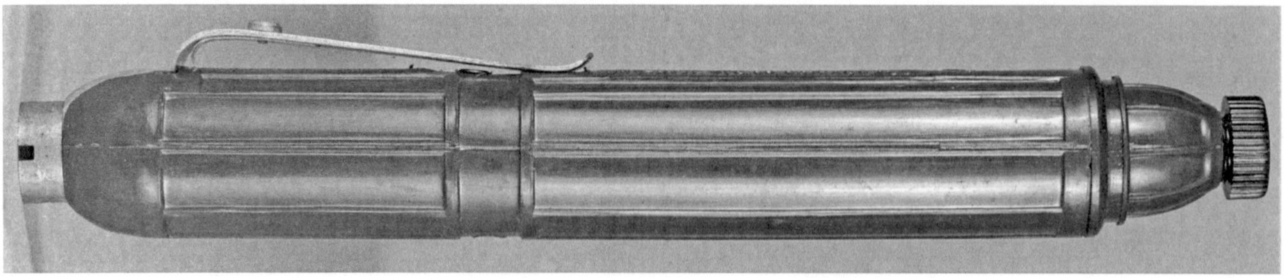

(Above and far right) Penlights for NavAer use were listed under the nomenclature of Flashlight-Penlight or Pencil, two-cell type. Four manufacturers are listed: Usalite Penlite, Eveready 210, Burgess 92, and Bright Star 100. Shown is a the Burgess 92 model requiring two AA batteries.

(Above and below) Revised Type A-6B penlight that included an integral red filtered lens in the nose cone. The body of the light remained consistent with that of the A-6A. (USAF)

(Left) 1944 contract USAAF Type A-6B penlight. Signal Corps contract numbers are stamped on clip. Further modifications were applied to these lights in the immediate post-war years. The three details to look for are: the specification number would reflect Amendment-C (94-32378-C), an updated switch that allowed the light to have a keying function for signaling and Morse code, and the property mark of Property US Air Force. WWII manufactured lights will have Property AF US Army. Production figures are unknown.

(Right) The BuAer procured the Type A-6B in August 1944 with a quantity of 125,600 under ASO contract 288s-24994. This contract would be stamped on the clip. The light is identical in all respects to the USAAF model, except the property mark reflects Property US Navy. These penlights were available from the ASO under stock number R17-F-13475. The only contractor for the A-6B penlights were the Justrite MFG. Co. In addition, and as part the request from the BuAer, the Justrite MFG. Co. supplied 50,000 17" lanyards under ASO contract 288s-25993 which were stocked separately under stock number R17-L-7840. No provision was applied to the contract lights for attachment of a lanyard, which was to be tied, preferably, to the clip.

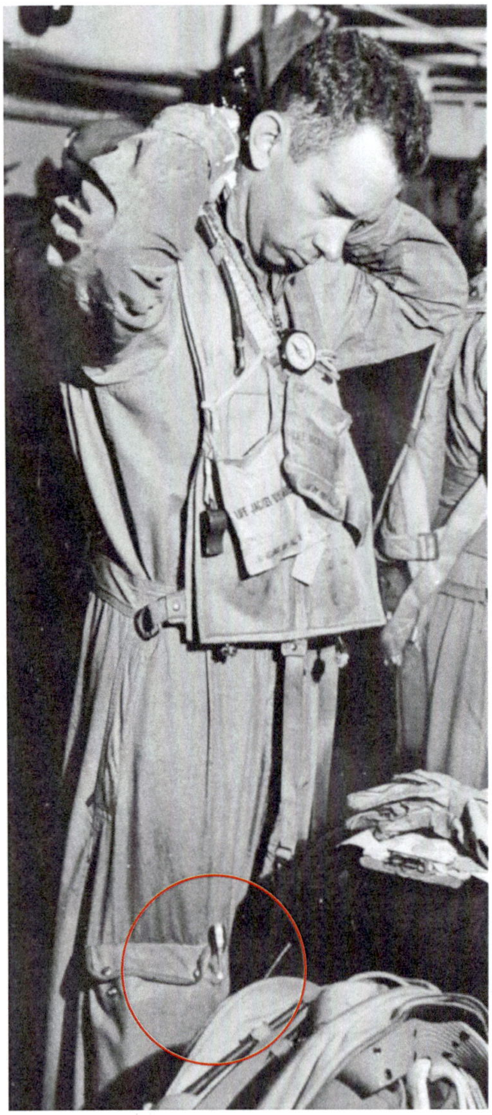

(Left) On the flight deck of the USS Core, pilot Lt jg. Howard Heard of VC-12 has a Type A-6B penlight secured to his life preserver utilizing the 17" lanyard, April 1945. (National Archives)

(Left) Pilot of VT-87 Lt. William Griese adjusts his life preserver in preparation for a strike on Japanese installations in the ready room of the USS Ticonderoga, July 1944. Located in the rigger-applied pocket on his flight suit is a Type A-6B penlight. (National Archives)

(Above and right) AN-L-30 penlight as procured under ASO contract 288s-33942, awarded in late summer 1945. Lanyards supplied with these lights were provided under the same contract, in 1944, under the Justrite catalog part No. 1110. These later production penlights would include a flat washer at the back for attachment of the lanyard. (Paul Faulkner Collection)

In late 1939, the USAAF investigated the idea of issuing a hand-energized flashlight to pilots. Several field tests were conducted in 1940, and it was concluded that a pilot could not devote time to operate the light while flying, giving no further consideration to generator flashlights. In 1942, flashlights were being added to emergency rescue equipment, first with life rafts, then sustenance kits. As these units entered service it was found that battery-operated flashlights were not dependable and that batteries deteriorated in storage, especially in extremely cold climates. In late 1942, the USAAF went back to the Viz Corporation, which had supplied their hand-energized light in 1939, for evaluation on possible adoption. The light evaluated was known by the commercial name of Vizlight.

By spring of 1943, the USAAF had determined that hand-energized lights were the best option to overcome the deficiencies of battery operated types, the most significant advantage being that they had an indefinite shelf life. Emergency or sustenance kits remained in storage for long durations of time, being inspected only on a semi-annual basis. The USAAF established an unofficial directive to develop a suitable model to be standardized for inclusion in life rafts and sustenance kits, as well as the vest kit that was in development at that time. The Vizlight and Niko models were not approved, but were utilized as an interim until a model could be standardized. In November 1943, the Dayton Acme Co. submitted their Model 163-B for committee evaluation.

(Left) Hand-energized flashlight manufactured by the Viz Corporation known as the Vizlight. The light proved unsatisfactory and was not recommended for standardization, but for lack of a better option it was incorporated into USAAF sustenance kits.

(Below) The Vizlight pictured with other recommended articles to be carried by individuals as personal emergency equipment, detailed in the manual "Dinghy! Dinghy! Prepare for Ditching!" compiled by Cpt. Donald Arlen for the 396th BG (H) AAB Drew Field, Florida dated June 1944.

(Below left) Coinciding with the evaluation of the Vizlight, the Nicoud MFG Co. submitted a model they had prepared under Signal Corps specifications drafted in 1942. The Niko Generator Emergency Flashlight was found to be superior to the Vizlight and was included in revised specifications for the Type's E-2, E-8 and E-10 Sustenance Kits in 1943.

(Below right) Detail from a display of components for the E-2 Emergency Kit including the box for the Niko generator.

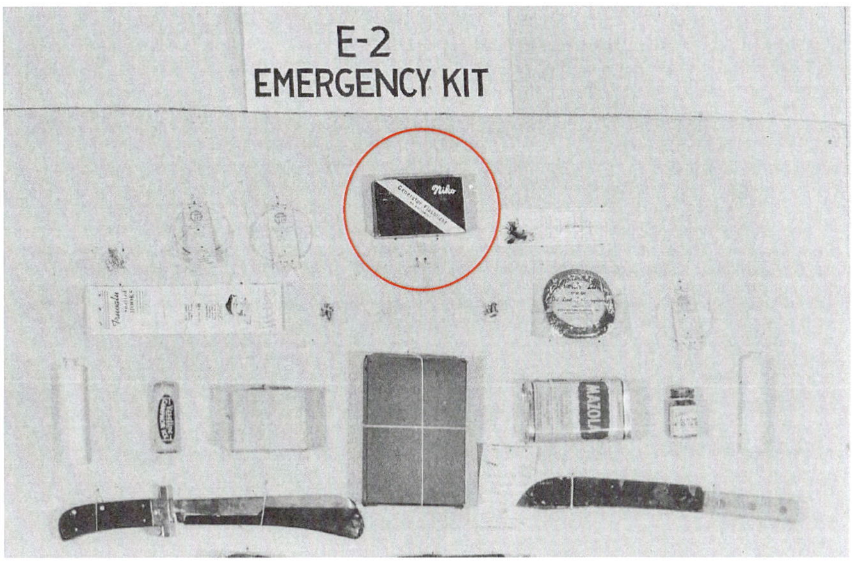

(Left) Dayton Acme Co. Model 163-B known as the Daco-Light. The model proved successful, prompting the USAAF to draft specifications and drawings based off this example. Procurements were initiated for the 163-B in February 1944 to equip life rafts and replace the Vizlight and Niko models for contractors of sustenance kits. Pictured is the 163-B as illustrated in the Air Sea Rescue Equipment Guide of 1945. The BuAer procured limited quantities of the 163-B in late 1944 for evaluation purposes for possible standardization.

(Right) Type A-9 flashlight with accessories for the USAAF multi-person life raft, c.1945.

(Bottom right) The Dayton Acme Co. revised their Model 163-B to conform to USAAF Specification No. 94-32454 Dwg. No. 44B9164, standardized 17 July 1944, as Lamp Assembly, Flashlight, Type A-9, and assigned stock number 7700-518591-65. Procurements were initiated at the time of standardization with limited deliveries by the close of the year. The Vizlight, Niko and 163-B were the three most predominant hand-energized flashlights used during the conflict, with the Type A-9 only available in limited quantities by V-J day. The principle purpose of the Type A-9 was to replace the Type A-7 floating lamp for life rafts, then for inclusion in sustenance kits, but it was never officially adopted as a component for the sustenance vest. The BuAer requisitioned limited quantities of the Type A-9 in January 1945, with further requisitions becoming available by August, through the ASO under stock number R17-F-13575.

(Below) Simultaneously adopted as standard was the Type A-10, identical to the A-9 in all respects except the diameter of the lens. Prior to their adoption, both styles were field tested, with the larger lens preferred. The type A-9 was the only model known to have been procured and appeared in supply catalogs and other literature. The A-10 flashlight illustrated is marked Specification No. 94-32465, and has a manufacture date of February 1953, by Tom Kincaid & Co. This example is an anomaly with no further background information.

Electric lanterns were standard issue for patrol type aircraft and a recommended accessory for rescue planes. It was important to have versatility for all types of conditions. Supplied aboard the aircraft would be a selection of droppable location markers including sea markers, pyrotechnics and radio buoys. Different types were employed by the US Navy and Merchant Marine which were approved by the USCG.

(Right) As detailed in Navy Department Specification 17-L-11. Lantern, Floating. Requisitioned utilizing stock number 17-L-7495. These three lanterns each weighed 3 pounds (with Type-C batteries installed) and had a continuous life of 22 hours. They featured a buoyant upper chamber with the lower shaft containing the batteries, which acted as a ballast to keep the light upright. These were classed as Standard. Far right is the Type-R, which had a dehydration unit to prevent internal condensation. Type R was assigned a different ASO stock number: R17-L-7496.

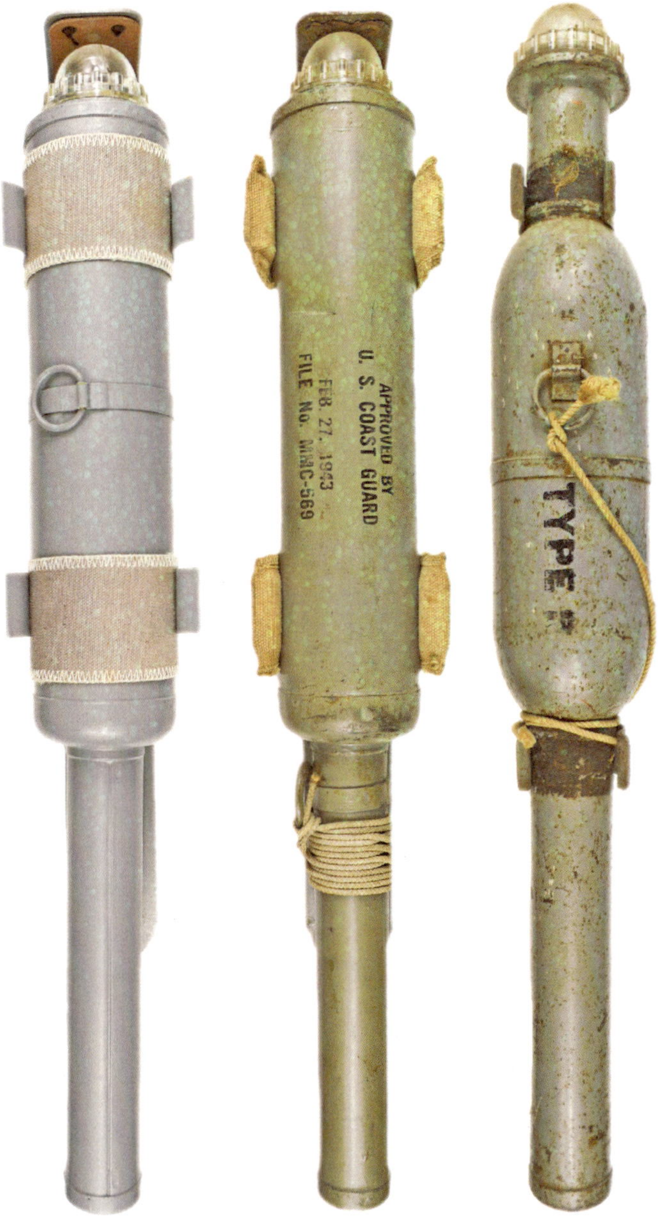

(Below) USCG approved types with celluloid bodies, principally used by the Merchant Marine.

(Below) Two images, c.1943, illustrating various types of floating lanterns used by nautical and aeronautical agencies. (National Archives)

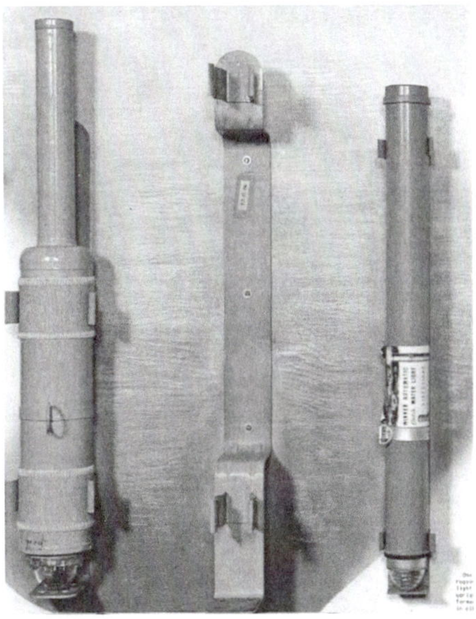

DISTRESS SIGNALS | Part VI. Whistles

Whistles, simple but effective, were incorporated into emergency rescue equipment used by both aeronautical services in WWII. A whistle could be used for communication and signaling, but most importantly of all as a distress signal.

The view towards whistles differed between the BuAer and the USAAF, with more value attached by the BuAer. The life raft and over-water kits developed by the USAAF included whistles, based more on maritime standards than any belief in actual practical use. The same could be said for the BuAer in the initial stages, but whistles would later prove to be vital location devices. For the BuAer, the whistle was first installed in life raft equipment, then back pad kits, then air-droppable units such as the Shipwreck kit. Gradually it became part of personal equipment.

Whistles helped attract searchers in low visibility situations like fog, cloud cover and low light. The whistle became of great importance within the BuAer during carrier operations at dusk and dawn. Aircraft were often required to operate in the dark, in the pre-dawn hours or after dusk upon returning from a mission. In either situation, aircraft could be forced to ditch for a variety of reasons. A survivor afloat could clearly see the spotlight of the searching craft, and by utilizing the whistle, could attract attention and create a definitive direction in which to search.

(Left) The BuAer adopted the whistle for life rafts in 1939. In 1940, the BuAer stipulated, by specification, that the whistle was to be "equal to that of the Official Scout Whistle as listed in the Local Council Exchange of the Boys Scouts of America." The scout whistle listed was nickel plated brass, often referenced as a Police Whistle. Pictured is a commercial police whistle made of nickel plated brass.

(Right) SBD pilot Lt. jg. Arnold Burrough of VB-16 has a "Police Special" whistle secured to the oral inflation tube of his life preserver, USS Lexington, December 1943. This is one of the earliest examples of whistles being worn as personal equipment and, like so many other articles covered in this study, it would become universal practice. (National Archives)

Due to wartime restrictions on the use of brass, plastic whistles were integrated to keep up with the demand and were supplied as CFE. For stock and replenishment purposes both metal and plastic types were available under ASO stock number R42-W-24000, titled Whistle, Police. The same was also true for the BuShips utilizing their stock number 74-W-70.

By 1944, plastic whistles had become the standard, with a bulk of CFE and other commercial types delivered directly to the services, but many were found to be deficient. What the services were receiving could be best viewed as cheap children's toys, often found to be dysfunctional and easily broken.

The BuShips began to draft a specification specifically for whistles, released early in 1944, as Navy Department Specification 74W1 (INT). The new whistles were to meet specific construction standards with the body made from either Bakelite or celluloid. Upon availability in mid-1944, each type was assigned separate stock numbers by the BuShips: 74-W-52 for Whistle, Bakelite, and 74-W-75 for Whistle, Celluloid. An additional type was available under stock number 74-W-54, simply titled Whistle, Signaling. These stock numbers, in conjunction with 74-W-70, caused confusion, with consistent requests for clarification within both the BuAer and BuShips. It is not clear which specific whistle was covered under stock number 74-W-54.

Specification 74W1 was officially standardized 1 September 1944. To further muddy the waters, by June 1945 the Bakelite whistle was re-designated under a new stock number 37-W-259, and the celluloid model under stock number 37-W-259-50. These new numbers officially superseded all previous types and stock numbers.

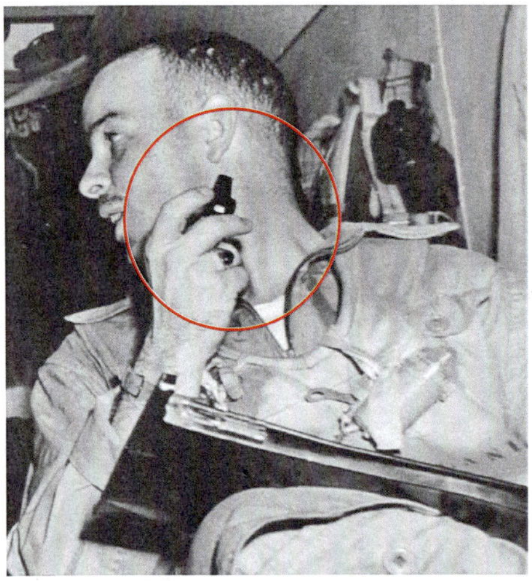

(Above left) As the war progressed, it became a recommended standard to have a whistle immediately available. The most common practice from late 1943 through 1944 was to remove the whistle from the emergency back pad kit and attach it to the life preserver. Pictured is a Naval aviator fidgeting with a whistle in the ready room of the USS Hornet, February 1945. (National Archives)

(Above right) The pattern is very much like that supplied by the New England Bedding Co. and issued in the M-592 back pad kits.

(Below right) Lt. Cdr. William B. Morton of VC-81 has a whistle tied to his life preserver aboard the USS Natoma Bay, December 1944.

(Below left) The whistle worn by Morton is of the same pattern as supplied in Belber Trunk & Bag M-592 back pad kits. (National Archives)

(Left) Specification 74W1 whistle fabricated from Bakelite or celluloid. Each whistle would be marked with the property mark USN, date of manufacture and manufacturer symbol. Dates of 1944 and 1945 can be found for both manufactures of this model. OD and yellow examples have been observed.

(Below) As supplies permitted, the 74W1 whistle would be adopted as standard by the BuAer for all emergency equipment and distributed as personal equipment.

(Left) Symbol for Eclipse Molded Products Co.

(Far left) Symbol for Plastic Engineering Inc.

(Above) An unknown TBF Avenger aircrew member of VC-96 on the flight deck of the USS Rudyard Bay has a Plastic Engineering Inc. whistle secured to his life preserver, April 1945. (National Archives)

(Above) Pilot Lt. jg Eugene Ternas of VT-16 has a black celluloid or Bakelite whistle secured to his oral inflation tube. USS Lexington, December 1943. (National Archives)

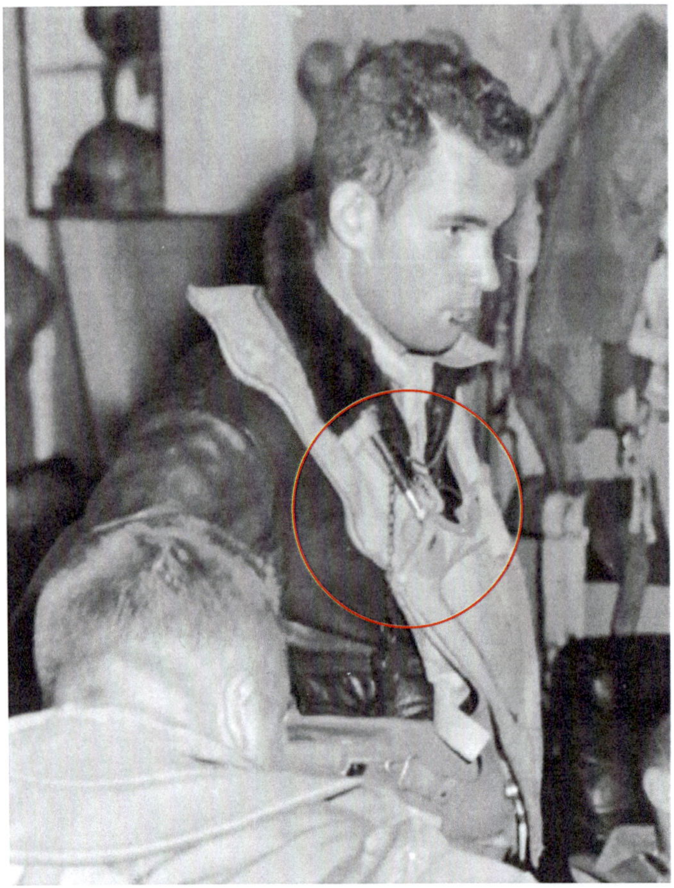

(Left) A pilot of VBF-83 has a 74W1 whistle secured to his life preserver by a chain lanyard. USS Essex, January 1945. (National Archives)

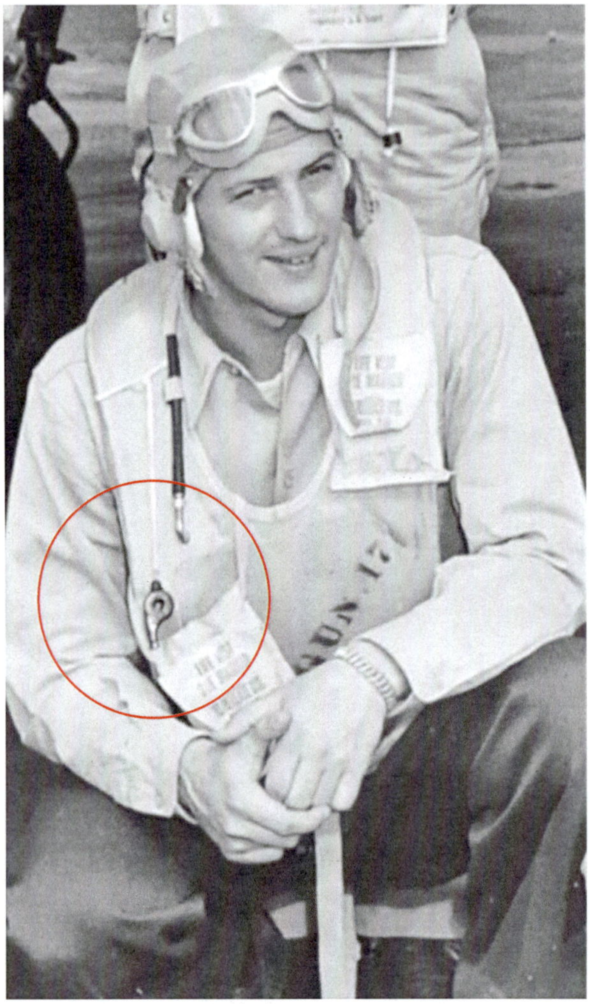

(Left) A third manufacture of the 74W1 of a different pattern. This whistle was manufactured by the Great American Plastic Co. and, as can be seen, it is only marked US NAVY.

(Above) This whistle being worn by Ens. Laush at NAS Melbourne, March 1945.

Whistles utilized by the USAAF would follow that of the BuAer, transitioning from metal to plastic, but only standard for life raft equipment. Whistles remained optional for personal use until the adoption of the Type B-5 life preserver, when they became a standard accessory.

It has proven difficult to find photographic evidence of whistles secured to life preservers of USAAF personnel operating in the Pacific Theater of Operations (PTO) and China Burma India theater (CBI), but it was a common occurrence in the European Theater of Operations (ETO). This fact is more related to operational situations; in the PTO it was found to have no real benefit, therefore no military requirement was established like that of the BuAer. However, a requirement was established in the ETO related to air sea rescue activities.

The principle workhorses of the Air Sea Rescue Service of the Royal Air Force were crash boats of RAF Coastal Command, which consistently patrolled the waters of the English Channel. Weather conditions in the channel often allowed only limited visibility, in which the whistle was best suited.

The whistle eventually became mandatory for all life preservers, effective 2 May 1946, being outlined in Technical Order 13-1-30. The Technical Order detailed that the lanyard of the whistle should be tied though the fabric loop used to retain the oral inflation tubes of the life vest.

(Above) The RAF provided a boat whistle as standard in their 1941 Pattern life preserver, which many American aircrew members used. This style whistle was available in plastic and metal. The USAAF was greatly influenced by equipment and procedures of the RAF, adopting many of their practices.

(Top right and right) RAF issue Air Ministry whistle usually worn attached to the flight suit or battle dress blouse.

(Above left) Somewhere in England, c.1944. Worn in a typical fashion on the zipper pull of his Type B-10 flight jacket, P-51 pilot Lt. Miles E. Elliot is utilizing an Air Ministry issue whistle. (National Archives)

(Above right) Pilot 1st Lt. James Hancock of the 38th FS has a typical pattern brass police type whistle secured to the zipper pull of his Type A-2 flight jacket in Nuthampstead, England. (National Archives)

(Left, upper and lower) Whistles stocked by the USAAF were requisitioned under stock number 6700-997970 titled simply Whistle, and described as "a generic plastic police type." Illustrated is one variant contained in life raft accessories for multi person USAAF life rafts c.1944. 1943 dated examples are seen which were procured by the QMC and referenced as the "Thunderer Model." USAAF life raft specifications required that they be marked "U.S. Army" only.

(Above) Commercial type plastic "Police" whistle which would have been utilized in emergency rescue equipment.

(Right) B-17 crew member of the 613th BS has a plastic police whistle secured to the zipper of his Type B-10 flight jacket, England, January 1945. (National Archives)

DISTRESS SIGNALS

Part VII.

Sails and Paulins

A protective cloth was one of the many items added to the compliment of equipment for life rafts, to protect the occupants in life rafts from the elements. Its inclusion was based on early experiences from survivors, and soon became a requirement.

The BuAer addressed the demand by standardizing a protective cloth 17 July 1942 titled Life Raft, Sail Fabric, under NavAer Specification M-551. The fabric measured 54" x 54" and colored orange-yellow on both sides for recognition purposes. It could also be used to catch water and provide shade. Though the vinyl resin fabric was called a "sail" by specification, it was more a general purpose paulin or tarpaulin, and officially classed as a cover. The cover was available under ASO stock number R83-C-76950, supplied as standard with the M-3 series life raft, 1 per MK-I and MK-II with 2 being provided as standard for the MK-IV and MK-VII. After January 1943 it was supplied with the pararaft as outlined in Specification AN-R-2A.

(Left) The M-551 sail fabric measuring 54" x 54" and colored orange-yellow on both sides. Fitted with grommets at each of the 4 corners for rigging as a sail or for other lashing purposes.

(Right) Two Naval personnel demonstrating the sailing capabilities of a life raft at NAS Norfolk, May 1943. Detailed sailing instructions were not provided with life rafts, so users were expected to improvise. In basic verbal instruction, one oar was to be wedged between the life raft bladder and the CO2 bottle. The oar, acting as the cross member, was to be lashed utilizing the 25 feet of 75# line provided in the supply pocket. In the revised specification M-3Q, a rubberized fabric socket was incorporated near the CO2 bottle for securing the butt of the oar. (National Archives)

By the summer of 1942 the USAAF was evaluating and revising their life raft equipment to include the standardization of paulins to suit their needs.

A feature the USAAF proposed was to include the color blue on one side of the cloth, adding camouflage and better signaling capabilities to the paulins.

This feature was incorporated into Amendment-1 of Specification M-551 for the addition of a non-specular blue to one side of the sail fabric 3 September 1942.

(Left) The revised sail M-551 with one side non-specular blue. All other construction details remained the same.

(Below)There was some variation with fabrics and markings relative to the contractor and supplier of the material. Shown is another variant.

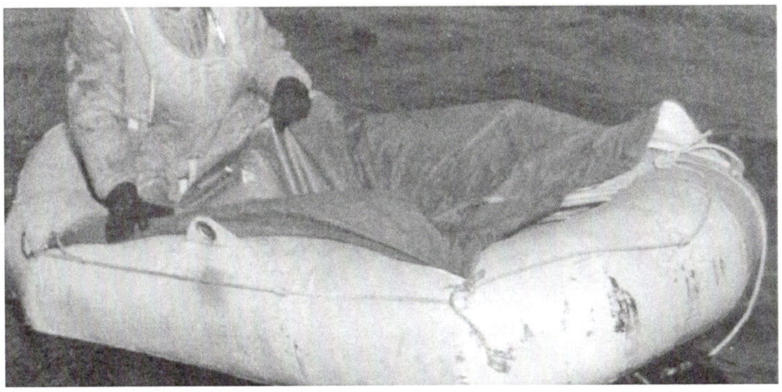

(Above) The cover M-551 shown lashed to the raft by utilizing the four grommets at the corners and being used to catch the rain for fresh drinking water, February 1944. (National Archives)

(Right) Two M-551 sails used together for shade over an M-3 series life raft. (National Archives)

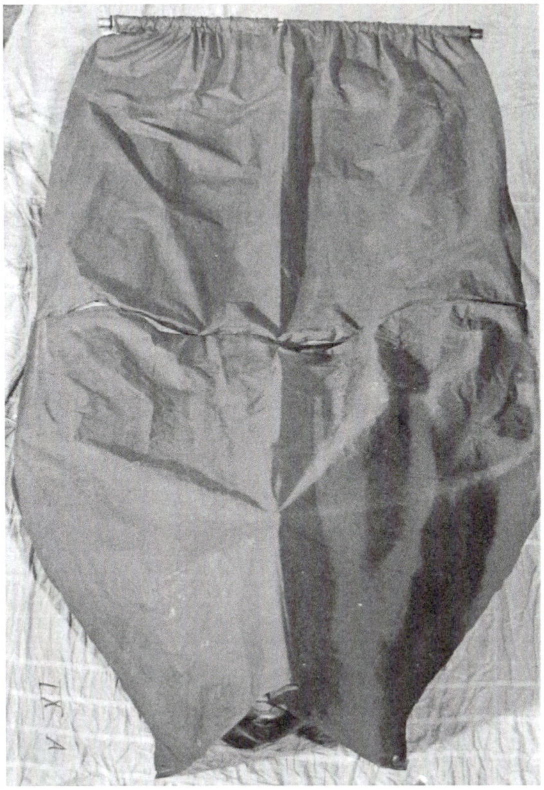

USAAF Specification No. 95-26756 of September 1942 was a general specification initially covering two sizes of paulins titled Paulin, Life Raft Utility, best describing them as large and small. The small paulin was almost a direct copy of the US Navy M-551, measuring 54" x 54", but with an added grommet in the middle, on opposite sides, making for three lashing points on two edges. Upon authority to procure the one-man parachute raft in August of 1942, the USAAF included the paulin as standard equipment prior to Specification AN-R-2A, additionally being manufactured with the non-specular blue side. One small paulin was standard for multi-person life rafts to be used as a combination water catcher, signal cloth, and sail, available under stock number 4500-743500 listed as Paulin, Sail, Size 18.

The M-551 and the USAAF Size 18 paulin were placed under joint Army-Navy form after the official inclusion of both in Specification AN-R-2A, and on 28 May 1943, the 54" x 54" sail fabric was standardized under joint Army-Navy Specification AN-S-34.

After May 1943, production of the sail fabric was executed under Part No. AN6521-1, being utilized by both aeronautical services in all life rafts. The AN-S-34 updated the sail by adding additional reinforced grommets, equally spaced along all four edges, retaining all other construction methods and materials.

Upon deliveries in late 1943, the AN-S-34 was placed under the respective service stock class for the M-551 and the Size 18 paulin, superseding these by specification only. By June 1944 the AN-S-34 was limited standard for Naval Aeronautical use, but remained in the supply system as optional equipment. Due to their elimination from one-man rafts under Specification AN-R-2B in February, the adoption of the new life raft paulin under NavAer Specification M-616 for multi-person life rafts took place in May. The USAAF were phasing out the small paulin from service, replacing it with the Size 35 sailing paulin and adopting spray shields for the one-man pararaft. The AN-S-34 remained in service for the duration of the war, being officially cancelled 20 November 1946.

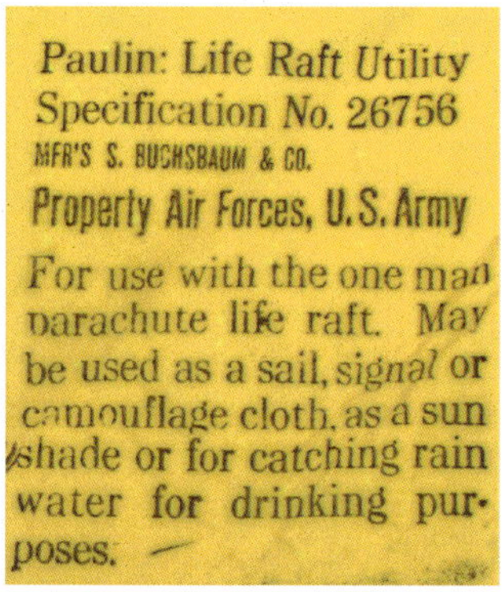

(Far left and left) Small utility paulin provided for one-man life rafts as referenced in the data printed on the fabric. The extra grommet can be seen in the middle of the top edge of the sail.

(Left) Printed data from an identical paulin. Though faded, the first sentence reads "For use with the five-man life raft". This is referencing the Type A-2 and early production Type A-3 life rafts. It also includes the contract and Size 18.

(Below) A variation of the USAAF small paulin marked with just the nomenclature, specification and property mark. As a standard stock item any variation would have been supplied to the one-man and multi-person life rafts.

(Left) The AN-S-34 Sail, Fabric, as titled by specification, is identical to the preceding patterns but with additional grommets for rigging.

(Above) The initial production included the drawing number AN-6521 in error.

(Below) The error was swiftly rectified by switching to the part number AN-6521-1 under which the contracts were actually awarded.

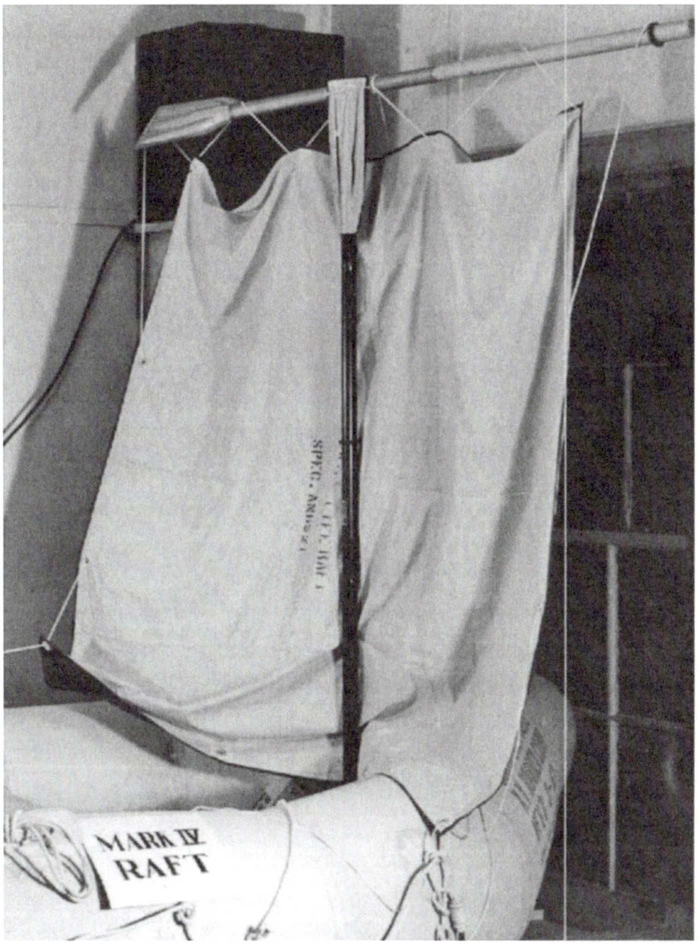

(Right) AN-S-34 sail rigged on a US Navy MK-IV M-3Q life raft. Unlike the rigging shown on page 374 with the paddle down for stability, the M-3Q raft provides a socket for the oar, placing the paddle up. The sail has two eyelets for securing the oar cross member and is laced to the horizontal oar by way of the four grommets. In addition, line is threaded down each side acting as toggles to manipulate the fabric while sailing. January 1945. (National Archives)

The USAAF large paulin was referenced as Size 77 under specification 26756, and was utilized in multi-person life rafts and sustenance kits as measuring 7'x 11.' It was to be used in conjunction with the Size 18 paulin as a shade and camouflage cover. Grommets were provided around the perimeter of the paulin for lashing purposes, and it was colored orange-yellow on one side and non-specular blue on the other. This paulin has remained as standard into the modern era, re-designated under MILSPECs.

(Above) The Size 77 Paulin shares all the same characterizes as the small paulins mentioned previously, other than its dimensions. Available under stock number 4500-635850 and listed as Paulin, Life Raft, Size 77. This paulin and the Size 18 were standardized prior to the adoption of the seven-man raft types E-1 and E-2.

(Below left) Examining surviving examples, it appears the contractors never updated the printed data to reflect the new person placement, the Size 77 retaining the "For use with the five-man life raft" label, which refer to the Types A-2 and A-3 life rafts.

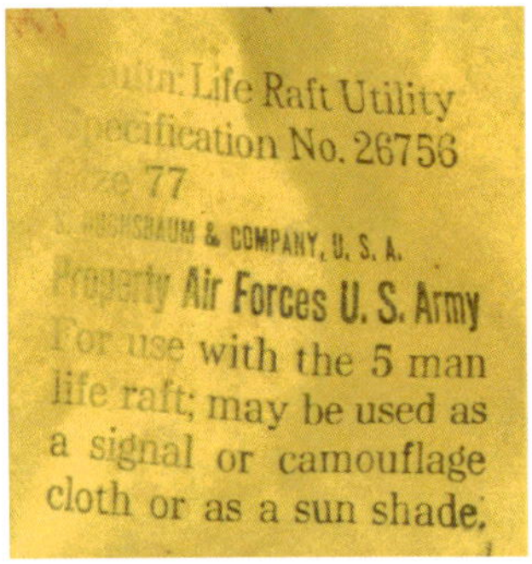

(Right) Size 77 Paulin is being demonstrated with the yellow side up during a field test of the Type E-1 seven-man raft on the Potomac River, January 1943. (National Archives)

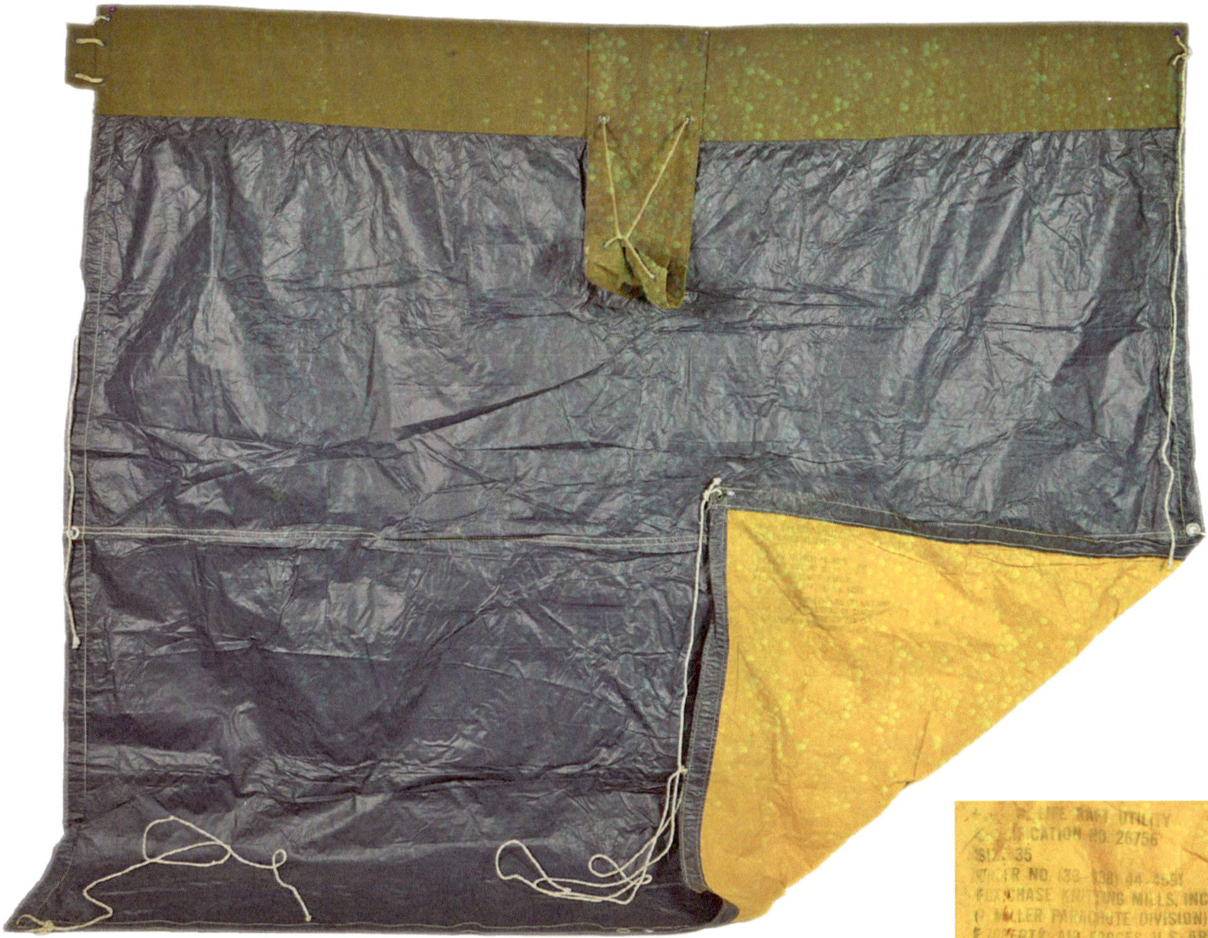

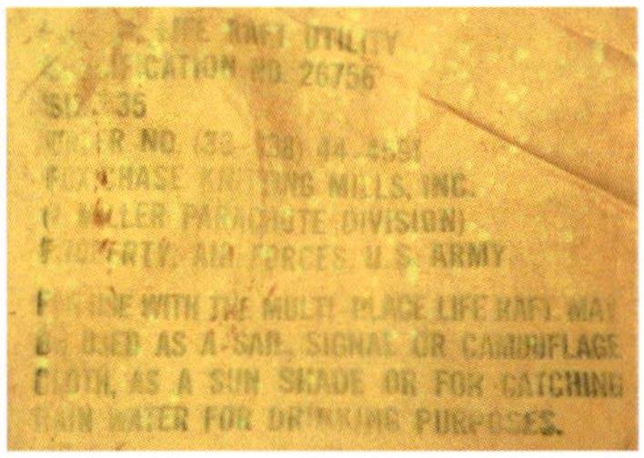

(Above and far right) An additional paulin was added to specification 26756 referenced as the Sail, Paulin Size 35 adopted in the spring of 1943. This paulin measured 6' x 5' and included sleeves for the oars to be used as a sailing rig. Orange-yellow on one side and non-specular blue on the other, this replaced the Size 18 for the Types A-3, E-1 and E-2 multi person life rafts available under stock number 4500-635825. The Size 18 or AN-S-34 still remained standard for pararafts and Type A-2 life rafts. All three paulins were supplied for multi-place rafts, however the Size 77 and Size 35 were the preferred types. Known contractors for all life raft paulins include Fox Chase Knitting Mills Inc., S Buchsbaum & Co., Buchman MFG., Troy Sunshade, and Scranton Love Lace.

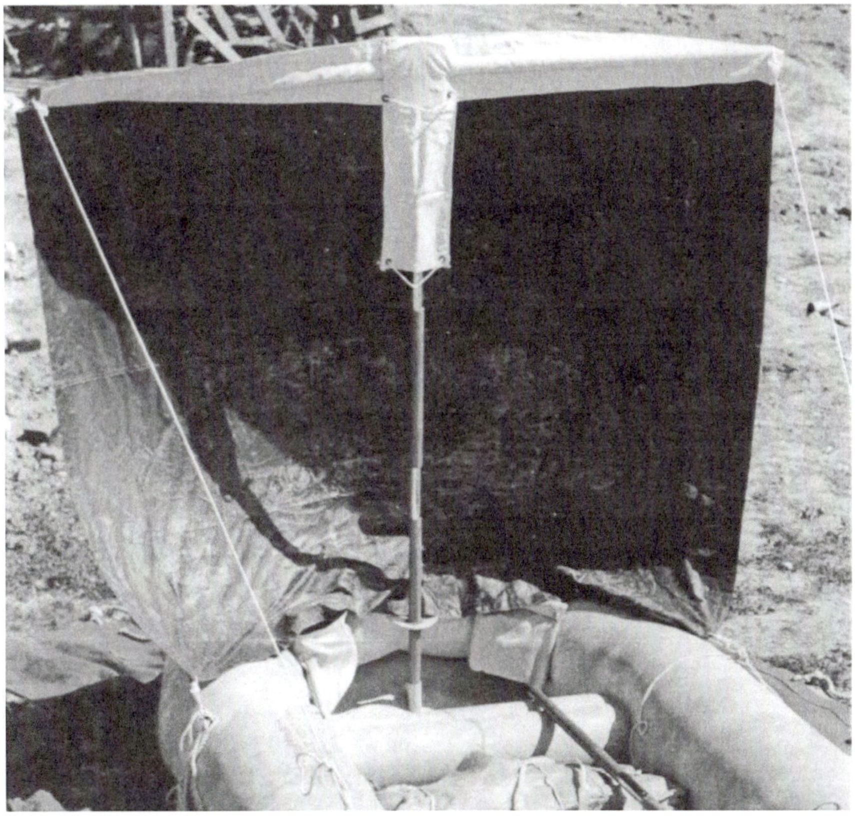

(Right) The size 35 sail demonstrated on an E-2 life raft, August 1943. Amended A-3, E-1, and E-2 rafts were equipped with a socket on the raft floor and a grommet cemented to the raft bladder to support the oar frame. The paddle of the vertical oar slipped into the sleeve along the top edge of the sail with an additional 6' oar in the horizontal sleeve. The third oar provided with the rafts was used as a rudder. (National Archives)

Several factors played a role in future development of life raft paulins within the BuAer. First to be established was that fresh drinking water was of vital importance to survivors. The second factor was protection of the survivors in life rafts from the elements – principally the sun – which accelerates dehydration and exhaustion. By late 1943, solar stills and chemical kits to supplement the water supply were in development, along with methods for storing the water collected.

The M-551 was found to be deficient for protection due to its relatively small size, but proved to be valuable in water collection from the rain. The BuAer developed an improved paulin for life rafts with emphasis on protection and water collection. This new paulin would be standardized 30 May 1944 titled as Paulin, Life Raft under NavAer Specification M-616.

There were three sizes of paulins covered under this specification, respective to which raft they were to be used with. Each was orange-yellow on one side and a non-specular blue on the other, and manufactured from a viscose rayon with nylon rip stop threads. This new paulin officially superseded specification M-551 and AN-S-34, and was incorporated into revised specification M-3R of May 1944.

There was a constant attempt to improve comfort aboard life rafts, being accomplished through various methods, as illustrated above with paulins providing protection from the elements. Other problems encountered were chaffing and salt water burns, and to address this issue sponges were adopted to help keep the raft dry. Hammocks were also introduced to suspend the survivor off the floor of the raft. Incorporated into the M-3R raft was a Hammock, Bed, and like that of the M-616 it was available in three sizes respective to the raft placement. The hammock was manufactured from a resin impregnated fabric that could be alternately used to collect water and provide shade. Wooden toggles secured the bed to web loops attached on the raft bladder.

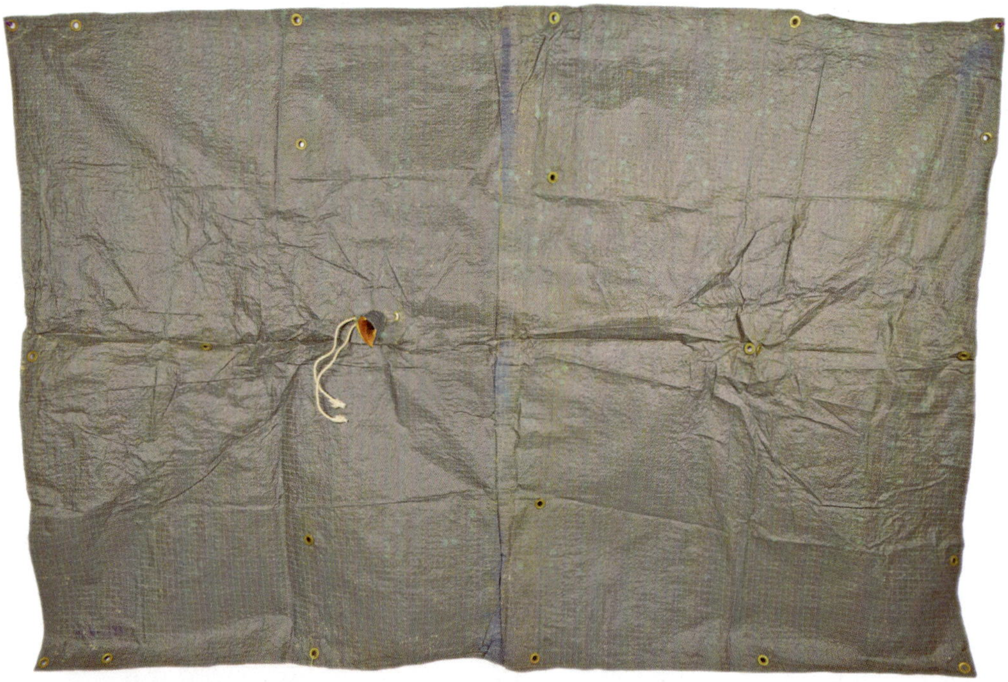

(Left) The M-616 was available in three sizes: 57" x 84" for the MK-II, stock number R83-P-92005; 72" x 108" for the MK-IV, stock number R83-P-92010; and 81" x 132" for the MK-VII, stock number R83-P-92020. Other features included specially located grommets along the edges and body of the paulin for multiple rigging purposes. Unfortunately no documentation has surfaced describing the installation methods or their specific intended purposes.

(Right) Incorporated into the paulin was a funnel to be used in conjunction with the five quart water bladder when collecting rain water.

(Left) Each paulin was stamped for the respective type raft with which it was to be used. The paulin pictured here is for the MK-II specification M-3R life raft.

(Left and below left) Front and back views of the Hammock, Bed which were available in three sizes: 33" x 45" for the MK-II, stock number R83-H-5000; 40" x 48" for the MK-IV, stock number R-83-H-5001; and 48" x 39" for the MK-VII, stock number R83-H-5002.

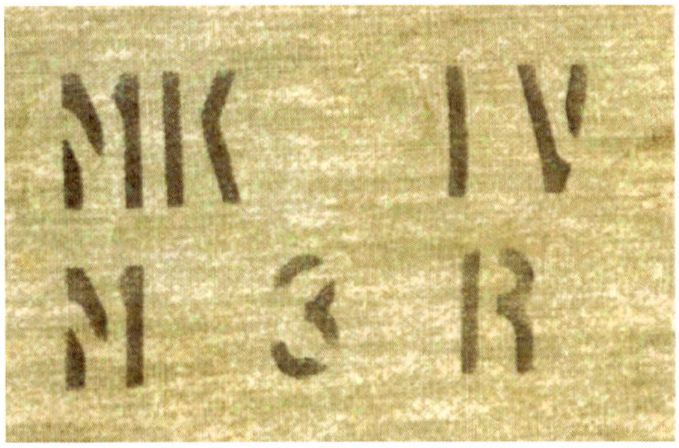

(Above) Each hammock was stamped with the corresponding type raft.

(Below) Detail of one of the 8 wooden toggles on the hammock.

(Below) The hammock was designed to turn half the raft into a bed, utilizing the bladders to increase overall square footage.

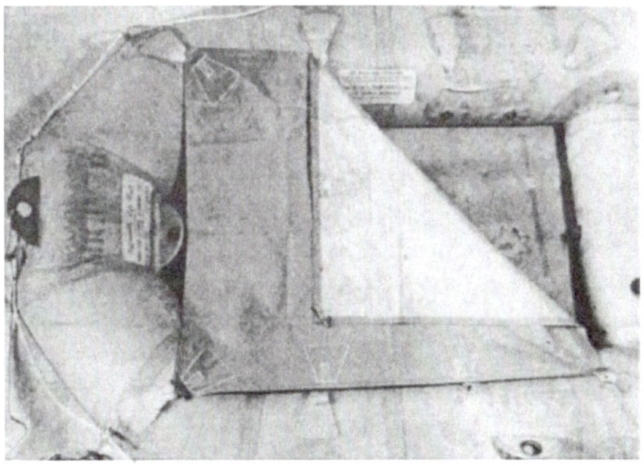

Life raft paulins were designed to serve multiple functions. Orange yellow was highly visible for signaling, and the blue side for camouflage. In December 1942, the Materiel Command recommended the distribution of 12 signals that were devised using the signal panel included in the B-4 parachute kit. Paulins could be folded to form various shapes, providing color contrasts, enabling communication between survivors and rescue sources.

The 12-panel communication system was first introduced and distributed in Arctic and Jungle and Desert Emergencies booklets, printed in the spring of 1943. Initially the messages were established at local levels and not universally with other units or services, particularly within the US Navy.

An improved joint Army/Navy message system

standardization was implemented in 1944, reflected through BuAer Technical Note 91-44 of 16 October 1944.

The new A/N system doubled most signals for translations "Over Land" and "Over Water," increasing the count to 20 possible messages. The messages were then suitable for joint operations over land or water. Upon adoption of the A/N system, the BuAer procured 300,000 rayon acetate charts to be supplied with all emergency equipment under contract 288s-26292, 11 October 1944, available under ASO stock number R83-C-22250, as Charts, Panel, Emergency Signaling.

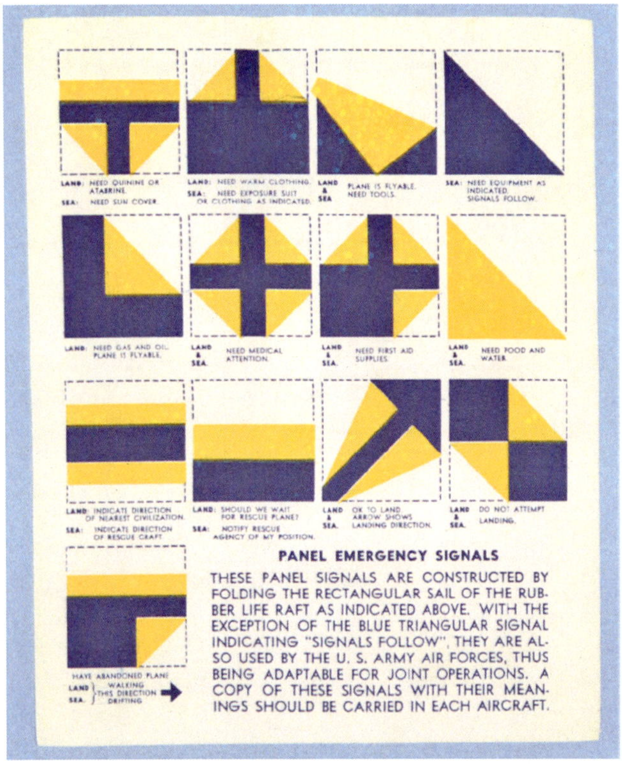

(Left) The single-sided rayon acetate chart, adopted for US Navy emergency equipment, illustrates the panels utilized in the joint A/N messaging system measuring 5" x 6 1/4". Minimal quantities of these charts were available by January 1945. T.N. 91-44 states that acknowledgement of the message is made by a series of slight zooms by the circling aircraft. Prior to the issuance of T.N. 91-44 the BuAer did not utilize a standard uniform system.

(Right) Two naval aircrew members demonstrate the universal panel for both land and sea which translates as "Need medical attention". (National Archives)

(Below) Two naval aircrew members are demonstrating a message over water which translates as "Notify rescue agency of my position", the same panel used over land would translate as "Should we wait for rescue plane?" (National Archives)

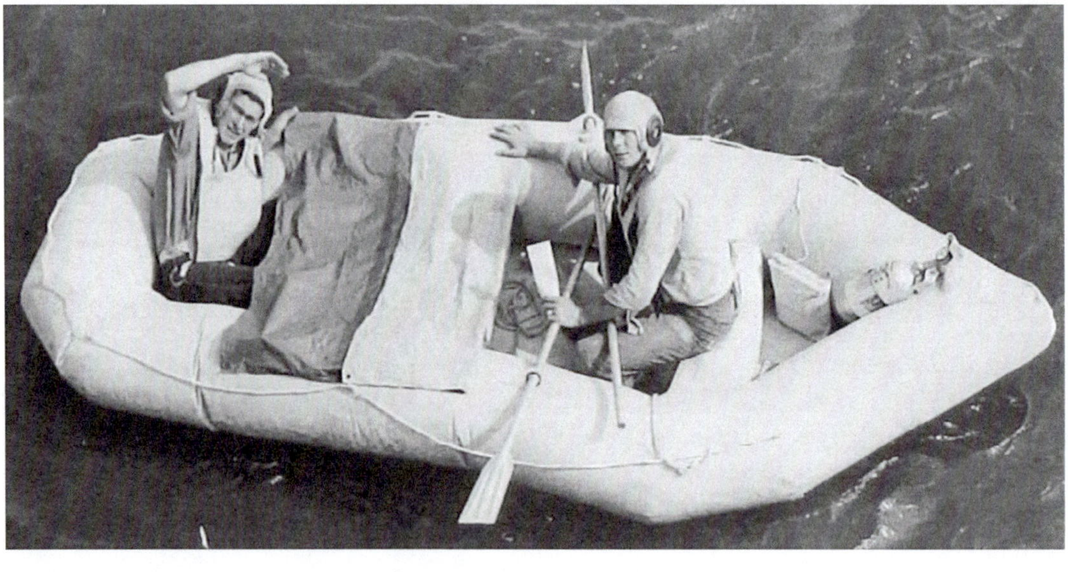

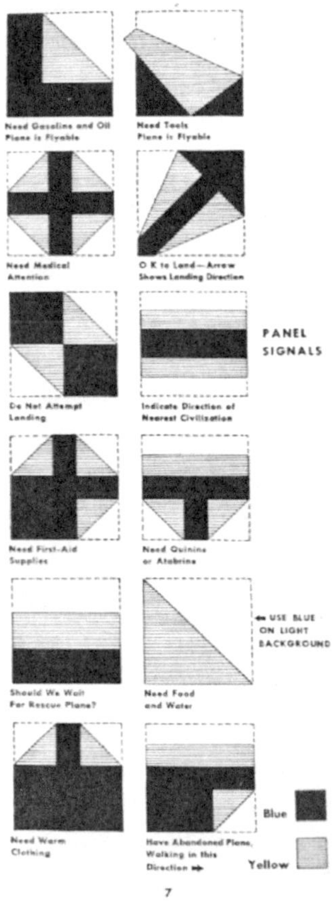

(Left and far left) The standard USAAF signaling system implemented in the spring of 1943, incorporating communication through both the paulins and hand signals. It also includes how the spotting aircraft is to signal for a "Yes" or "No". These signals would be formalized under a joint A/N system in 1944, adding alternate paulin communication for "Over Sea". The pages shown are from AAF manual 21W. The new A/N system was illustrated in the revised Manual for Personal Equipment Officers printed in June 1945.

(Below) A Size 77 paulin being utilized in this simulated jungle survival situation for signaling a searching aircraft, stating they need quinine or Atabrine. In the left hand of one of the survivors is the emergencies booklet he referenced to fold the panel. (National Archives)

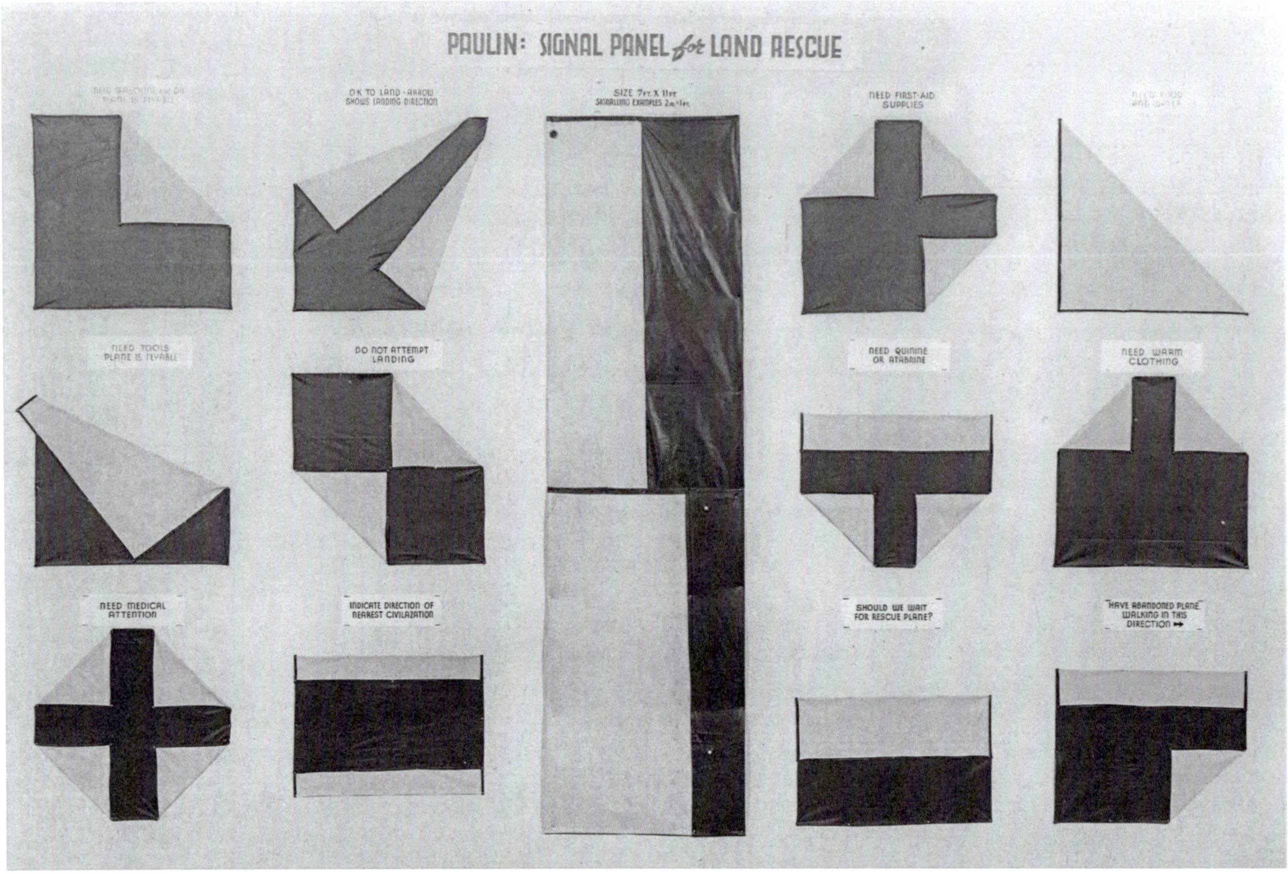

(Above) Wartime full-size display board illustrating signal panels for land rescue. The same panels were used for messages for water rescue, but of differing translations. (USAF)

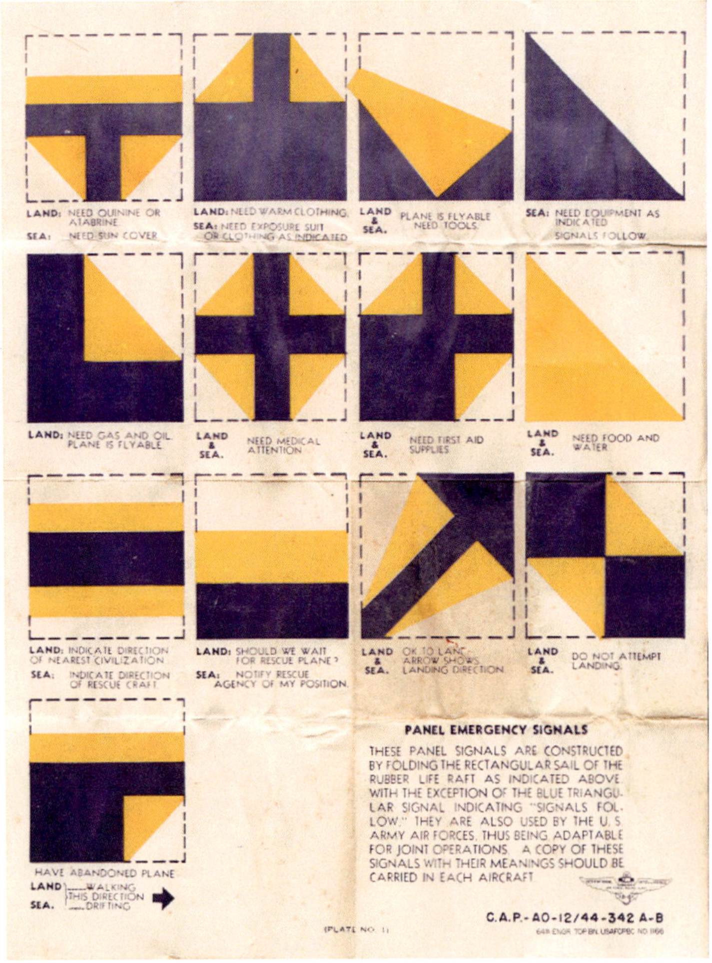

(Left) 8" x 11" paper sheet printed in Hawaii by the 64th Engineer Topographic Battalion in December 1944. It's a direct copy of Plate No. 1 included in T.N. 91-44. It was found folded and stuffed in a pocket of an M-592 back pad kit.

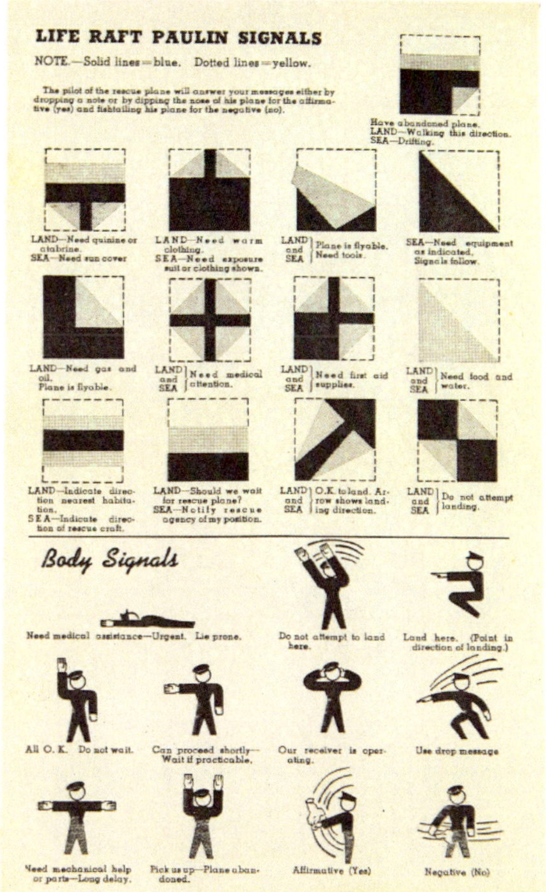

LIFE RAFT PAULIN SIGNALS

NOTE.—Solid lines = blue. Dotted lines = yellow.

The pilot of the rescue plane will answer your messages either by dropping a note or by dipping the nose of his plane for the affirmative (yes) and fishtailing his plane for the negative (no).

Have abandoned plane. LAND—Walking this direction. SEA—Drifting.

LAND—Need quinine or atabrine. SEA—Need sun cover

LAND—Need warm clothing. SEA—Need exposure suit or clothing shown.

LAND and SEA — Plane is flyable. Need tools.

SEA—Need equipment as indicated. Signals follow.

LAND—Need gas and oil. Plane is flyable.

LAND and SEA — Need medical attention.

LAND and SEA — Need first aid supplies.

LAND and SEA — Need food and water.

LAND—Indicate direction nearest habitation. SEA—Indicate direction of rescue craft.

LAND—Should we wait for rescue plane? SEA—Notify rescue agency of my position.

LAND and SEA — O.K. to land. Arrow shows landing direction.

LAND and SEA — Do not attempt landing.

Body Signals

Need medical assistance—Urgent. Lie prone.

Do not attempt to land here.

Land here. (Point in direction of landing.)

All O. K. Do not wait.

Can proceed shortly—Wait if practicable.

Our receiver is operating.

Use drop message

Need mechanical help or parts—Long delay.

Pick us up—Plane abandoned.

Affirmative (Yes)

Negative (No)

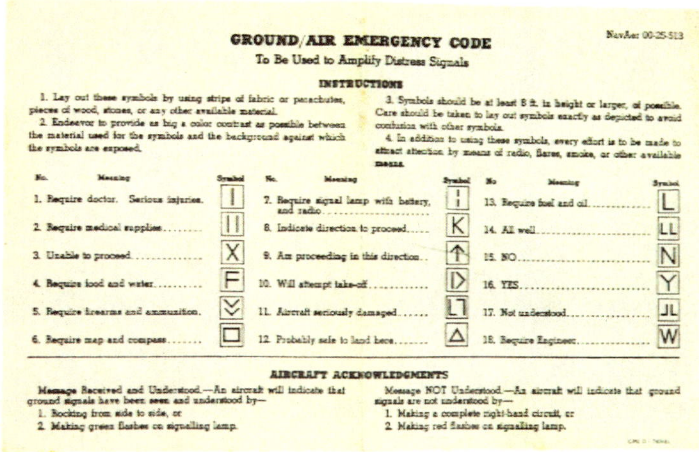

NavAer 00-25-513

GROUND/AIR EMERGENCY CODE

To Be Used to Amplify Distress Signals

INSTRUCTIONS

1. Lay out these symbols by using strips of fabric or parachutes, pieces of wood, stones, or any other available material.
2. Endeavor to provide as big a color contrast as possible between the material used for the symbols and the background against which the symbols are exposed.
3. Symbols should be at least 8 ft. in height or larger, if possible. Care should be taken to lay out symbols exactly as depicted to avoid confusion with other symbols.
4. In addition to using these symbols, every effort is to be made to attract attention by means of radio, flares, smoke, or other available means.

No.	Meaning	Symbol	No.	Meaning	Symbol	No.	Meaning	Symbol
1.	Require doctor. Serious injuries.	I	7.	Require signal lamp with battery, and radio	¦	13.	Require fuel and oil	L
2.	Require medical supplies	II	8.	Indicate direction to proceed	K	14.	All well	LL
3.	Unable to proceed	X	9.	Am proceeding in this direction	↑	15.	NO	N
4.	Require food and water	F	10.	Will attempt take-off	|▷	16.	YES	Y
5.	Require firearms and ammunition.	≚	11.	Aircraft seriously damaged	⌈⌉	17.	Not understood	JL
6.	Require map and compass	□	12.	Probably safe to land here	△	18.	Require Engineer	W

AIRCRAFT ACKNOWLEDGMENTS

Message Received and Understood.—An aircraft will indicate that ground signals have been seen and understood by—
1. Rocking from side to side, or
2. Making green flashes on signalling lamp.

Message NOT Understood.—An aircraft will indicate that ground signals are not understood by—
1. Making a complete right-hand circuit, or
2. Making red flashes on signalling lamp.

(Left and above) The A/N systems were again revised to a joint Army/Navy/British Ground/Air Emergency Code in 1946. This system incorporated the USAAF hand signals and new navy ground signal. The new ground signals were merely a series of shapes that could be constructed by any contrasting method, such as scratching them in the sand or using logs or branches. A two-sided paper chart was provided for life rafts and emergency equipment measuring 5"x 6" as Naval Aeronautical Publication 00-25-513.

(Below) Paulins manufactured post-war included a signal chart permanently laminated in one corner.

DISTRESS SIGNALS

Part VIII.
Transmitters

Two fundamental operations are necessary for a successful rescue: firstly the initial alert must be sent from the distressed aircraft in order for air sea rescue organizations to act. Secondly, the position of the survivors must be determined by the rescue party.

Aircraft were provided with electronic radio and radar systems for a multitude of functions, including emergency and distress communication. The first two direction-finding networks employed were a high frequency (HF) and medium frequency networks, then integration of a very high frequency (VHF) during the war. VHF had the advantage of a longer range and was limited to line of sight, which allowed the use of radio communication when it was necessary for radio silence on HF systems.

When an aircraft was in distress, they would first turn on their IFF (Identification Friend or Foe) transmitter. This system communicated with direct finding stations via radar with a distress code. They would then verbally transmit their position over designated HF/VHF frequencies. If this proved unsuccessful, a broadcast would be transmitted on the frequencies of 500 kilocycles and 8280 kcs. 500 kcs had been used for many years before the war as an international distress frequency, intended as a navigational aid to surface vessels.

Since the transmission of medium frequencies over long distances required power not usually available in aircraft transmitters, the range, under optimal conditions, was within 300 miles. This frequency was reserved and continuously monitored by watch stations. In 1944, the United States Joint Chiefs of Staff established a VHF emergency and safety frequency of 8280 kcs, which effectively doubled the range of the 500 kcs. With HF and VHF in continued use in an emergency, ground stations and searching aircraft or vessels were able to triangulate and home in on the position of the distressed aircraft.

Before WWII there was little need for an extensive air sea rescue organization and, as mentioned previously, there were very few planes attempting long over water flights. If an aircraft was forced to ditch, international merchant shipping was considered sufficient to lend aid. With an increase in ocean traffic, both on the surface and in the air, demand for rescue coordination grew greater.

Great Britain was the first to establish dedicated resources towards rescue, in May 1940, in the Dover area. Due to their success, a deputy directorate was established as a branch of the office of the British Directorate General of Aircraft Safety, unifying Fighter, Coastal, and Naval Commands for air sea rescue duties. As part of this organization a system of communication networks (or chains) were established along the coastlines of Great Britain. The British developed, and were the first to implement, the IFF system, which allowed radar watch stations to establish bearing and range. In basic terms, radar, which is the interrogator, communicated with the aircraft transponder in code. Through these frequencies, identification of friend or foe could be determined. Updated systems would allow limited communication capabilities using IFF. When using HF and later VHF radio transmissions, Direction Finding (DF) stations could triangulate to determine the location of the aircraft. As long as either or both of these systems remained in operation, rescue resources could home in on survivors. In reality, aircraft quickly lost power transmission, could not transmit, and were no longer able to be located. Battle damaged planes often lost the ability to transmit in flight, or their flight path may carry them some distance from the last known location. The emphasis shifted towards survivors being able to communicate once the aircraft was abandoned.

In spring 1941, a captured German emergency transmitter was evaluated by British and USAAF representatives in Britain. Of immediate interest was its intended purpose of being a life raft transmitter, which included a hand-generated power supply. This was recognized as having benefits for locating survivors at sea. Since multiple samples were available, one was sent to the USAAF Aircraft Radio Laboratory for evaluation. The US Army Signal Corps and Bendix Aviation Limited became involved with reverse engineering of the German emergency transmitter, as well as with extensive testing of the existing model. Through a correlation of multiple branches and services it was decided to adopt a radio set for emergency life rafts carried in aircraft. By the close of 1941, Bendix had submitted a suitable transmitter, which they had designated as the Model T.1333. In January 1942, the Chief of the Army Air Forces recommended the emergency transmitter for standardization.

A Signal Corps Technical Committee then drafted preliminary guidelines for military characteristics for a standardized model, assigning the designation as the Radio Set SCR-578. In addition, the British negotiated procurement of 15,000 units through a reverse lend lease contract to Bendix Aviation.

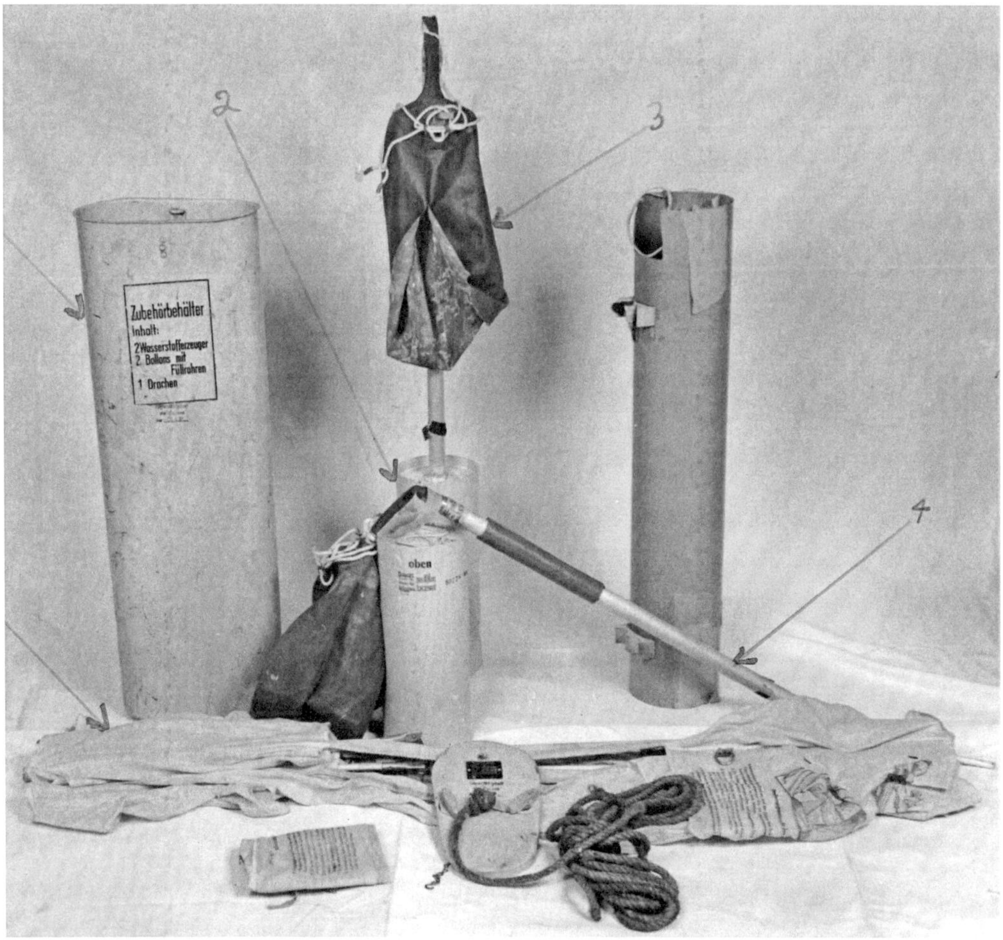

(Left and below) German NS2 (Notsender) Emergency Transmitter. The complete assembly was known as the NSG2 Not Sende Gerat 2. It operated at a frequency of 500 kHz. Accessories were stowed in a metal container and comprised two balloons, two hydrogen generators with inflator tubes, a box kite and instruction booklet. The transmitter included at 260' long, stainless steel wire antenna on a reel with a ground wire. The British T.1333 closely copied the design of this transmitter. (National Archives)

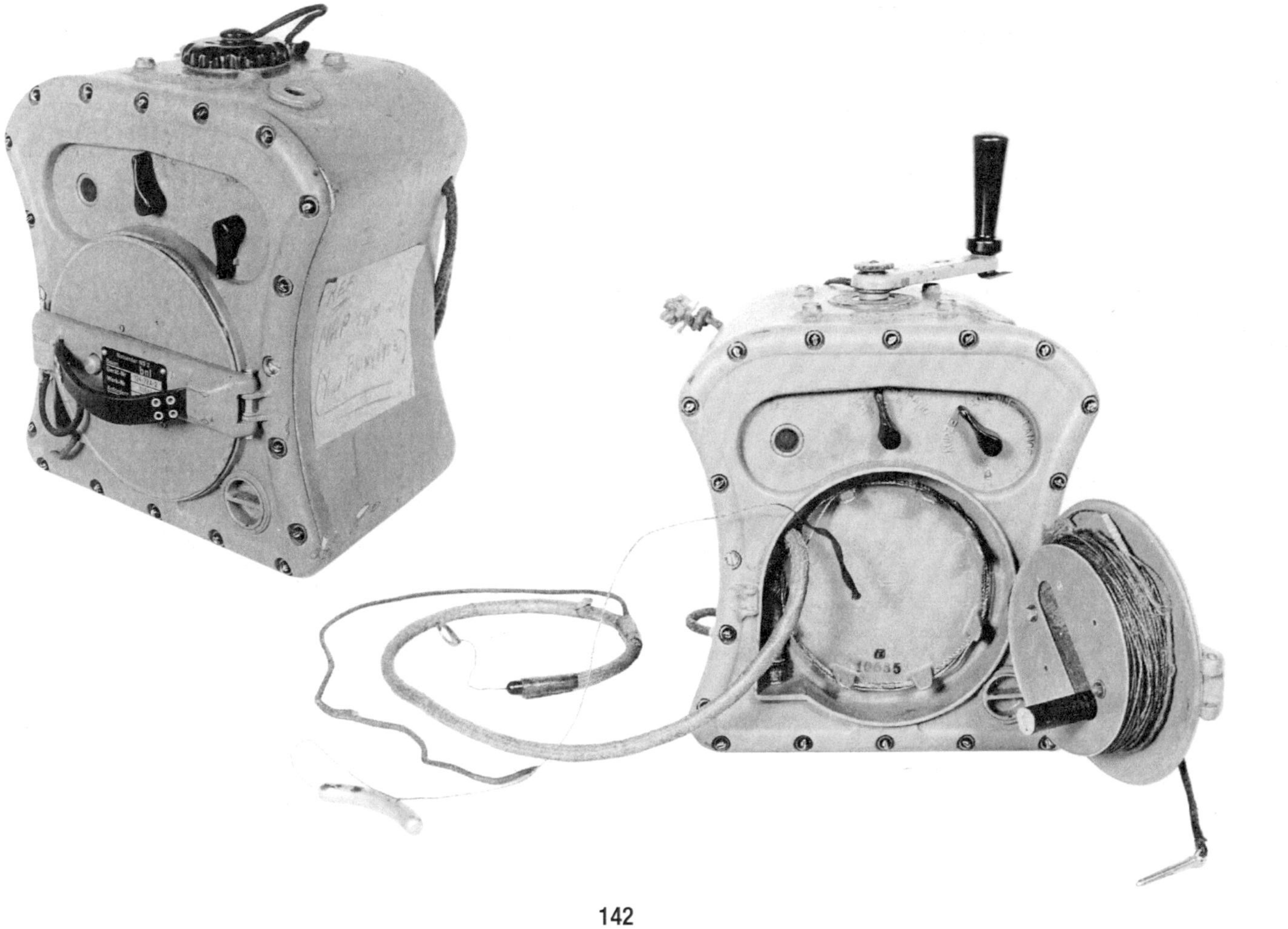

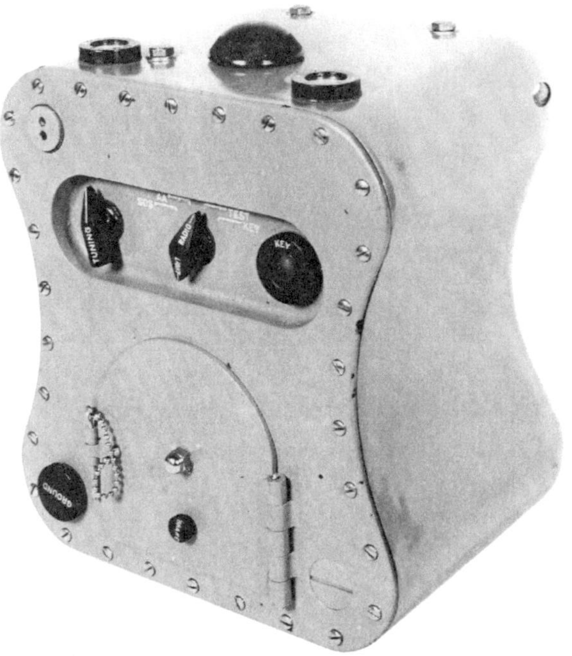

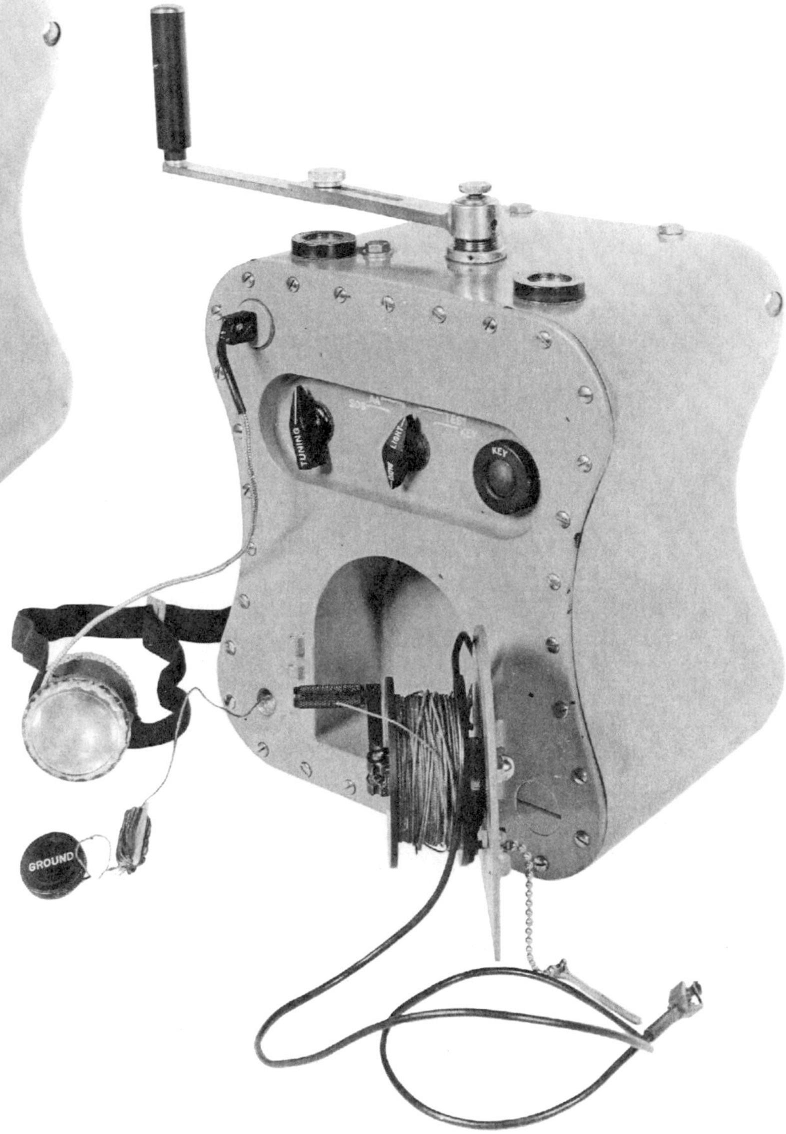

(Above and right) The first generation transmitter developed by Bendix Aviation Limited, designated as Type 3011 (known by the Signal Corps as the Radio Set). The general appearance of the radio is much like the NS2, with improvements. This unit is a portable radio transmitter, powered by an integral hand driven generator. It is capable of remaining afloat in sea water and includes automatic and manual keying of radio and signal light. It was designed to be carried aboard all types of bombers and transport aircraft, to be used in case of forced landings at sea by personnel forced to take to rubber boats. In addition, it was designed to be air droppable. Originally intended to become part of the life raft accessories, it added too much bulk for that application. Instead it was to be stored in a convenient location, specifically in the radio compartment aboard bombers. The Chief of the Air Corps requested the procurement of 1,000 sets as of January 1942, and for them to be placed into service as soon as practical. The transmitter, now designated BC-778, was not suitable for mass production at that time, as it still did not meet established requirements, principally remaining under development for the remainder of 1942.

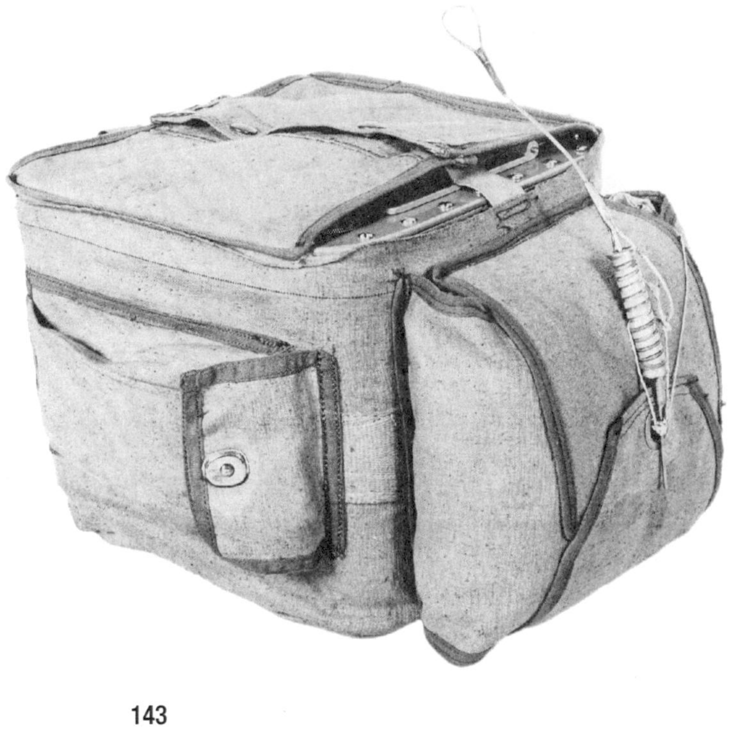

(Right) The BC-778 was packed in a fitted, padded case which protected the device but added extra bulk.

(Right and below) By June 1942, further refinements were integrated into the radio set, primarily in the internal components for power and durability purposes. Other external differences can be noted as well. The complete assembly with accessories was designated as the SCR-578, and the radio itself was designated as the BC-778. Production of the set was limited, with a minimal quantity available by the end of 1942. The SCR-578 was adopted as standard 11 November 1942, under Signal Corps Specification No. 27-1198 and officially designated as the SCR-578-A. The USAAF was the only using branch with control of distribution by the Signal Section, Air Service Command. For replenishment of parts the complete assembly was maintained under Signal Corps stock. (National Archives)

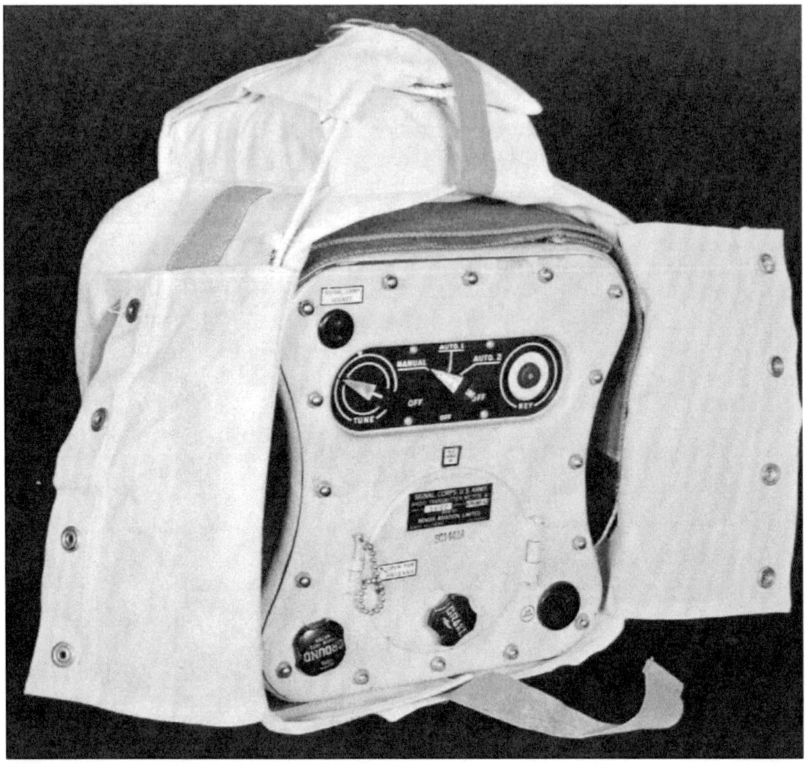

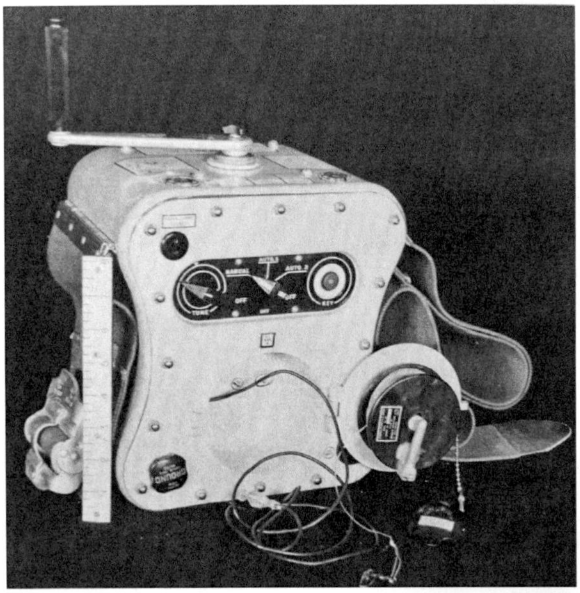

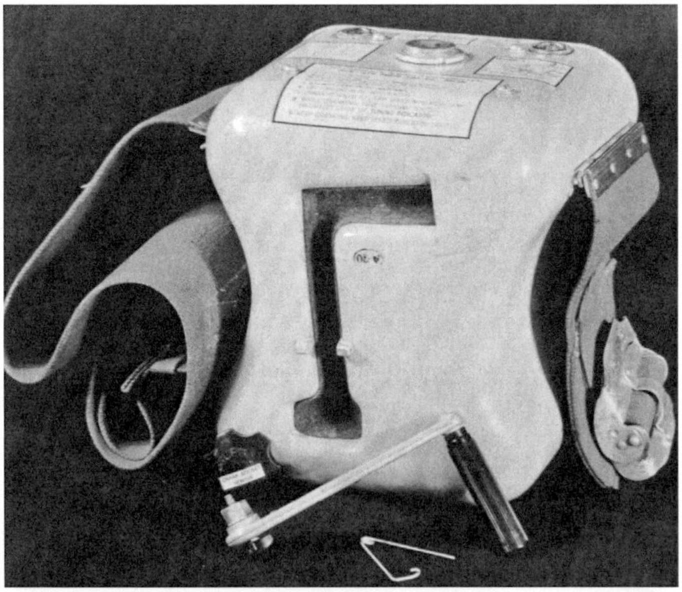

(Right) The complete assembly was stowed in two orange/yellow waterproof bags designated as the BG-109-A for the accessories and BG-110-A for the radio. (National Archives)

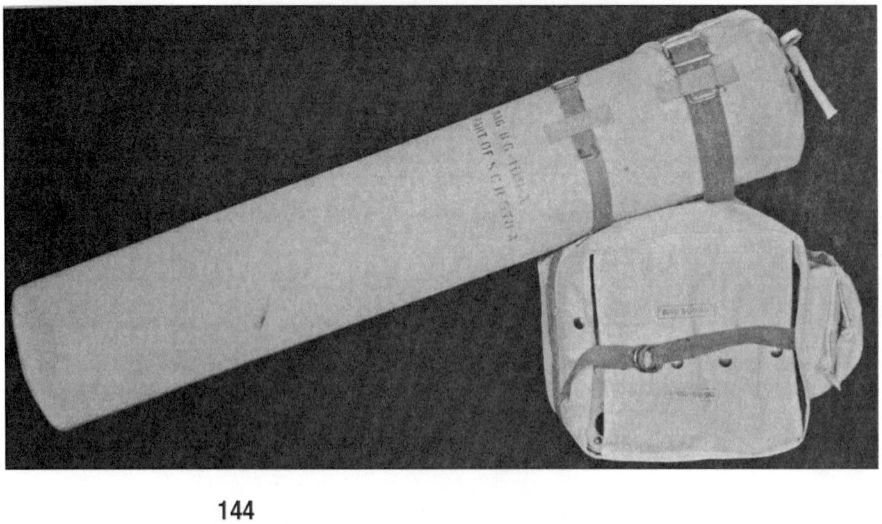

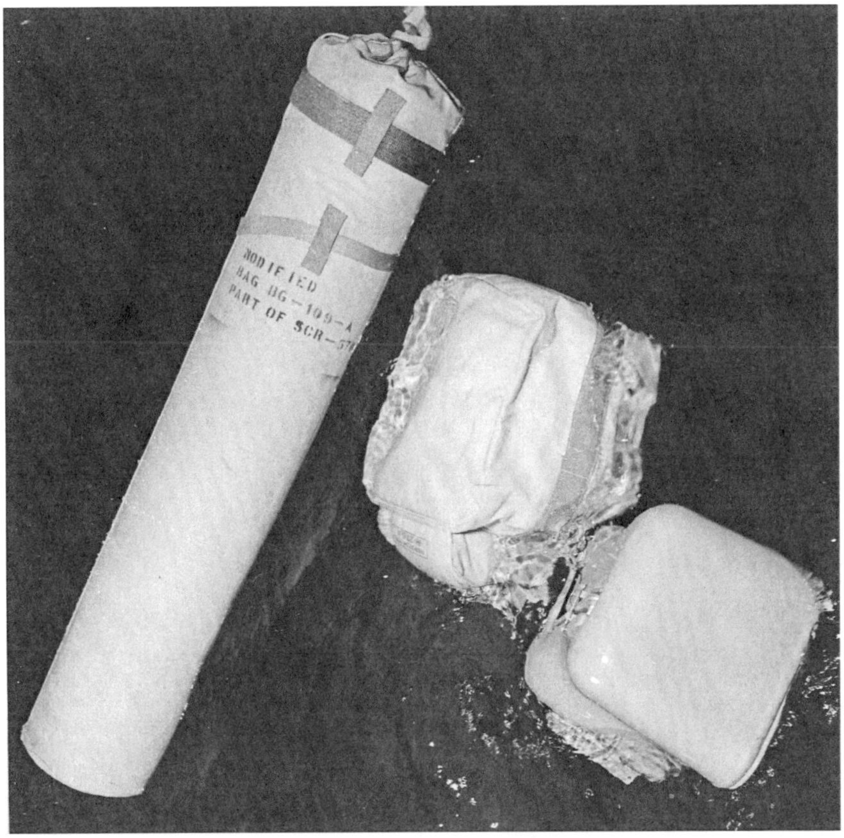

(Left) The two containers were designed to float as a unit or individually. (National Archives)

(Below) The compliment of accessories for the SCR-578-A included two balloons, two hydrogen generators with inflation tubes, signal lamp, spare wire, box kite, and instruction booklet-all contained in BG-109-A. (National Archives)

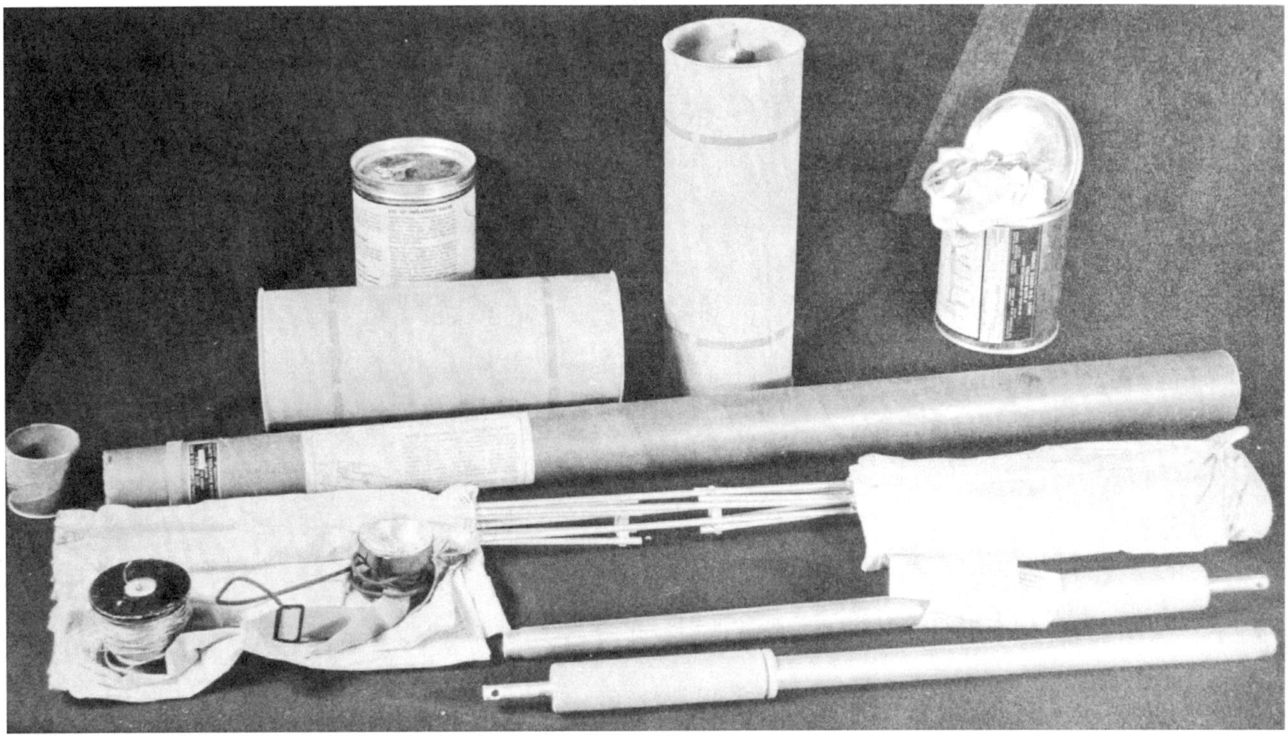

Procurement of the radio set was executed through the Wright Field Signal Corps Procurement Division, maintained and distributed through the Air Service Command. Due to the limited initial supply available, its distribution was controlled, being supplied to aircraft at departure points, staging areas, or embarkation points in conjunction with life rafts for overseas service. Coastal patrol aircraft would be equipped with the radio set when supply permitted.

The SCR-578-A was considered a component to life rafts and was to be used in conjunction with them. Being a Signal Corps division, it was assigned SC stock number 2Z578A. Upon adoption of the SCR-578-B the letter designation would be changed, and on 1 January 1943, radio sets would begin to be allocated to the BuAer available from the ASO under stock number R16-T-9169.

The international frequency of 500 kcs by international law was to be watched during three minutes following the quarter hour (15 minutes) and three-quarter hour (45 minutes) after the hour. All ships, as well as patrol and transport aircraft, had equipment for keeping watch on this frequency.

For operation of the BC-778-A, the selector switch marked RADIO and LIGHT were set to the AUTO 1 or AUTO 2 positions. The crank was rotated clockwise at 80 RPM, and when sufficient speed had been obtained an indicator light on the top would illuminate. The TUNE dial was adjusted to provide optimal transmission, which was indicated by the brilliance of the indicator on the top.

In Auto 1 position, the transmitter would automatically alternate between SOS for 20 seconds and a continuous dash for 20 seconds. In Auto 2, it would automatically alternate between AA for 20 seconds and a continuous dash for 20 seconds. In the MANUAL position the operator could send Morse code messages by the button marked KEY.

For use of the signal lamp, the selector switch marked LIGHT was placed in the AUTO 1, AUTO 2, or MANUAL positions. No radio transmission occurred when the signal light was used. Through IFF, HF, and VHF radio and radar transmissions, a distressed aircraft could communicate and be located. With loss of these resources, the signal could be regained by rescue parties through the automatic distress signal by the SCR-578-A to permit bearings.

(Right) Together with BG-109-A, the BG-110-A comprised the Radio Set SCR-578-A weighing 34 pounds. The bag contained the radio transmitter BC-778, instruction booklet AN 08-10-94, and permanently attached parachute M-276-A.

(Above and right) Details of the bag. Note loops for the attachment of BG-109-A.

(Below left) Stowage of the unit would be relative to the aircraft. For transport types the typical application was to secure it to the cargo door for easy access in the event of an emergency, as seen on this C-47 in the South Pacific. It was prescribed that the unit be stowed in the radio compartment in bombardment aircraft, or close to an emergency exit.

(Below right) A SCR-578-A can be seen secured to the fuselage near the waist gunner position of a B-17 in the 388th BG operating from England. (National Archives)

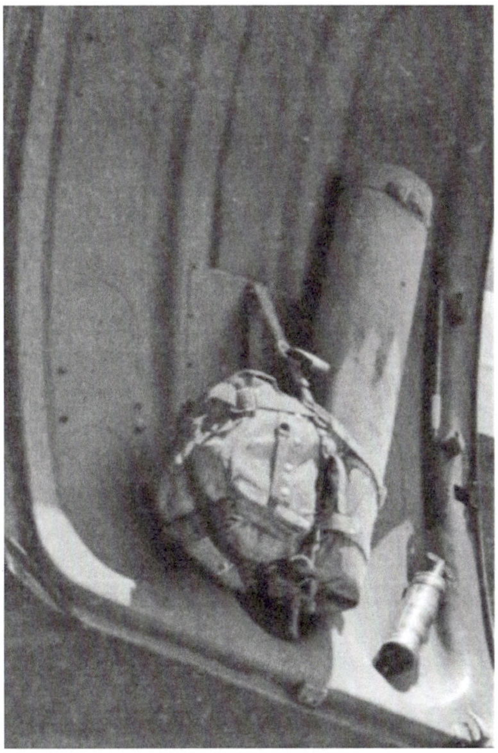

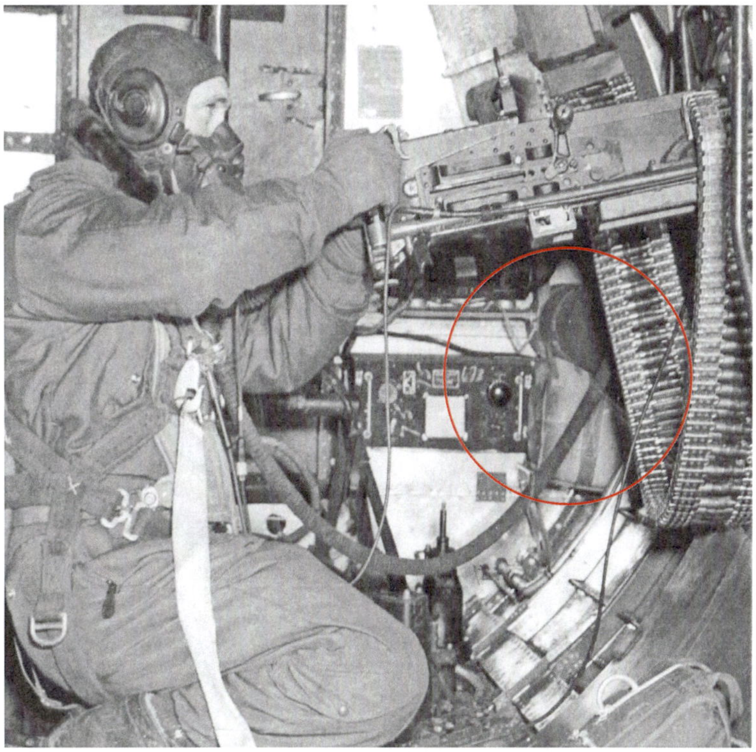

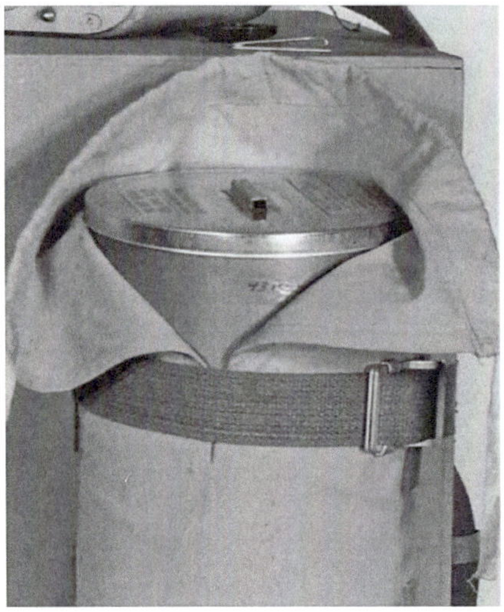

(Far left) The BG-109-A consisted of an orange/yellow canvas draw string bag with a cardboard inner sleeve measuring 39.5" x 7" in diameter. It contained the kite, balloons, generators, signal lamp, spare wire and an additional instruction booklet AN 08-10-94. Two straps with buckles enabled attachment to the BG-110-A.

(Left) The cardboard tube had a thin sheet metal bottom and a removable cap.

(Below) Drawing from the instruction booklet showing the proper packing of accessories in the tube.

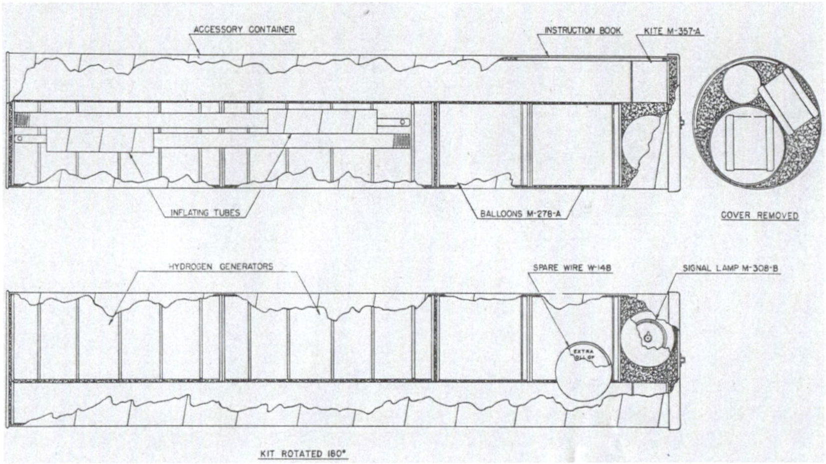

(Below left) From late 1942 through spring of 1943 there was a series of progressive improvements in design and accessories, including the BC-778-D or –E and the new signal lamp M308-B. (National Archives)

(Below right) Equipment laid out with the addition of the new hydrogen generators M-315-B. These new components were could have been replacements after various inspections. (National Archives)

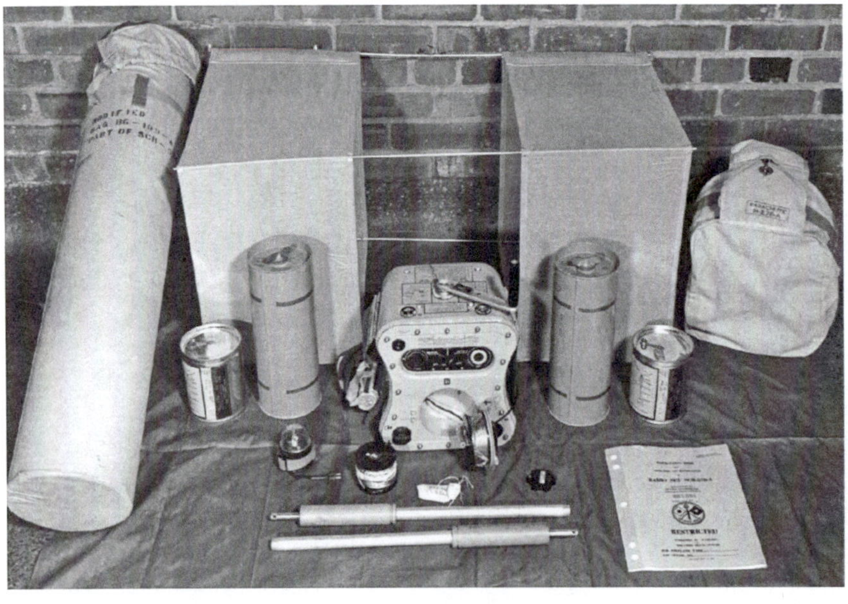

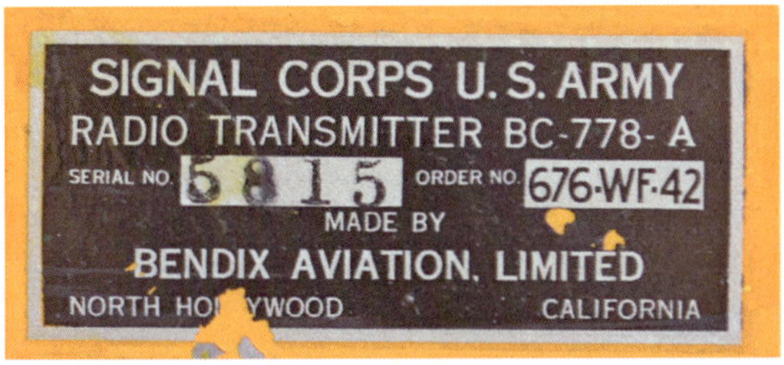

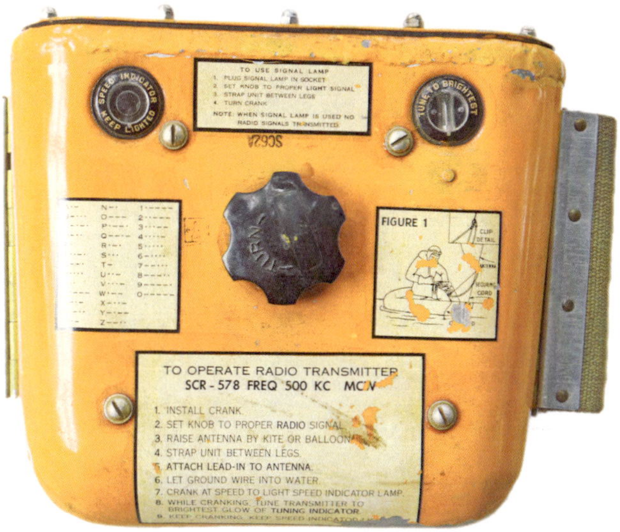

The radio transmitter was self-contained in a cast shell designed for buoyancy, being air tight and waterproof. The shell was molded to the contour of the operator's legs. For operation, the radio was held between the thighs secured in place buy the 5 wide strap provided. The radio measured 13" x 10" x 9" and weighed 15.8 pounds. There were two principle variations of the BC-778, The first model was designated by the suffix "–A" with subsequent modifications labeled "–B" and "–C" and so on. They all shared the same external characteristics as the unit illustrated here. Instructions and Morse code were provided, printed on decals or data plates on the face plate and top of the radio. Some transmitters would be modified according to revision "-F" as can be seen by the over-stamp on the lower data plate pictured. The modifications were internal to the power and gear trains. Radios that required updates or repairs were sent to modification centers.

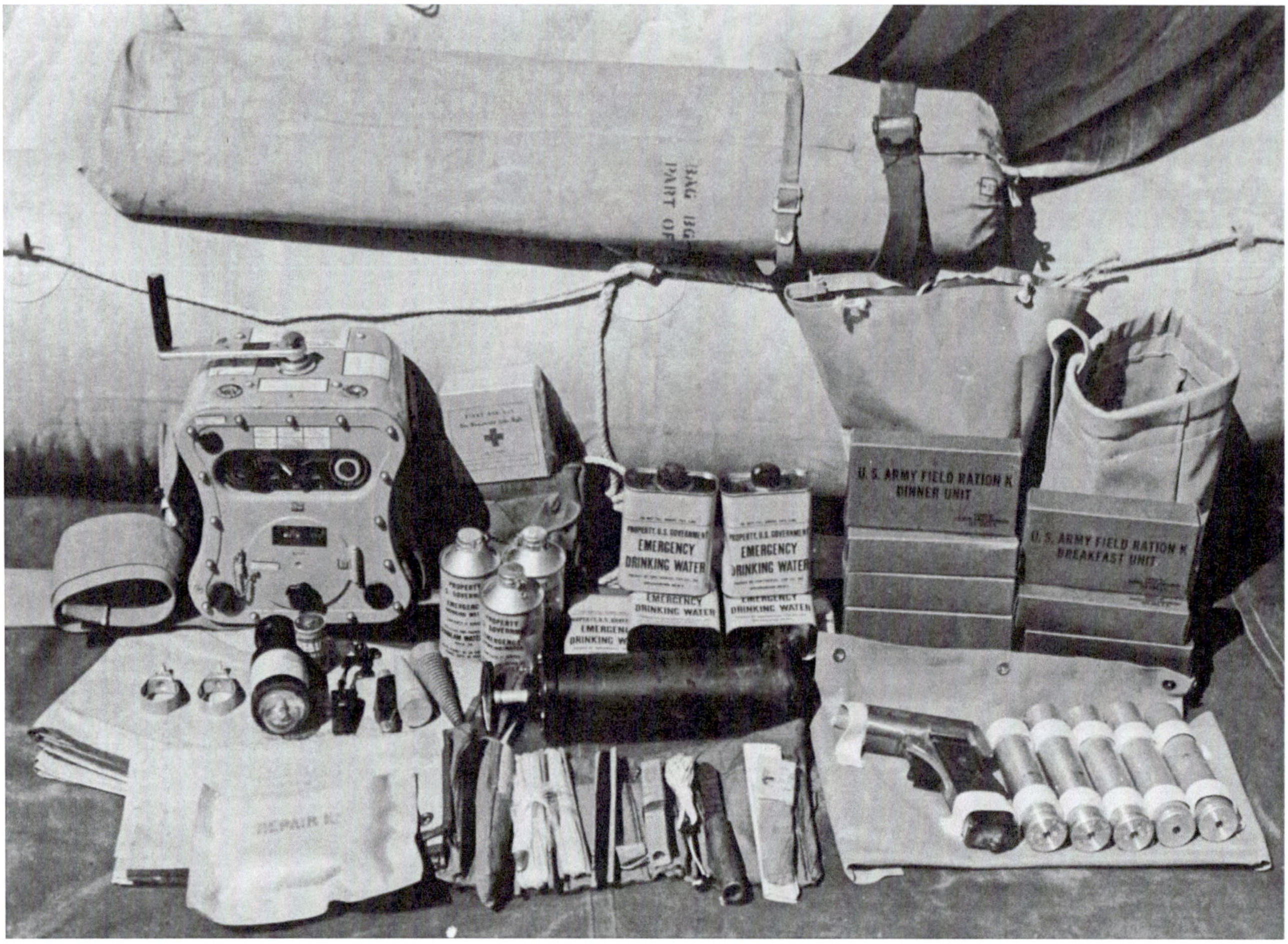

(Above) The SCR-578 radio set was designed to be a standard item for all multi-person life rafts, but too large to be directly incorporated, being stored separately. It was however alternately designed to be droppable by aircraft to survivors. This display by the 386th BG in England illustrates a typical compliment of emergency equipment carried aboard bombardment aircraft. (National Archives)

(Right and below) At some point in its development, the radio was nicknamed "Gibson Girl," a term coined at the turn of the century by artist Charles Gibson. The shape was not unlike the feminine waistlines depicted in Gibson's drawings. However, it was not an official designation, and was used loosely in some period documentation much like the term "Mae West" for life preservers. This test sample of a BC-778-D, provided in an exhibit at the equipment display room for the Liaison Committee for Emergency Rescue Equipment, November 1943, has the unique screen printing of "Gibson Girl." (National Archives)

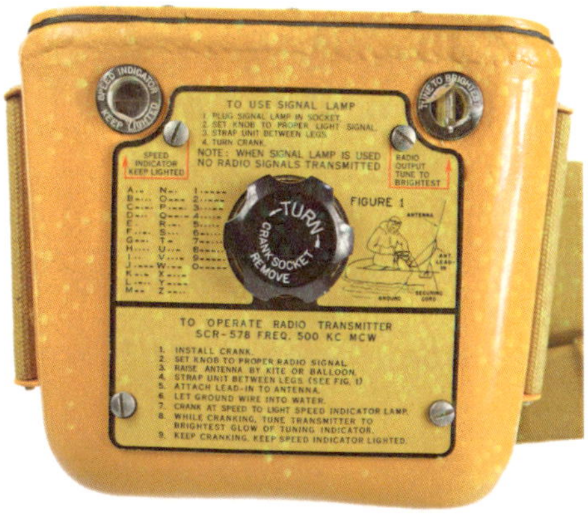

(Left and below) The BC-778 was an item pressed into service which remained under development before and after standardization. From December 1942 until April 1943 there was a continuous series of improvements applied to the complete set. New features would be applied under revision -D that would change the external appearance. A new type of brake was incorporated, centered in the antenna door, and the provision to lock the door open (which wasn't provided on the -A through -C models). New types of instruction and data plates were also applied. Bendix Aviation Limited was the prime contractor for the BC-778 but sub-contracted to the Kingston Products Corp, noted by the red stamped "K."

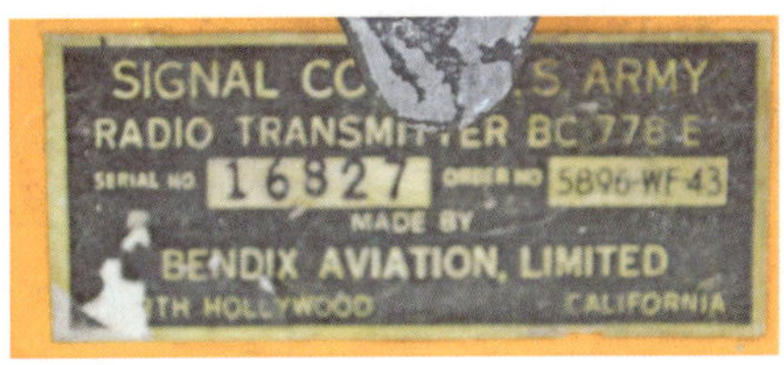

(Above) Another example of a variation of screen printed instructions and data on the face and top of a BC-778-E. There are multiple subtle variations in the radios as they consistently underwent repairs and modifications, with only the principle types illustrated here.

Upon standardization, an instruction booklet titled *Preliminary Instructions for Radio Set SCR-578-A* was printed. Formalized as *Instruction Book for Operation & Maintenance of Radio Set SCR-578-A* and Technical Order 08-10-94, dated 27 November 1942. This was updated and retitled *Handbook of Maintenance Instructions for Radio Set SCR-578-A or SCR-578-B* and approved as publication number AN 08-10-94 on 19 August 1943. Two booklets would be included with the Radio Set SCR-578-A, one in BG-110-A, one in BG-109-A, and one copy in SCR-578-B.

The booklet of August 1943 detailed models A through E, with the revised booklet of 23 March 1944 detailing revision F. Revision F would be the last modification before the switch to the manufacture of the AN/CRT-3. The total quantity of SCR-578-A and -B radio sets manufactured was 117,809.

The quantity per known contracts is unknown, but breakdown is as follows: 13,720 in 1942, 48,040 in 1943, 46,525 in 1944 and 9,524 in 1945 (up to March when the manufacture of the AN/CRT-3 commenced).

Known contracts are: 676-WF-42, 1152-WF-42, 5896-WF-43, 8127-WF-43, 11953-WF-43, 391-DAY-44, 1510-DAY-44. WF signifies Wright Field and after the fiscal year 1943 that designation was changed to DAY signifying Dayton, Wright Field, Dayton, Ohio. As of May 1942, only 103 radio sets had been delivered, steadily increasing each month.

(Below) Breakdown of the revisions of the BC-778. Dates of changes unknown.

BC-778-A	Original transmitter supplied with Radio Set SCR-578-A.
BC-778-B	Is an electrically modified version using a transformer-coupled audio oscillator instead of the resistance-coupled oscillator.
BC-778-C	Modified version of BC-778-B differs only in that no automatic keying of the signal lamp is provided in "AUTO 1-LIGHT" and "AUTO 2-LIGHT" selector switch positions, the signal lamp is continuously illuminated.
BC-778-D	Is a mechanically modified BC-778-C. Improvements include a new type lock-open antenna reel door, a new type of antenna-reel brake, improved instruction name plates and longer life gears.
BC-778-E	Same as BC-778-D with the exception of a new type of antenna lead-in fastening.
BC-778-F	Is a modified BC-778-C or –A or –B consisting of replacement of the main-power and governor gear trains. Name plates are to be changed to an over stamp of –F to signify modifications.

(Above left and right) Service testing had proven a variety of different deficiencies in the BG-110 and BG-109. Some of the inherent reported issues were the weakness of the cardboard tube, allowing damage to the components, and waterproofing for the BG-109. The parachute assembly of the BG-110 was found to fail in repeated cases. In December 1942, a new configuration was tested at Eglin Field, comprising of a single felt padded bag and new type of parachute. (National Archives)

(Below) Testing continued for three months looking at various concepts. In March 1943, a new single unit was adopted, designated as the Radio Set SCR-578-B, which included a new bag BG-155-A and parachute assembly M-390-A. The SCR-578-B comprised of all the same components as the -A unit with the only change in the kite. Existing contracts were amended to the fabrication of the new configuration, becoming available by the summer of 1943. In March 1944, the USAAF Aircraft Radio Laboratory directed that the older models be converted to the use of the BG-155-A upon their availability. When the BC-778 was used in conjunction with BG-110 & 109 it was referenced as the SCR-578-A, and when used in conjunction with the BG-155 it was referenced as the SCR-578-B. Illustrated on this page is the concept SCR-578-B with BG-155-A tested at Eglin Field. (National Archives)

(Above left and right) The production of BG-155-A would include four parachute webbing grab handles, one on each side, sharing all the other same characteristics of the test model. The complete packaged unit would weigh 38 pounds.

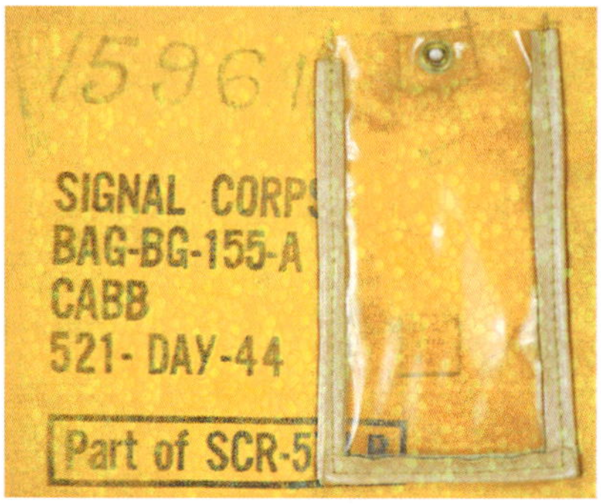

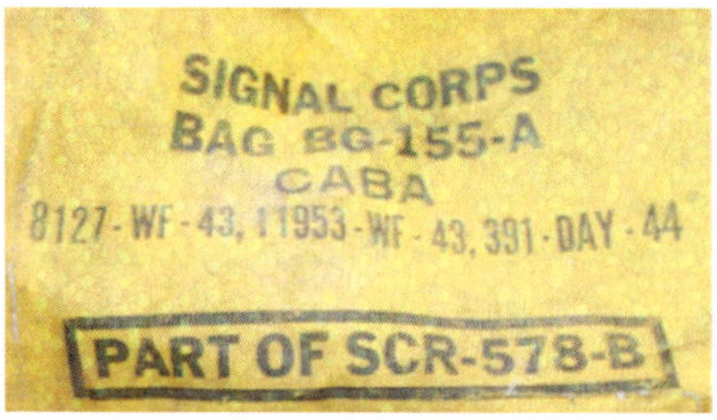

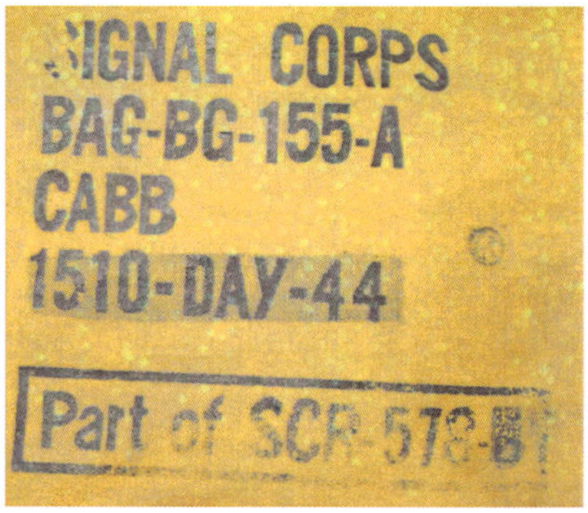

(Top left and center) End views of the packed BG-155-A

(Above) Various nomenclature

(Left) As part of the assembly of the BG-155-A, a set of pads would be included to protect the transmitter.

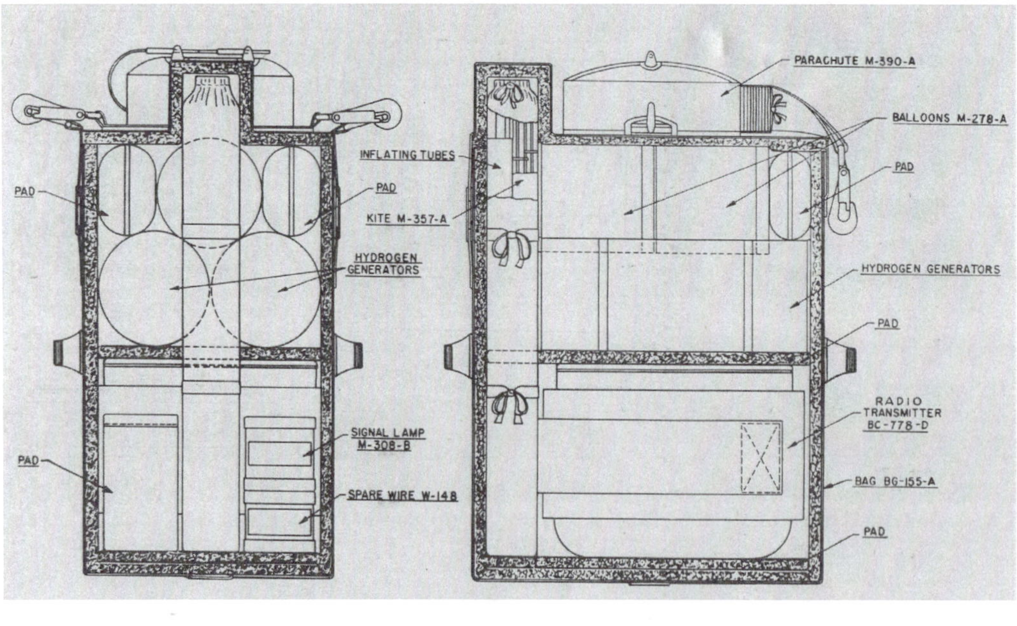

(Left) A schematic showing the proper packing of accessories.

(Right) A hydrogen generator was used to inflate the balloon. The original type M-315-A was painted grey with two red lines and contained one pound of Lithium hydride. For use, the two plugs were unscrewed, an inflation tube was installed at one end with the other hole placed in the water, allowing water to enter. Upon contact with water the balloon inflated. This model is depth sensitive (hence the red lines) having an operating time of about 40 minutes.

(Far right) Each of the four pieces of the apparatus would have an instruction tag.

(Right center) Some examples have been found with an outer cardboard sleeve printed with instructions.

(Below) Inflation tube provided with Radio Sets SCR-578-A or SCR-578-B. Two generators and two inflation tubes were provided with each complete assembly.

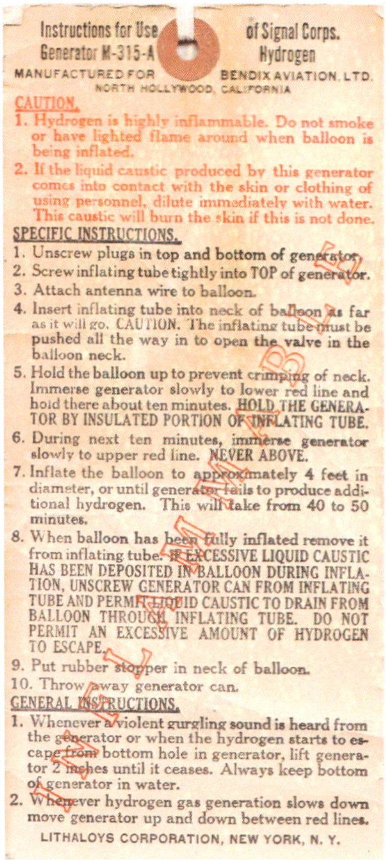

Instructions for Use of Signal Corps. Generator M-315-A Hydrogen

MANUFACTURED FOR BENDIX AVIATION, LTD. NORTH HOLLYWOOD, CALIFORNIA

CAUTION.

1. Hydrogen is highly inflammable. Do not smoke or have lighted flame around when balloon is being inflated.
2. If the liquid caustic produced by this generator comes into contact with the skin or clothing of using personnel, dilute immediately with water. This caustic will burn the skin if this is not done.

SPECIFIC INSTRUCTIONS.

1. Unscrew plugs in top and bottom of generator.
2. Screw inflating tube tightly into TOP of generator.
3. Attach antenna wire to balloon.
4. Insert inflating tube into neck of balloon as far as it will go. CAUTION. The inflating tube must be pushed all the way in to open the valve in the balloon neck.
5. Hold the balloon up to prevent crimping of neck. Immerse generator slowly to lower red line and hold there about ten minutes. HOLD THE GENERATOR BY INSULATED PORTION OF INFLATING TUBE.
6. During next ten minutes, immerse generator slowly to upper red line. NEVER ABOVE.
7. Inflate the balloon to approximately 4 feet in diameter, or until generator fails to produce additional hydrogen. This will take from 40 to 50 minutes.
8. When balloon has been fully inflated remove it from inflating tube. IF EXCESSIVE LIQUID CAUSTIC HAS BEEN DEPOSITED IN BALLOON DURING INFLATION, UNSCREW GENERATOR CAN FROM INFLATING TUBE AND PERMIT LIQUID CAUSTIC TO DRAIN FROM BALLOON THROUGH INFLATING TUBE. DO NOT PERMIT AN EXCESSIVE AMOUNT OF HYDROGEN TO ESCAPE.
9. Put rubber stopper in neck of balloon.
10. Throw away generator can.

GENERAL INSTRUCTIONS.

1. Whenever a violent gurgling sound is heard from the generator or when the hydrogen starts to escape from bottom hole in generator, lift generator 2 inches until it ceases. Always keep bottom of generator in water.
2. Whenever hydrogen gas generation slows down move generator up and down between red lines.

LITHALOYS CORPORATION, NEW YORK, N. Y.

(Right) In March 1943, a new generator was adopted as standard, designated as the M-315-B. The principle reason for the change was the cost of manufacture, and the new generator utilized typical commercial can fabrication methods. The redesign of the internal components meant it was no longer depth sensitive and reduced the operating time in half from the M-315-A. The assembly Included one pound of lithium hydride. By March 1944 the unit was updated to include three pounds of calcium hydride and redesignated as the M-315-C, with no external changes. Keys on the bottom and top were provided for opening the can for operation, functioning under the same principle as the M-315-A.

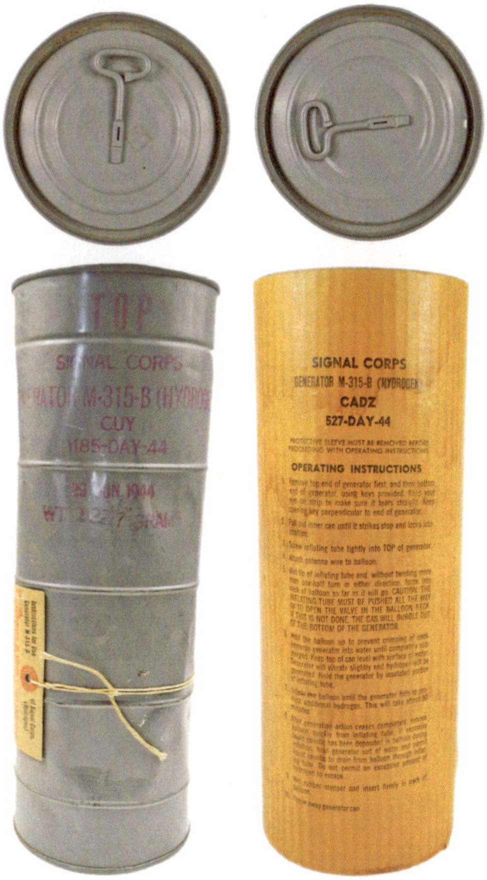

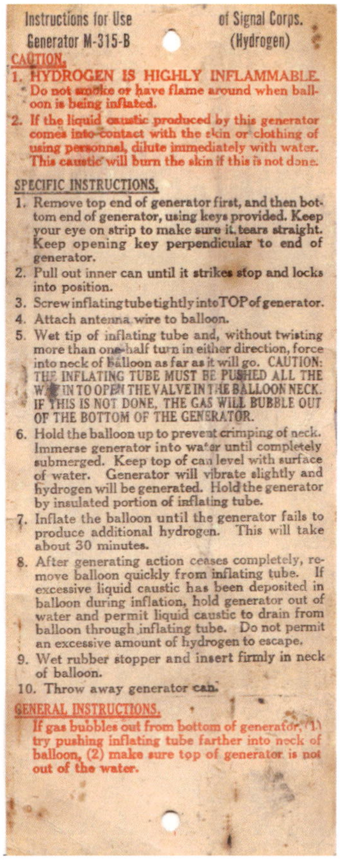

Instructions for Use of Signal Corps. Generator M-315-B (Hydrogen)

CAUTION.

1. HYDROGEN IS HIGHLY INFLAMMABLE. Do not smoke or have flame around when balloon is being inflated.
2. If the liquid caustic produced by this generator comes into contact with the skin or clothing of using personnel, dilute immediately with water. This caustic will burn the skin if this is not done.

SPECIFIC INSTRUCTIONS.

1. Remove top end of generator first, and then bottom end of generator, using keys provided. Keep your eye on strip to make sure it tears straight. Keep opening key perpendicular to end of generator.
2. Pull out inner can until it strikes stop and locks into position.
3. Screw inflating tube tightly into TOP of generator.
4. Attach antenna wire to balloon.
5. Wet tip of inflating tube and, without twisting more than one-half turn in either direction, force into neck of balloon as far as it will go. CAUTION: THE INFLATING TUBE MUST BE PUSHED ALL THE WAY IN TO OPEN THE VALVE IN THE BALLOON NECK. IF THIS IS NOT DONE, THE GAS WILL BUBBLE OUT OF THE BOTTOM OF THE GENERATOR.
6. Hold the balloon up to prevent crimping of neck. Immerse generator into water until completely submerged. Keep top of can level with surface of water. Generator will vibrate slightly and hydrogen will be generated. Hold the generator by insulated portion of inflating tube.
7. Inflate the balloon until the generator fails to produce additional hydrogen. This will take about 30 minutes.
8. After generating action ceases completely, remove balloon quickly from inflating tube. If excessive liquid caustic has been deposited in balloon during inflation, hold generator out of water and permit liquid caustic to drain from balloon through inflating tube. Do not permit an excessive amount of hydrogen to escape.
9. Wet rubber stopper and insert firmly in neck of balloon.
10. Throw away generator can.

GENERAL INSTRUCTIONS.

If gas bubbles out from bottom of generator, (1) try pushing inflating tube farther into neck of balloon, (2) make sure top of generator is not out of the water.

(Below) Breakdown of all the parts that consist of the M-315-B and -C.

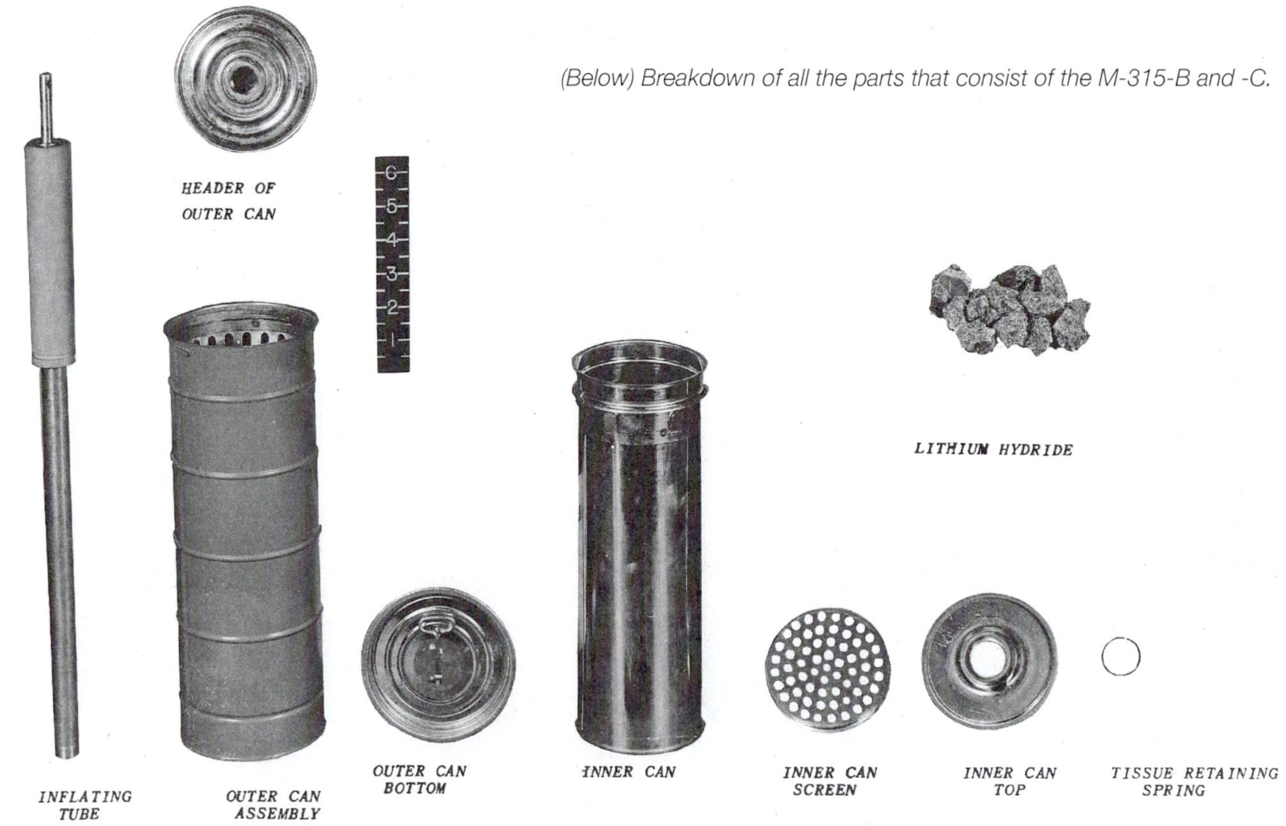

(Above and right) The process of inflation. The antenna is connected to the balloon and paid out from the transmitter, Two 4' diameter latex rubber balloons designated as M-278-A were provided for raising the antenna when there was no wind for use of the kite. (National Archives)

(Above) Each balloon was stored in a sealed can, 5-1/8" x 4-1/4" in diameter and provided with a key for opening. Most production cans for the SCR-578-A & -B were provided with paper labels (left and center), later changing to a painted panel for the AN-CRT-3 (right).

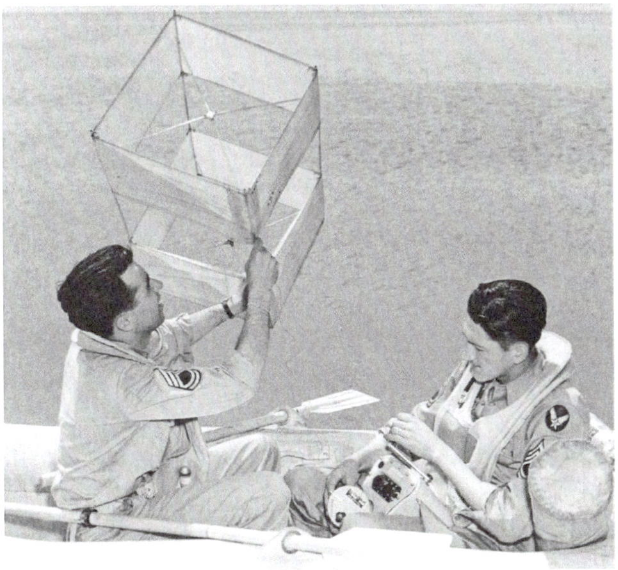

(Above and right) Erection and use of the box kite, provided for raising the antenna (National Archives)

(Below) The SCR-578-A utilized a cardboard tube, 36-1/2" long x 3-1/8" diameter for kite stowage. This complete assembly was designated M-277-A. The original specification of November 1942 designated the kite itself as the M-357-A.

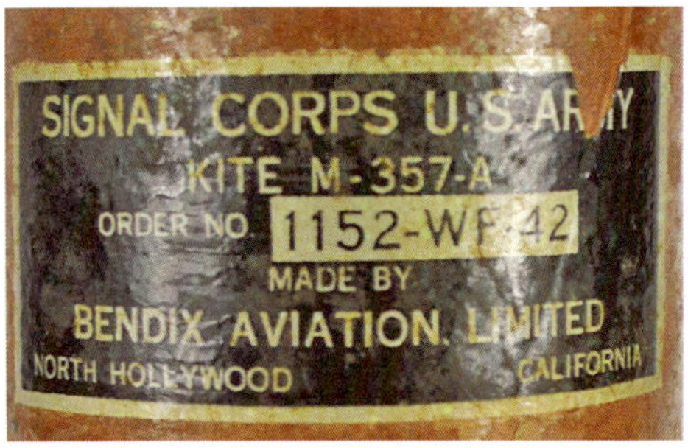

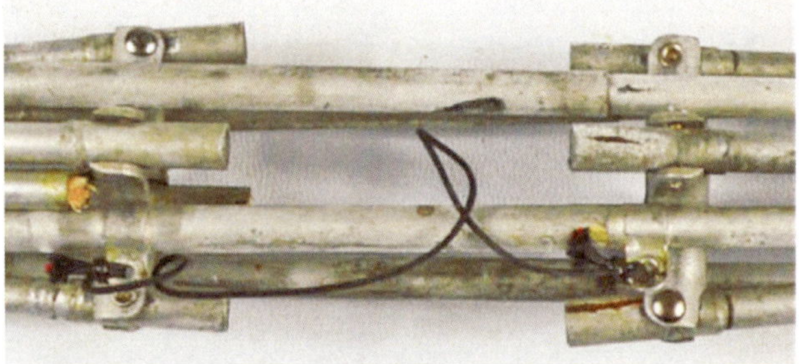

(Above and left) The kite was modified for inclusion in the BG-155-A by making two sectional halves. (The original design did not allow the kite to hinge or separate in half).

(Below) The kite, with the two inflation tubes, would be stowed in a cotton draw-string bag for the SCR-578-B assembly.

(Right) Detail of list on the label of the bag

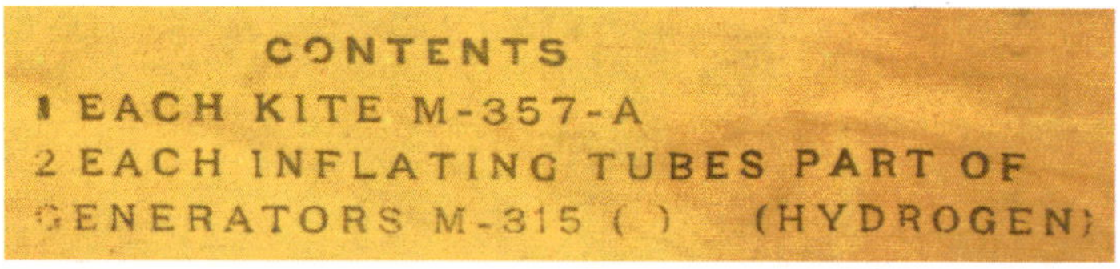

(Top left and right) One signal lamp was provided, used for low-light distress signaling purposes. The original lamp included was designated as the M-308-A. A strap fastened it around the operator's head and a 5' cord connected it to a socket on the BC-778.

(Center left and right) In January 1943, a new modified light was adopted designated as the M-308-B. The principle change was a domed lens, allowing the light to be non-directional, to be worn on the top of the head with the strap under the chin.

(Above left) Stowage containers for the lamps. The designation M-308-C is printed in error.

(Right) Testing the M-308-A.

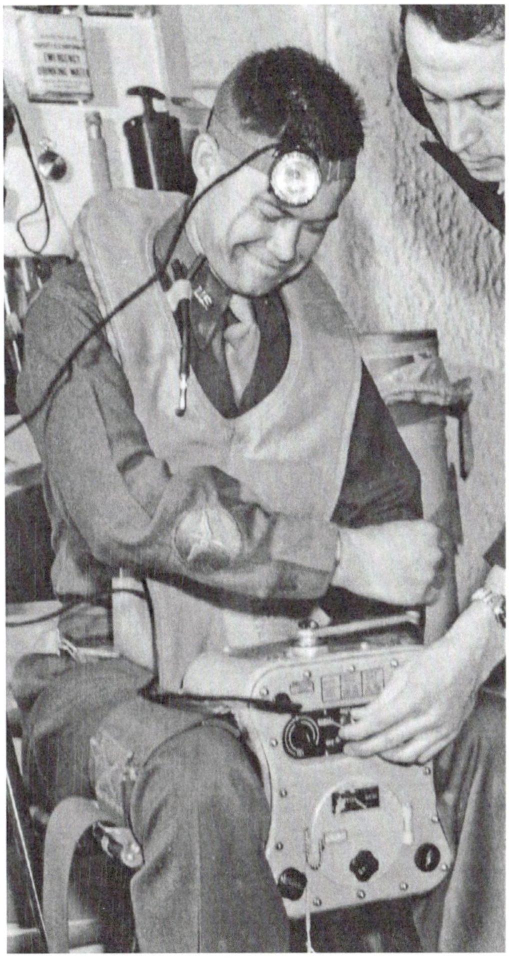

(Above) Antenna wire was provided on Reel RL-48 located on the door of the BC-778. The reel allowed the operator to pay out the wire or reel it back in, as needed. A clip was provided on the end for attachment to the kite or balloon. Inside the compartment was the Lead-In wire which attached to the extended antenna. Note additional length of copper alloy wire stowed in compartment to the bottom left for use a ground wire which was dropped over the side of the raft and anchored by a weighed cap.

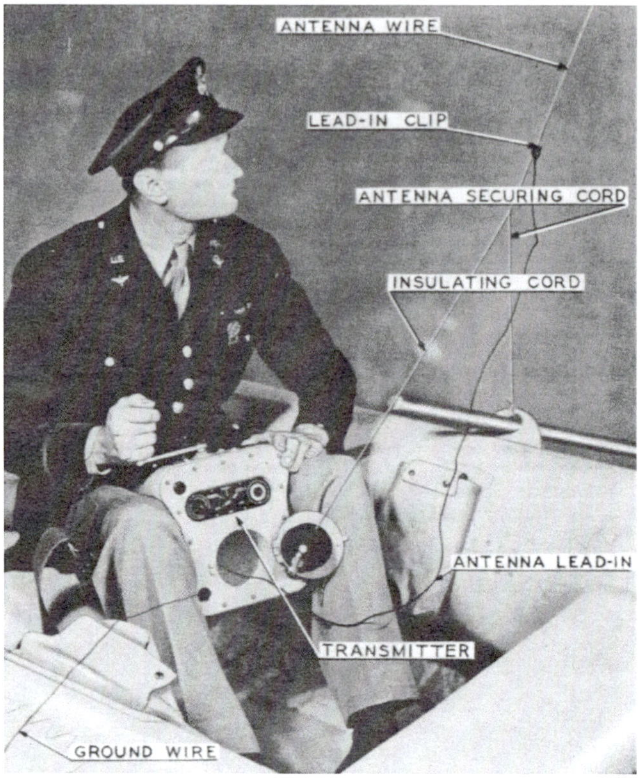

(Above) Configuration of the copper wiring for proper operation.

(Below) Various containers for the 300' of spare, copper alloy antenna wire provided for each unit designated as W-147 or W-148.

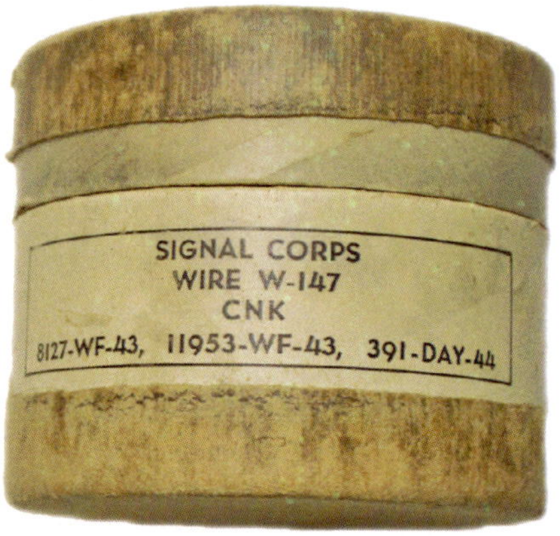

(Below) One hand crank GC-18-A is provided for the BC-778. A slot is provided on the back of the radio for stowage, a cabled lanyard is provided to prevent loss. A black phenolic cap protects the crank receptacle on the top of the radio.

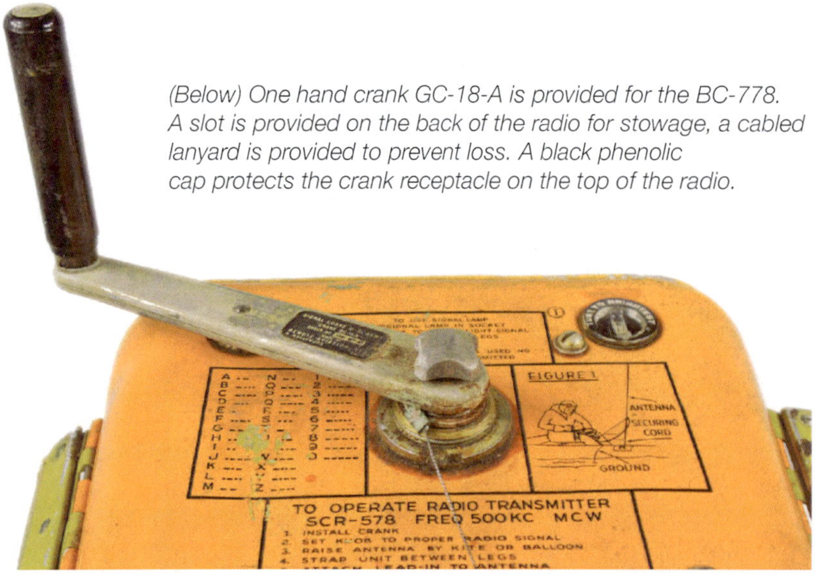

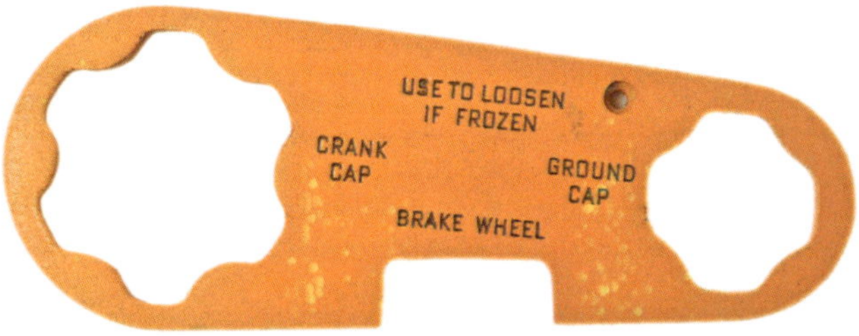

(Above) Integrated into the compliment was a wrench for leveraging the knobs of the brake, crank and ground cap. The first generation were wood for flotation, colored orange yellow, but since they were prone to breakage it was later revised to an aluminum type. These were introduced with the revision -D and not provided with the -A through -C BC-778 radios.

(Left and above) A permanent feature on the BG-110-A was the integral parachute M-276-A. The parachute was folded accordingly and stowed in a pocket. A cotton static cord was provided with snap hook for deployment.

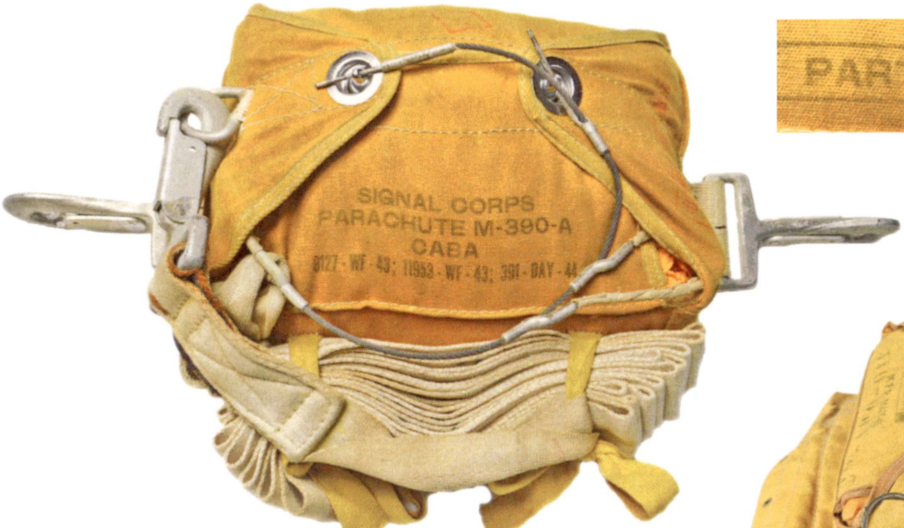

(Above and right) Due to durability issues with the M-276-A a new parachute system was developed for the BG-155-A. The new parachute designated as the M-390-A conformed to the conventional design of typical parachute systems, the parachute and static line were now stronger. D rings were provided on the BG-155-A for attachment.

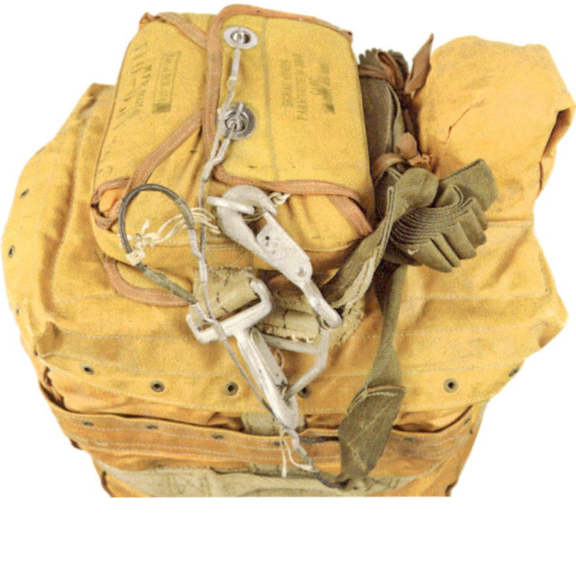

Page from the handbook illustrating the complete assemblies for both the SCR-578-A and SCR-578-B.

AN 08-10-94

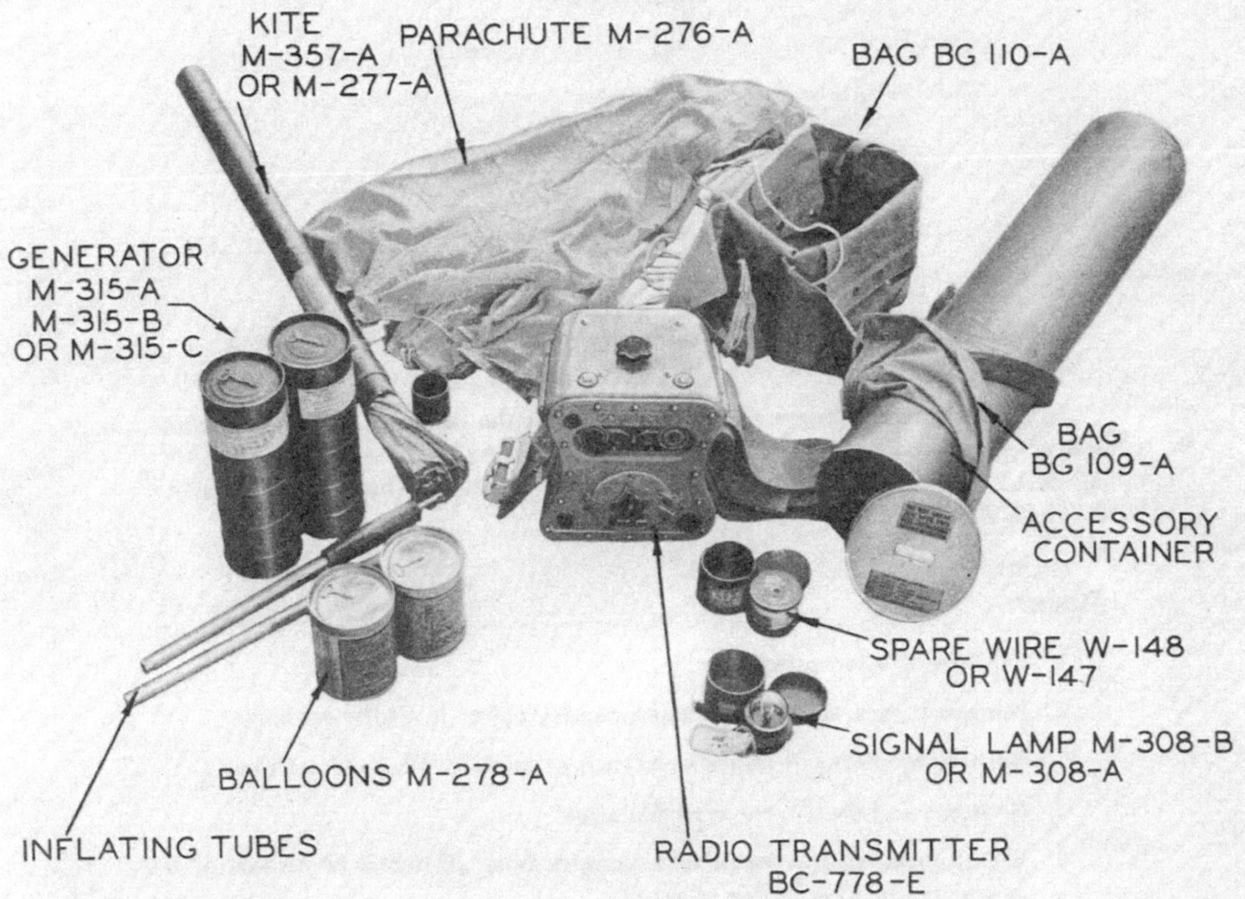

Radio Set SCR-578-A

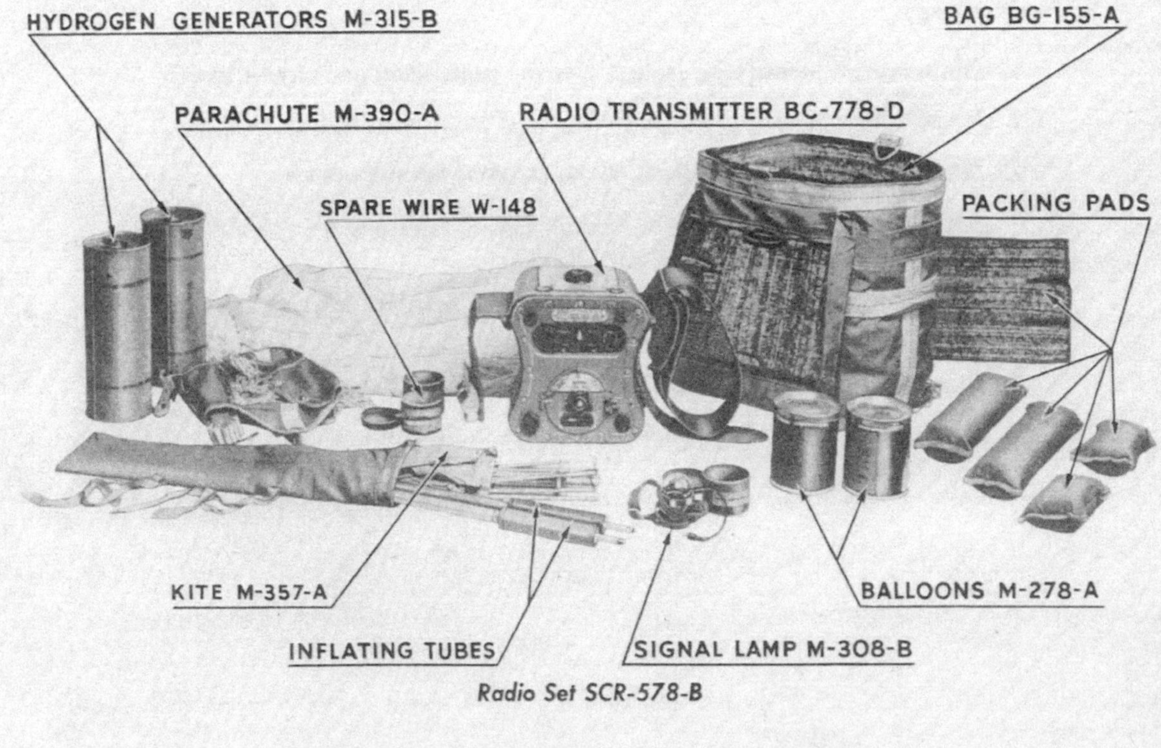

Radio Set SCR-578-B

Figure 1—Radio Set SCR-578-A or SCR-578-B, Component Parts

(Left) The BuAer requisitioned radio sets from the USAAF for equipping their large patrol aircraft. An SCR-578-B has been removed from a US Navy PBY after a mission stationed in New Guinea, September 1944.

(Above) An SCR-578-B being stored in a U.S. Navy PB4Y Liberator near the ball turret position. In the event of an emergency, retaining the transmitter would be the waist gunner's responsibility.

(Left) Both models of SCR-578 radio sets can be seen in the storage room at USAAF Burtonwood Airfield, England c.1944. The older SCR-578-A remained in service and were issued until exhausted, or until suitable quantities of the -B became available (or replacement BG-155-As).

(Right) Inflated USAAF Type A-2 life raft shown with the SCR-578A radio compliment. (National Archives)

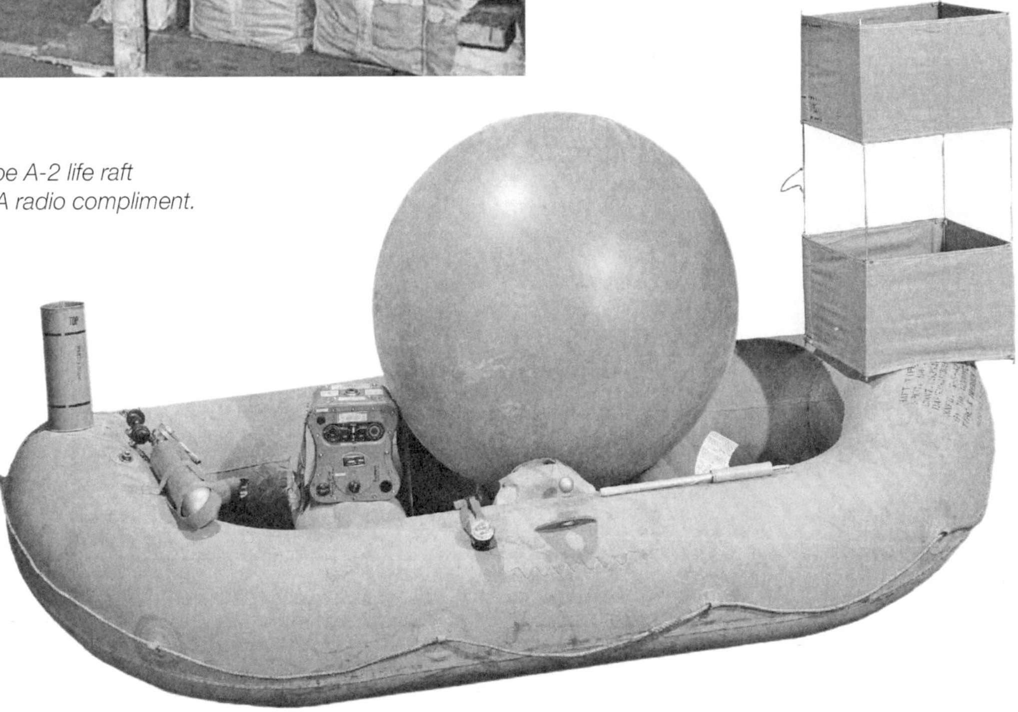

(Right) The SCR-578B in an actual rescue being conducted. USAAF P-38 pilot Lt. Joseph Devona was forced to ditch after he lost engine power near Truk Island in the eastern Carolines while escorting Navy PB4Y reconnaissance planes, January 1945. He was spotted in his pararaft after a brief search by one of the PB4Ys. As part of a standard operating procedure when spotting survivors, a large life raft and a SCR-578B were dropped on his position. The PB4Y was able to continue circling for ten hours. Devona used the transmitter from which two destroyers picked up the signal, arriving on scene at dusk and picking him up. PBYs were dispatched but were unable to land. (National Archives)

(Left) Five airmen of the 6th Ferrying Group were forced to ditch 800 miles off Long Beach, California, c.1945. They were located by the distress signal of the SCR-578, then spotted by a USCG float plane, which directed surface craft to the location. The balloon can be seen above their position. Though not noted in this instance, the balloon was also effective as a line of sight spotting device. (National Archives)

Through extended service use by 1944, the deficiencies of the BC-778 were noted. It worked efficiently in the ETO, principally due to the short ranges in which it needed to communicate, but also because of the extensive watch and direction finding stations available. However, over water flying distances in the PTO were much greater than can be covered by the range from the transmission of 500 kcs, far from range of ground stations.

As early as March 1944 it was recommended to convert the BC-778 to a two frequency transmitter. Modified sets were field tested through the remainder of the year finishing with final approval in January 1945 for the production of a new radio to operate on both frequencies of 500 and 8280 kcs.

The new assembly was designated as the AN/CRT-3, with the transmitter designated as the T-74/CRT-3, entering production in March 1945. Projected production schedules detailed that deliveries would commence by August of that year. Due to the late delivery, this radio set would not have been used in the combat zone before VJ-Day.

The T-74/CRT-3 was a modified BC-778 which would transmit signals on 500 and 8280 kcs.

Transmission would be automatically shifted every 40 seconds from one frequency to the other. 8280 kcs provided the means for long range bearing from land based DF stations, allowing them to direct ships and aircraft to general locations. 500 kcs would be used by ships and aircraft to home to the survivors.

During testing, the transmission utilizing the 8280 kcs was reported have been received at distances of up to 1,600 miles.

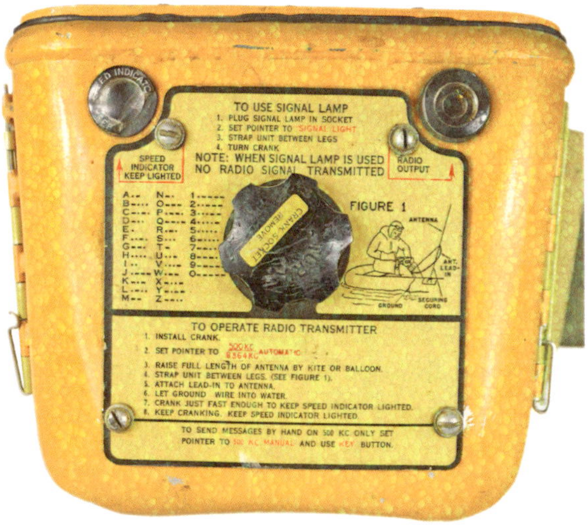

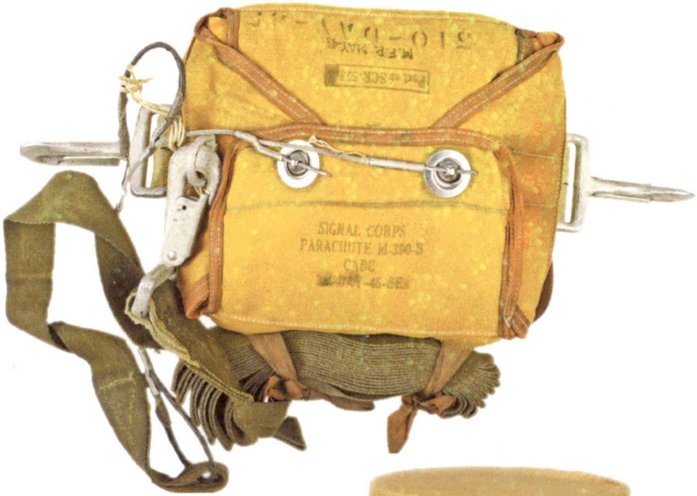

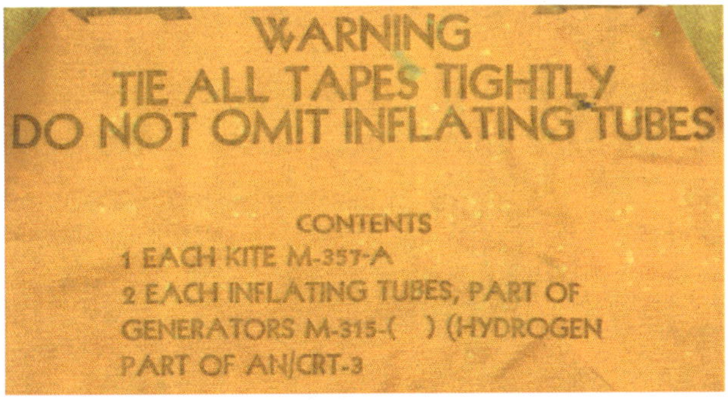

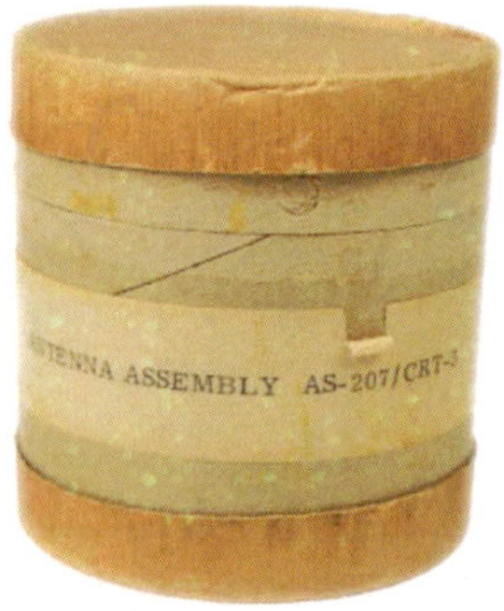

(Above) Initial assemblies of the AN/CRT-3 comprised components from the SCR-578. BG-155-As were over-stamped with the new designation. For the radio T-74, overall features remained the same except for the operation control panel and the instruction data plate on the top. Improvements were adopted and integrated into the set to include a new parachute, M-390-B and spare antenna assembly, AS-207/CRT-3. Once available it was requisitioned under SC stock number 2S1501-1. The example shown is a wartime contract, but has undergone updates to CRT-3A which changed the frequency from 8280 to 8364 kcs.

Many technological advancements were made during the war, one being the capability of radar.

Tests began in December 1942 on radar reflective devices at Langley Field to test the feasibility of fixed stations on land, and floating apparatus at sea, for aircraft to home in on. The final report from Langley was released in August 1943 with positive results.

The concept of a radar reflector piqued the interest of the BuAer soon after the release of the report, as any aircraft equipped with radar could detect a survivor, night or day. Much of the design work was conducted during tests at Langley, with the BuAer refining the apparatus to incorporate into pneumatic life rafts.

Two sizes were adopted and standardized, one for the multi-person life raft and one for the pararaft. Each corner reflector consisted of an approximate 4 sq. ft. knitted mesh on three perpendicular and intersecting planes, having an effective range of 7–12 miles.

(Above) The first corner reflector to be adopted was for multi-person life rafts, standardized 16 June 1944, as MX-138/A under NavAer Specification M-619 nicknamed "Emily." In the designation breakdown, the M signifies the device as Ground Mobile and the X signifies it as Identification Recognition with the 138/A being the model number. The packaged reflector measured 26" long and 2" square, and was stored in the oar pocket. Available from the ASO under stock number R83-R-AN-MX-138/A, it was initially proprietary to the BuAer, but soon adopted by the USAAF and assigned its Army-Navy designation of US Army Signal Corps stock number 2Z756-138A. The Vendo Co. was the only contractor for the MX-138/A. It was not available until early 1945, with limited actual use during the war, however several rescues were executed by utilizing the device.

(Right) The life raft oars acted as the mast, with a socket provided on the reflector for securing the device to them.

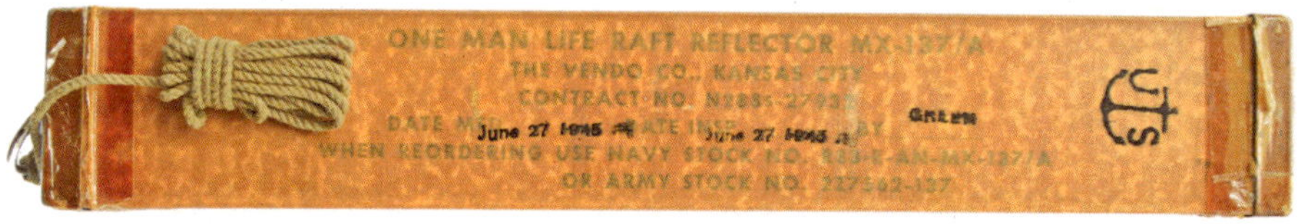

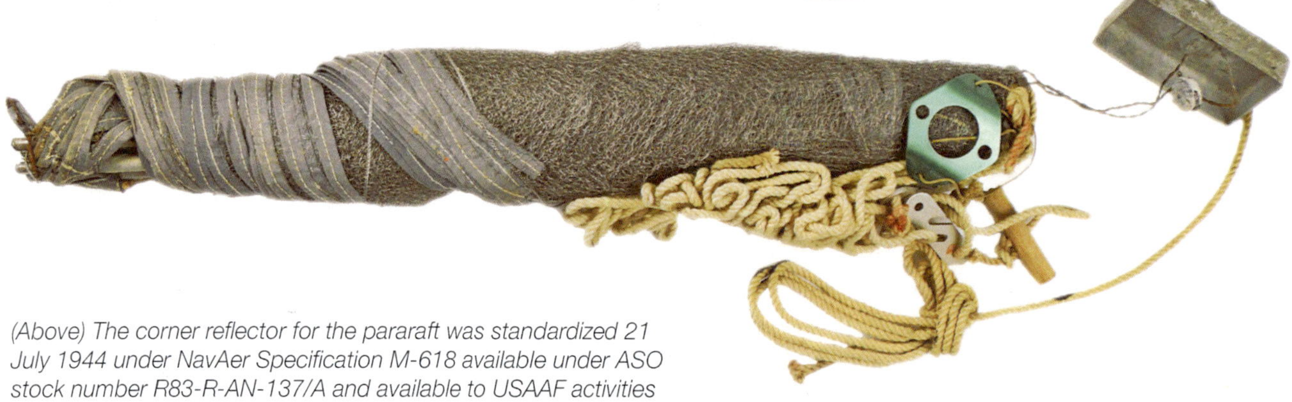

(Above) The corner reflector for the pararaft was standardized 21 July 1944 under NavAer Specification M-618 available under ASO stock number R83-R-AN-137/A and available to USAAF activities under SC stock number 2Z756-137/A. The packaging for the pararaft was reduced to fit in the seat packs measuring 14-1/2" x 2-1/4" x 1-1/2", with a crush-proof fiber container. This reflector was provided with a telescoping mast for installation. The Vendo Co. and Reed & Barton Co. were the two contractors for the MX-137/A. The corner reflectors had a very short life span, and as technological advancements were made in electronic radios, they were superseded in the immediate post-war years. In 1946, as detailed in BuAer Technical Order 29-46 of 29 August 1946 (addressing both multi-person and one-man life rafts) the corner reflector was to be replaced by the CRC-7 transmitter, with priority to pararafts. Other CRN radio beacons in both aeronautical services were developed in the immediate post war years rendering the radar reflector obsolete. Both the M-618 and M-619 were finally cancelled and removed from service in November 1954.

(Above left) A corner reflector being tested in a pneumatic life raft by the BuAer at NAS Key West, February 1944. (National Archives)

(Above right) An MX-138/A corner reflector being demonstrated by USAAF personnel in a Type A-3A life raft at Wright Field, Ohio. (National Archives)

(Left) One type of radar scanning device that was used to locate survivors utilizing a corner reflector. The SCR-521 could be used for contact up to 12 miles. (National Archives)

(Below left) A USAAF aircrew member aboard a PBY-5A checks the screen on the SCR-521. (National Archives)

Below right) In late 1943, the British developed a lightweight beacon for use in the RAF K Dinghy that would transmit an osculating pulse that could be picked up by radar with a range of 5 miles. Designated as the T.3180 (nicknamed "Walter"), this device compensated for the lack of the SCR-578-A in small aircraft, performing the same function. By October 1944 a limited quantity of 5,000 American copies was procured under the designation of Radar Set AN/CPT-2, and nicknamed "Walter," with an effective range up to 25 miles. Shown is the Walter beacon in use with a USAAF Type C-2 pararaft.

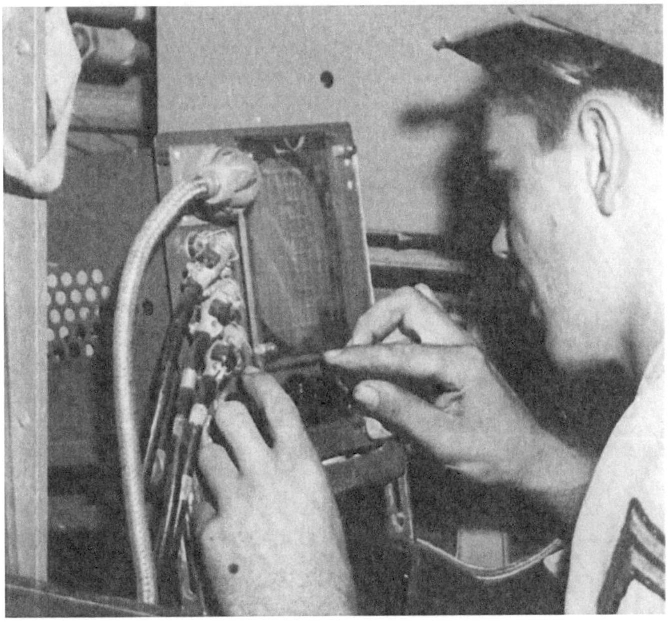

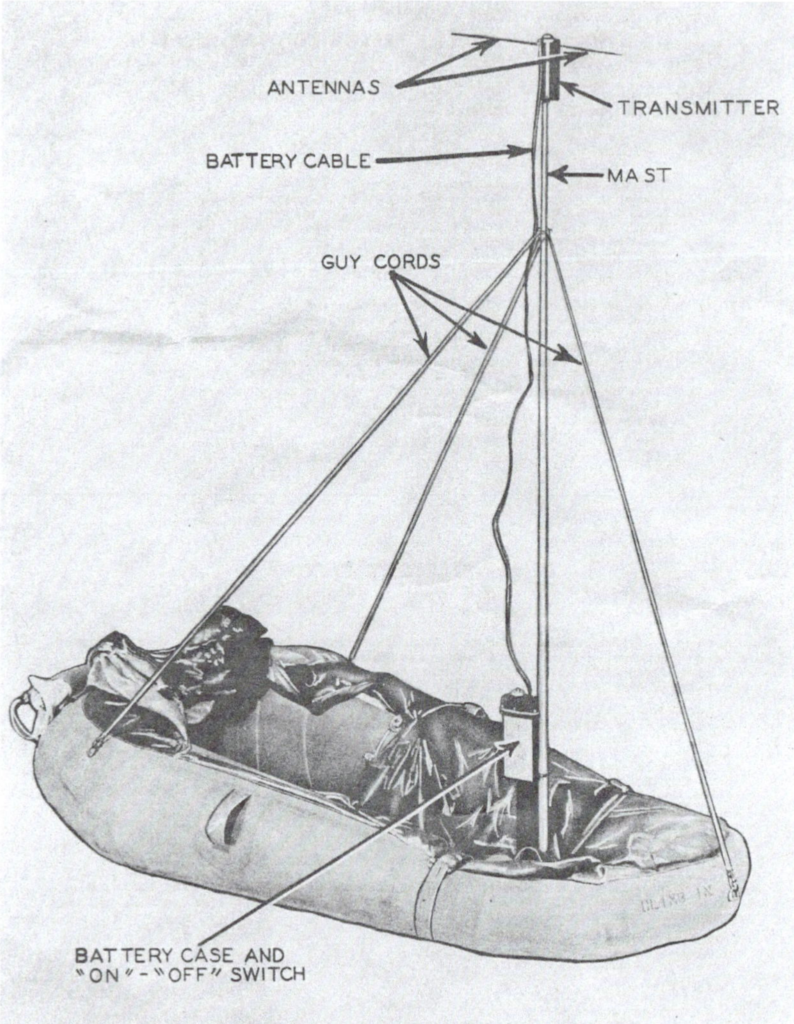

(Left) Installation of the The AN/CPT-2 "Walter" transmitter on a Type C-2 pararaft. The British Air Ministry began procurement of its "Walter" transmitters in early 1944, but did not receive significant deliveries until the end of the year. The 8th Air Force Command was working in coordination with the Air Ministry, testing the Walter beacon alongside radar reflectors for air sea rescue work into spring 1944. The results were favorable, with recommendations for procurement and supply of a similar beacon for US pararafts. When available it would replace the sail and mast in the kit, stowed in its place available under SC stock No. 2S1514-2. The Walter beacon was found to be more reliable than the radar reflector.

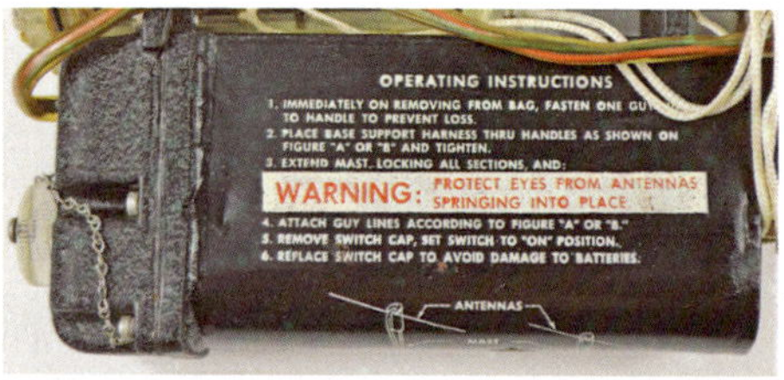

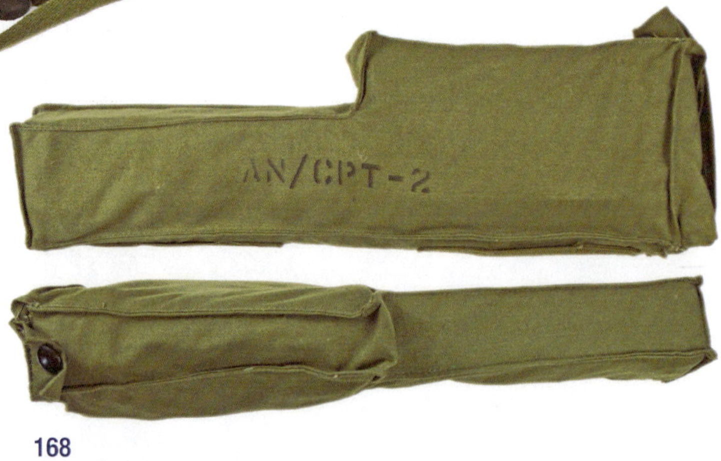

(Above and right) The AN/CPT-2 "Walter" shown ith its fitted case.

The adoption of the SCR-578-A temporarily satisfied the requirement for a transportable emergency radio, but lacked the capabilities for receiving messages. With advancements in radar and radio technologies and the creation of air sea rescue organizations, a requirement was established for a compact, lightweight, portable transmitter. The first device to help meet this demand was the AN/CPT-2 but, just like the SCR-578, it could not receive messages. Range of signal actually became less of a concern considering the air sea rescue nets operating in the combat zone.

Building from the principle of the Walter beacon and the Handie Talkie BC-611, a VHF emergency transmitter-receiver was developed in 1945, designated as the AN/CRC-7. It was designed to provide a tone and voice transmission, voice reception, and a homing beacon with an effective range of 45 miles. The unit did not enter production until the spring of 1945, becoming available in 1946, effectively replacing the Walter and radar reflectors for joint Army-Navy use. Like all electronic devices, the AN/CRC-7 seemed to have a short life span, remaining in use for a few years before being replaced by a better model, progressing into the present day.

The AN/CRT-3 remained in the arsenal until the 1980s because of its actual range capabilities with additional subsequent revisions.

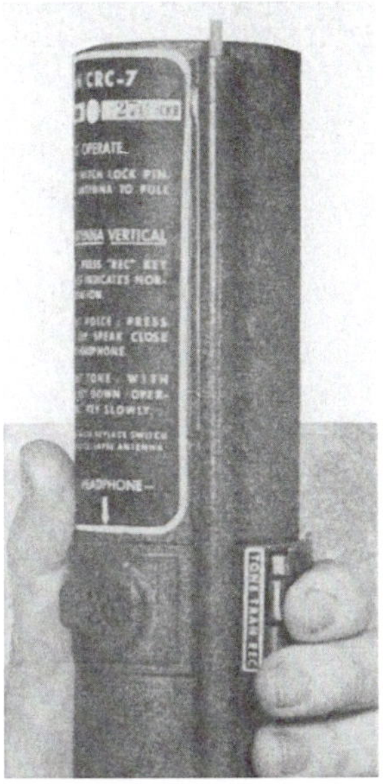

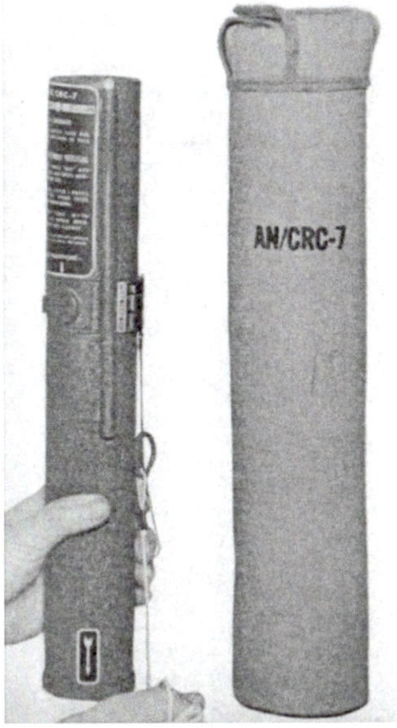

(Left) The CRC-7 measured 15" x 2-1/2" in diameter and weighed 3 pounds 12 ounces. This unit was revolutionary in that it allowed two-way communication with rescue craft while simultaneously transmitting a homing beacon on a very high frequency.

(Above) Another electronic beacon employed for air sea rescue purposes was the AN/CRN-1 buoy radio, which was used for marking locations of life rafts. The buoy would be dropped into the water by parachute from patrolling aircraft, providing an automatic transmission for homing purposes. It had an affective range of 50 miles and a battery life of 12 hours. A patrolling aircraft dropped one of these buoys near a TBF aircrew on a rubber raft, May 1944. One hour and 45 minutes later they were picked up by escort destroyer USS Loy. The Buoy can be seen in the bow of the life raft. (National Archives)

Homing pigeons were used by the USAAF and US Navy during WWII to supplement other forms of communication as a back up. The USAAF Pigeon Service was established 15 February 1943, organized into companies, divided into four platoons or six sections. Each platoon or section was divided into three-man detachments handling 120 to 240 birds. Airfields were assigned one detachment. A typical compliment per plane comprised two sets of two birds under the responsibility of the radio operator.

On an operational flight the aircraft took off from point A with the destination as point B, one bird in a set was homed for point A and the other for point B. In an emergency situation, the two pigeons were released with a message, each flying in a different direction. Standard operating procedure outlined that one pigeon was to be released while in flight and the other after ditching. The use of pigeons in the combat theater comprised one homing station, with aircraft each having a compliment of two, four or more birds.

(Above) Located at each station or airfield would be a loft for the homing pigeons, cared for by the detachment of Pigeoneers. This was the location the bird would return to with the message. A homing pigeon could fly for a steady 12 to 15 hours comfortably up to distances of 500 miles or more. (National Archives)

(Left) A RAF SOS message being opened and read from a returning bird in North Africa. (National Archives)

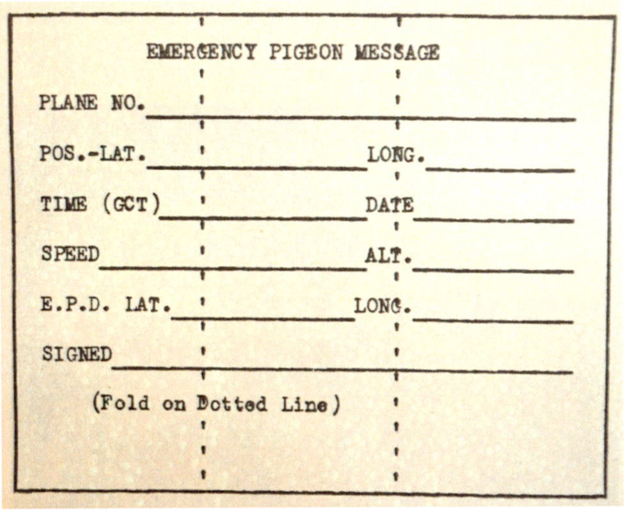

EMERGENCY PIGEON MESSAGE

PLANE NO.

POS.-LAT. LONG.

TIME (GCT) DATE

SPEED ALT.

E.P.D. LAT. LONG.

SIGNED

(Fold on Dotted Line)

(Above) The messages carried by the bird would be very basic in order to be stowed in a capsule strapped to its leg. There was no standard form, most being locally printed.

(Above) Examples of types of capsule.

(Left and above) RAF pigeon carriers were often used by the USAAF in Europe. Note the distinctive yellow-orange color. Photographs: Steve Milnthorpe

(Below) A standard USAAF two-bird carrier. It contained a message book, map overlays, message holders, pencils, and food. The US Army Signal Corps had standardized carriers of several sizes for two, four, or more birds.

(Below) A locally fabricated two-bird carrier. It would contain similar articles as the standard types in the top.

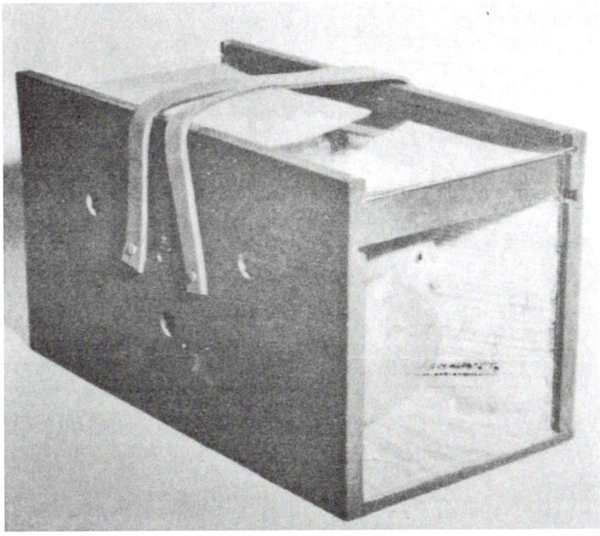

(Below left) USAAF aircrew prepare a message during a simulated ditching scenario. When the crew evacuates the aircraft the pigeons are brought aboard the raft. The locally fabricated pigeon carrier sits on top the SCR-578-A radio set. (National Archives)

(Below right) A B-24 waist gunner of the 9th AF brings aboard two sets of pigeons in RAF carriers, August 1943 North Africa. (National Archives)

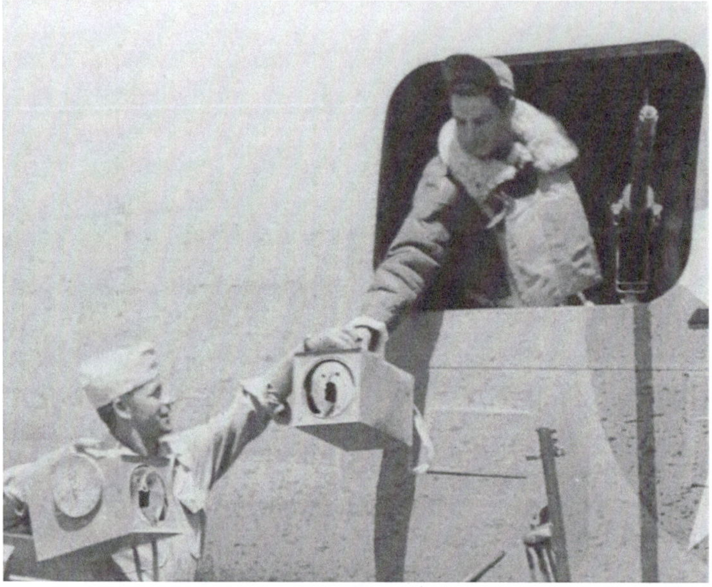

(Above left) A pigeon service was also maintained by the US Navy, instituted just prior to the turn of the 20th century. Pigeon trainers or Pigeoneers were part of the Quartermaster rating. Use of carrier pigeons within the fleet declined by the 1920s with advancements in radio technology, but remained in use primarily with aviation activities. Lofts were maintained until a need for their use arose with the outbreak of war. The birds were used with continental United States coastal patrols, namely by airship or ZP squadrons. A series of lofts was established at naval air stations, with the birds being used for communication when radio silence was in effect. A specialist pigeon trainer, rating of Specialist-X, was instituted in 1943. WAVE Specialist-X 2nd Class Marcelle Whitman takes care of a pigeon at NAS Santa Ana.

(Above right) WAVE Specialist-X 3rd class Sidney Cazort monitors a loft at NAS Moffet Field.

(Above) WAVE pigeoneers undergo training, in the back ground is the pigeon loft.

(Above) Carrier pigeon indexed as a Hall of Fame Bird at an unknown NAS, with a map of the continental United States as the back drop. It's likely it flew from coast to coast earning its entry into the Hall of Fame. (National Archives)

(Left) Airship aviation cadets undergo training on the use and care of using carrier pigeons at NAS Moffet Field, 1943. (National Archives)

(Right) A pigeon carrier being loaded prior to a patrol. (National Archives)

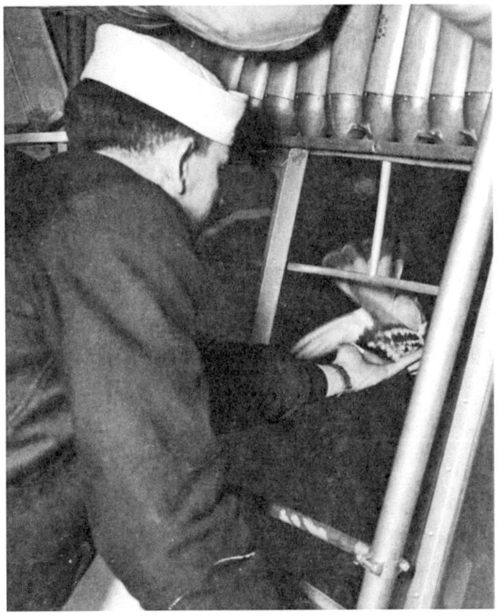

(Left) A pigeon is released with a message from the airship. (National Archives)

(Right) Message is removed from the capsule on its leg by pigeoneers after it has landed at the homing loft. (National Archives)

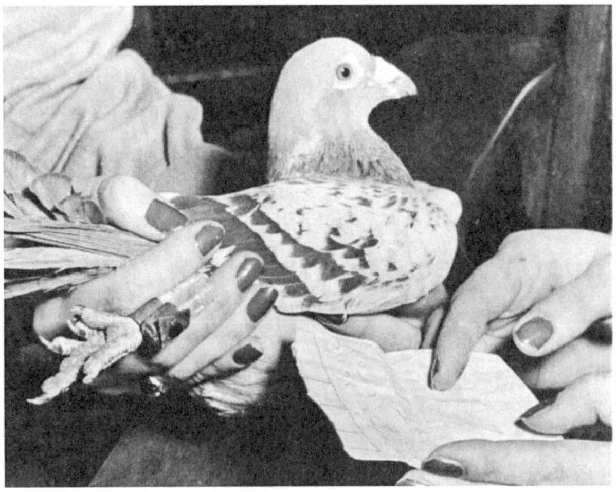

Air sea rescue organizations employed alternate means of communication with survivors by dropping message bags or message blocks from an aircraft into the water. They were locally fabricated and provided with a message carrier made from either cork or a soft wood.

One report detailed the block to be 6" x 6" x 3" with a drilled hole in which to place the message, which was then sealed by a plug. A piece of colored bunting about 4" long would be attached to the block to increase visibility both in the air and after it landed in the water.

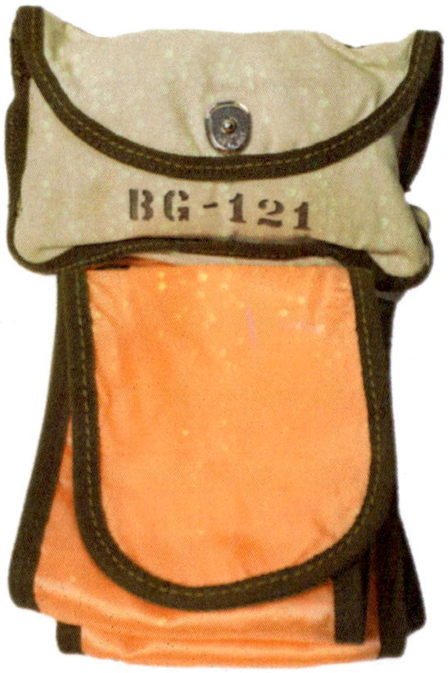

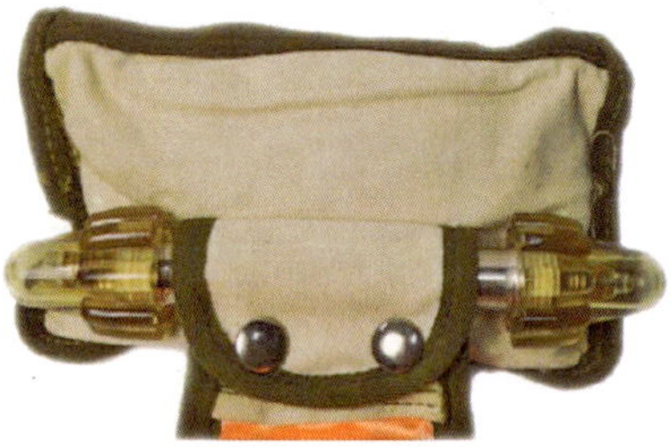

(Above) Message bags or blocks were not a new innovation, having been used since the introduction of aircraft for communication dating back to WWI. They did fall from use with the introduction of radio transmitters and receivers, and were merely a rediscovered method repurposed for air sea rescue. Shown is a message bag, standardized by the US Army Signal Corps, designated BG–121 with locating light TL-214. Based on descriptions, locally fabricated message blocks closely resemble this unit. This message bag and others were employed by USAAF search and rescue activities.

(Top right and right) Copy of a message utilized by US Navy Rescue Squadron 3 (VH-3) printed on waterproof material 10"x 12-1/2." This message would be dropped to survivors giving detailed instructions.

UNITED STATES PACIFIC FLEET
AIR FORCE
RESCUE SQUADRON THREE

IMPORTANT

YOU HAVE BEEN SIGHTED BY RESCUE SQUADRON THREE. FOLLOW THESE INSTRUCTIONS AND YOUR RESCUE WILL BE EXPEDITED THROUGH YOUR **COOPERATION.**

1. **MAINTAIN IDENTIFICATION:** If you will trail dye markers, show smoke, use mirror, it will enable rescue craft to keep trace of your position.
2. **SIGNAL IF INJURED BADLY:** By crossing arms across body. If uninjured, hold arms straight out at side. Rescue plane will acknowledge by dipping wings.
3. **DO NOT USE EMERGENCY EQUIPMENT NOT REQUIRED:** It may not be possible to pick you up at once so conserve your gear, particularly pyrotechnics, which are of little value by day.
4. **PLAN FOR RESCUE:**
   You will be picked up by:

   CIRCLE ONE — This Airplane — Submarine — Seaplane — Surface Craft

   In approximately.................Minutes. Hours — Approximate bearing..................°.

   Detailed instructions regarding cooperation expected of you for each type of rescue following. In each type of rescue remember: **REMOVE HELMET AS RESCUE CRAFT APPROACHES AND SECURE ALL FIREARMS.**

RESCUE

1. **FLYING BOAT.**
   A. Favorable Sea.
      1. Airplane will land; remember it is hard to keep you in sight when plane is on the water so **CONTINUE TO SIGNAL** by mirror or smoke.
      2. **STAY CLEAR OF PROPELLERS** as a swell may wash you into them. It is recommended that you remain outboard of the wing tip floats. If one of the engines is cut, approach the plane between the wing tip float and the hull on the side of the cut engine.
      3. A life preserver with a line will be thrown to you, or if you are incapacitated a man in a life raft with a line attached will guide you to the plane.

   B. Unfavorable Sea.
      1. This plane will report your position and arrange for you to be picked up by a submarine or surface craft.
      2. This plane will drop provisions as required for your rescue. The equipment or supplies dropped will depend on the circumstances. Most of this equipment will be dropped with lines having floats to assist you in recovery. Check out in the use of all of your equipment.
      3. Rescue Squadron THREE will stand by as necessary to direct your rescue by leading pick up craft to your position.
2. **SEAPLANE** (Cruiser or Battleship type)
   A. **CONTINUE TO SIGNAL** when plane is on water.
   B. If **EXHAUSTED:** swim or paddle outboard of wing tip float if forward of wing as swell may wash you into propellers. Crew will try to get you to the main float. **DO NOT HOLD ONTO WING TIP FLOAT—DANGER OF CAPSIZING PLANE.**
   C. If capable, swim or paddle to main float aft of wing.
   D. Plane may or may not have two (2) man crew. If only pilot aboard, you must be prepared to get in plane with little help.
   E. **DO NOT ATTEMPT TO CLIMB FROM FLOATS TO WING UNTIL** pilot or radioman instructs you or you may capsize plane. Crew will give you verbal or hand signals.
   F. If seaplane is equipped with small life preserver and line, crew will show this so stand by for line throw.
3. **SUBMARINE** (Day)
   1. **CONTINUE TO SIGNAL** preferably with mirror or smoke. Also dye marker can be seen.
   2. If **SUBMARINE SURFACES** look for life preserver and line to be thrown.
   3. If **PERISCOPE TOW** is desirable due to proximity of shore batteries.
      (a) Submarine will make pass at raft with periscope showing.
      (b) Have line ready to throw around periscope (remember if you miss on first pass, submarine cannot reverse course submerged and must make another circle). Best tow length 20 feet aft.
      (c) Do not secure line to periscope and raft unless it can be slipped quickly in case submarine dives.
      (d) Maintain surface and aircraft watch for submarine and signal to periscope if unfriendly aircraft or surface craft in vicinity.
      (e) Submarine will tow until safe to surface; **MAINTAIN CONTACT THROUGH PERISCOPE.**

   SUBMARINE (Night)
   1. **WATCH** for flares from submarine, answer them with flares, flashlight; **LISTEN** for whistle, answer with whistle, flares or flashlight.
   2. **SHIELD FLASHLIGHT** to maintain contact with approaching submarines without giving away operation to enemy.
4. **SURFACE SHIP.**
   A. **DAY:** Best signal is smoke, then mirror, less dye marker.
   B. **NIGHT:** Best, signal flares, then flashlight.

In 1943, both aeronautical services considered the concept of developing a personal balloon locating device to aid in locating survivors.

The BuAer spent a relatively short time investigating the idea, concluding by August 1943 that it would be necessary to have two pounds of equipment to fill either an 18" helium balloon or a 3' hydrogen balloon, and gave the matter no further consideration.

The USAAF considered including a balloon with the parachute first aid kit, but it was soon realized that the inflation equipment required made the concept impracticable.

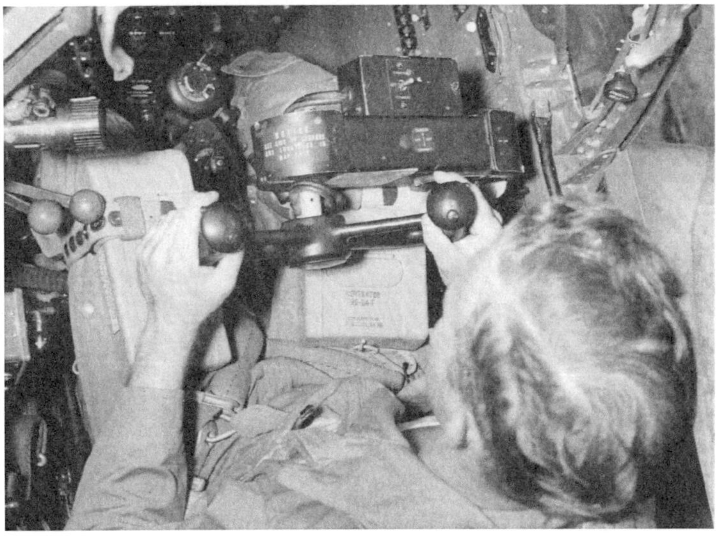

(Above) The concept field tested by the USAAF in July 1944. The complete unit was contained in two canisters, one for the balloon and the other for the gas, designed to strap to the user's legs. (National Archives)

(Top right) The balloon was found to function satisfactorily. (National Archives)

(Right) In conclusion, the idea was judged completely impractical. Note how the canisters could easily hamper the aircraft controls. (National Archives)

PERSONAL AID KITS

The USAAF had four principle classification groups for sustenance and survival equipment kits: Parachute Emergency Kits, Airplane Kits, Droppable Kits and Personal Aids Kits.

This section details Personal Aids Kits, which are described as units attachable to pilots and crew members of an airplane for the USAAF and units adopted by the BuAer.

There was an almost endless variety of personal aids kits utilized by the aeronautical services, if units assembled in theater are included. These were associated with unit level assemblies and by intelligence sections such as MIS-X operating in all theaters of war.

However, there was a series of standardized models adopted by the USAAF and BuAer to suit service requirements as established by regional operations.

The RAF would initiate the first influence in the development of personal aid kits for US issue. Prior to the United States entry into WWII, the British Military Intelligence Service, MI-9, had begun producing a series of what they referenced as Ration Boxes or Aids Boxes. These were intended to be carried by aircraft personnel and used if forced to bail out over enemy territory as an escape and evasion aid.

Several variations of the RAF Aids Box were developed from the original concept by Christopher Hutton of MI-9, each identified by a Mark designation. All Mark series boxes contained a principle compliment of components to include: rubber water bottle, Horlicks malted milk tablets, Benzedrine sulfate, halazone, packets of chewing gum, bar of chocolate, box of matches, tubes of condensed milk, hacksaw blade and compass.

(Left) The Mark II Aids Box. Components from the Mark II to Mark VI were housed in a sealed cellulose acetate type container, providing basic sustenance to an individual on the ground in enemy territory attempting to evade or escape to friendly lines. The unit was designed to be compact to fit in the flight suit pocket, measuring 6" x 4-1/2" x 1" respectively. The Mark II could also contain currency and maps, not included in the remaining production of the Mark series. The later series would be complimented with money and map purses issued separately.

(Left) Components of the Mark III aids box.

Escape & Evasion was new to the United States armed forces and, fortunately, units arriving in Britain in early 1942 had the benefit of guidance from the RAF. Their practices and equipment would influence future developments, especially within the USAAF.

In May 1942, the 8th Air Force recommended the development of a ration or aids box like that of the RAF, initiated by Maj. Gen. Carl Spaatz, who had recently returned from England and been briefed on MI-9 activities. The USAAF Materiel Command began the development program the following month, establishing preliminary guidelines for a suitable design, based on the British version. The American copy was assigned the tentative nomenclature of Kit, Emergency Sustenance (Escape), Type E-3, under USAAF Specification No.94-40441. It would comprise: one box of matches, one pocket compass, one small hacksaw blade, 16 Horlicks malted milk tablets, one bar sweet chocolate and a chemical to purify water. All the components would then be stored in one large drawstring, rubberized bag that doubled as water storage.

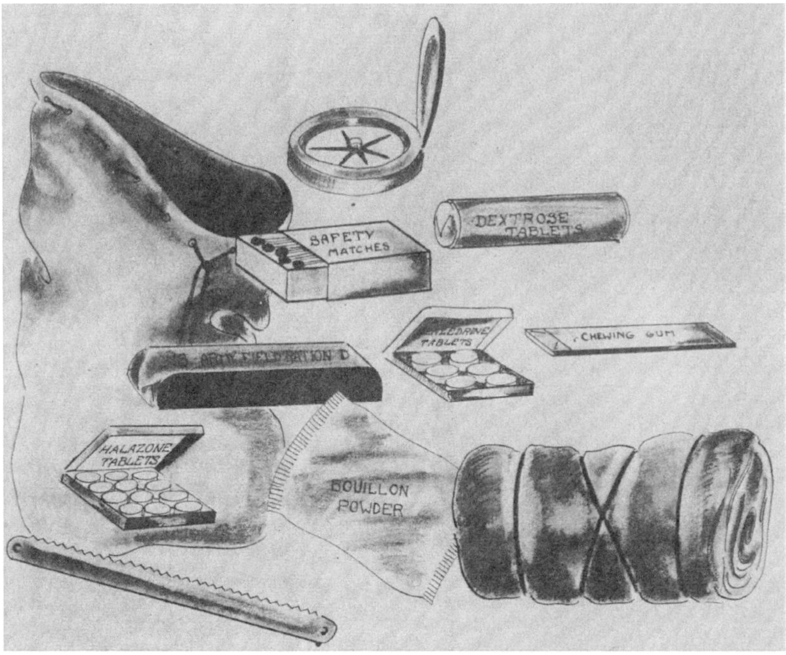

(Right) Drawing attached to preliminary Specification No. 94-40441, illustrating the original concept of the Type E-3 sustenance kit. It has been slightly revised from the original list of components by adding a package of bouillon powder and one stick of chewing gum. Additionally, it shows halazone as the water purifying agent. In the bottom right is the complete assembly bundled into a roll shaped package.

Approval for release of the specification was withheld pending a discussion regarding combining the unit with the parachute emergency kit. At the time, a new concept of a parachute kit was under development, which would later be standardized as the Type B-4, and the intention may have been to include the Type E-3 with this unit.

The evolution into a ration pack was to replace the standard U.S. Army Field Ration D and combine components from the Bail Out Ration.

The engineering section was operating under the premise that all personnel would be equipped with the parachute kit on operational flights and, in the event of a bailout, all the necessary components would be available from the kit for escape and evasion purposes.

It was not Materiel Command's sole decision to adopt the Type E-3. As an escape aid, approval of standardization involved the Assistant Chief of Air Staff/ Intelligence (ACAS). Through lack of communication, ACAS was unaware of the withheld submission of specifications. The concept of the combination was dropped after it was determined to repackage the Type K ration for inclusion in emergency parachute kits. A formal specification was then drafted and submitted for approval 24 August 1942 to ACAS.

In the interim, the unit was revised by adding components from the Bail Out Ration as follows: one box safety matches, one pocket compass, one hacksaw blade, 12 halazone tablets, six Benzedrine tablets, one U.S Army Field Ration D, 4" x 2-1/4" x 7/8," two packages of 15 flavored dextrose tablets (six lemon and nine malted milk), one package of bouillon powder and one stick of chewing gum.

The dextrose, bouillon and gum were to be contained in a waterproof package, 4" x 2-1/4" x 1-1/4," with the entire assembly stored in a resin-coated fabric drawstring bag, 12-1/2" x 6." Upon review, and from service recommendations, it was advised to increase the count to four sticks of gum and add a spool of 3/4" adhesive tape, 24" long. Specifications were then revised to include the recommendations being re-submitted, and approved as standard, 8 October 1942, as Kit, Emergency

Sustenance, Type E-3, USAAF Specification No. 94-40441, with the term "Escape" removed.

The Materiel Command had full responsibility over all aspects except final approval and distribution of escape and evasion aids, including responsibility for procurement utilizing their funds.

At this point, approval for procurement was not authorized, pending review from overseas stations. It was not until February, 1943, that responses to the design were received in Washington, stating that they were under the assumption the kit was already under procurement and requesting it be placed into production immediately, as this item was urgently needed.

As reports from the combat theaters were being reviewed, the Type E-3 underwent an additional amendment: Amendment-1, dated 31 March 1943, revised the components, dropping the pocket compass and replacing it with the match safe compass BQD 82, as well as reducing the dextrose tablets to one box and increasing the bouillon powder to two packages. Everything was to be packed in a cellulose acetate container.

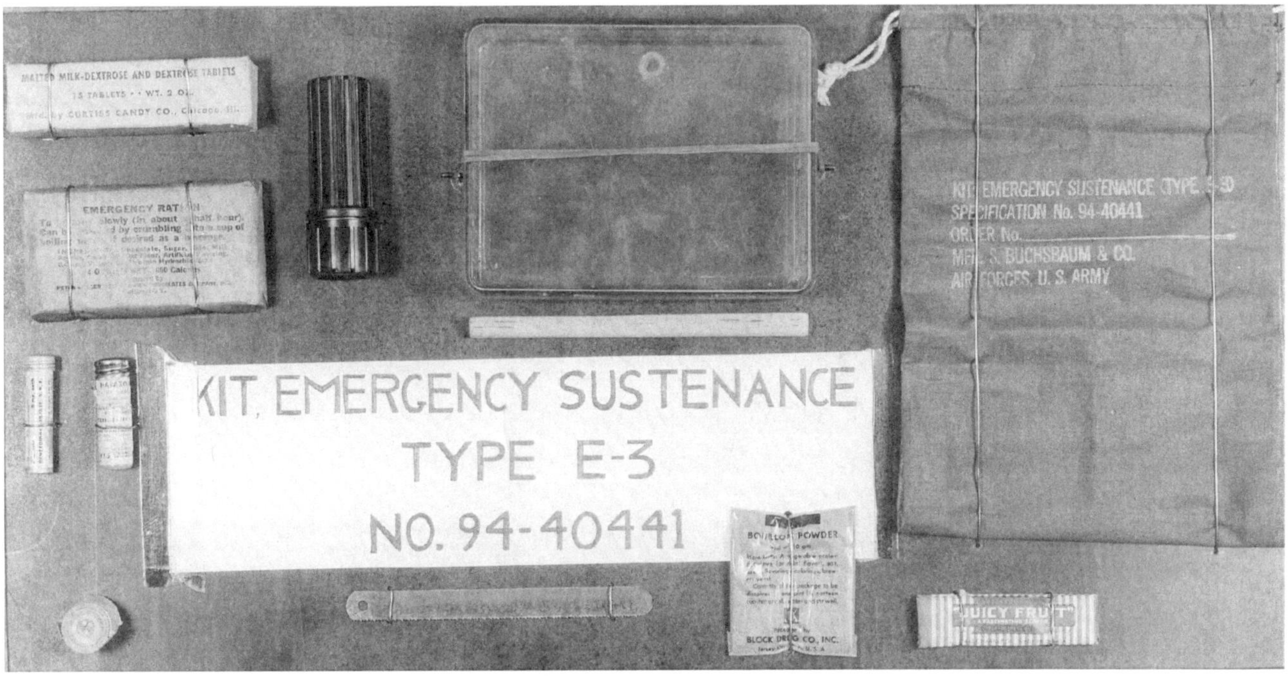

(Above) Conceptual design assembled as per Amendment-1. The cellulose acetate container was originally intended to have an interlocking rim to snap closed. This was dropped on the production model. The components seen here were principally carried over to the contracted type. The resin water bag was manufactured by the S. Buchsbaum Co. and, as per specification, the outer container was legibly marked with nomenclature, specification, contract number, contractor and property marks. Being a test sample, the Order No. was left blank.

Escape aids were an untried concept at the time, and as part of typical procedure for the Materiel Command, the Type E-3 kit, like other equipment, was being drafted for inclusion in technical orders and other literature to be distributed to service activities.

ACAS approved the Type E-3 for procurement, 25 May 1943, and immediately ordered that the kit be removed from any descriptive literature such as stock lists, Status of Equipment, Pilot Information Files, etc. These were for special issue only, distributed through S-2 intelligence sections of the Assistant Chief of Air Staff/Intelligence and therefore could not be issued without approval.

The final recommended change to the Type E-3, prior to production, was that the water bag be reduced in size and packed with the components in the cellulose container detailed in Specification 94-40441-A, dated 1 June 1943. The new nomenclature was Kit, Emergency Sustenance, Type E-3A.

The original intention was for the procurement of 80,000 kits, but this was reduced to 40,000, awarded in two purchase orders dated 25 June 1943: 20,000 to the Nathan Products Co. under P.O. 43-29112, and 20,000 to the Mintz Co. under P.O. 43-29113.

Shipments of the sustenance kits commenced by August, with all being distributed fairly evenly to Air Force intelligence sections in all combat theaters.

These would be the only two procurements awarded for quantity production of the Type E-3A sustenance kit.

(Left) Type E-3A sustenance kit as contracted and assembled by the Nathan Products Co. and the Mintz Co. The unit measured 5-3/8" x 4-1/4" x 1-3/4" and consisted of two halves taped together.

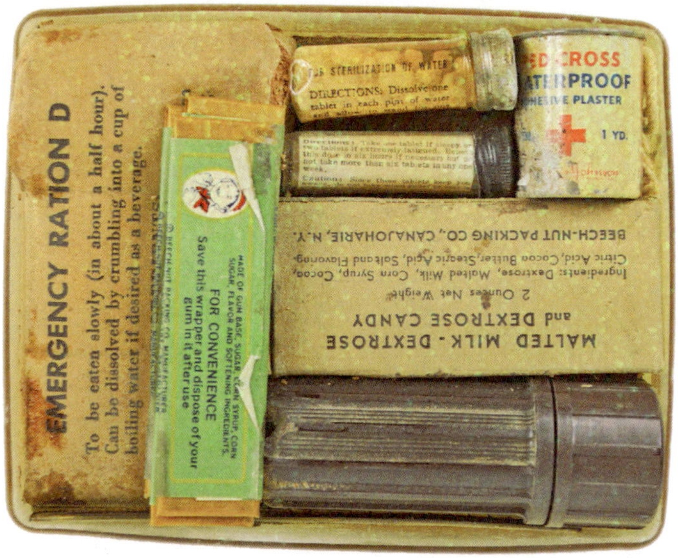

(Left) The unit with its lid removed to show contents more clearly.

(Top right) In keeping with typical Materiel Command administrative procedure, nomenclature, type designation, specification, stock number, contract, contractor, and property mark were required to be placed on the item. This was not required to be removed by ACAS. Initially the engineering section believed the Type E-3 to be organic to supply and assigned it stock number 8300-559158. Unusually, the nomenclature also reflected the quantity of the contract

(Left) 1" wide friction tape was utilized to seal the kit, and each unit was marked on the seal with an acceptance stamp.

(Above and right) The resin bag measured 12" x 4-3/4" and was typically folded and stored in the bottom of one of the halves of the kit as shown.

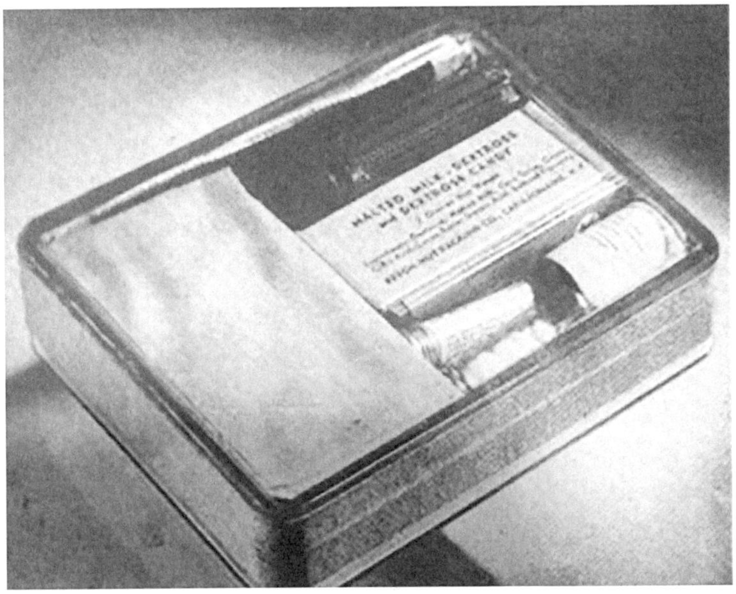

(Left) Wartime image of the Type E-3A. Like its MI-9 counterpart, it was intended to be carried in a convenient location by the user, typically in a pocket of the flight clothing. It was not an uncommon occurrence for them to be kept in the parachute emergency kit, if worn.

It was the responsibility of Materiel Command to procure and maintain equipment peculiar to the Army Air Forces, with the development of the Type E-3 placed under its cognizant.

ACAS was an agency oriented towards strategic dissemination, having only minimal influence at the time from MI-9 activities.

In October 1942, the Military Intelligence Service Section-X was established and would take over responsibility of all aspects orientated towards escape and evasion, working in coordination with MI-9 in all theaters of war. As the United States intelligence sections were being established, technical requirements and departmental guidelines were being refined. By spring, 1943, there had been sufficient data accumulated from survival accounts and Escape & Evasion reports emphasizing the needs of service activities. The Type E-3A was developed based on the requirements of the ETO with sole emphasis on escape and evasion, but was soon proven deficient for the needs of those in the PTO and CBI. The requirement in those theaters transitioned towards "Survival," with suggestions of development of aids kits to supplement the parachute emergency kits.

Soon afterwards, revisions began on the Type E-3A. One service requirement was the development of a suitable canteen to replace the M-1910 for issue to aircrew members operating in a confined space. It was reported that aircrew were reluctant to carry the canteen and, in the event of an emergency, they were not being retained. A container for potable water was a valuable survival aid in arid and unpopulated areas. Possibly influenced by the MI-9 development of the Mark-IV & V canteen aids boxes, a new container was developed to fill this much needed gap.

(Left) With assistance from Master Plastics Molding Corp., who fabricated the container for the Type E-3A, a new canteen type flask was developed, molded from an ethyl cellulose. In this design concept, the components were stored in the flask with a removable telescoping top. After the components were removed, the lid was sealed shut with tape provided in the unit, then used to hold water.

Another complaint of the Type E-3A was its size, which created discomfort. The new flask container was slightly streamlined to 4-1/4" x 5-1/2" x 1-1/4."

One feature was a small hexagon shaped screw top for drinking purposes. To further meet service requirements components were added and, to better utilize the limited space, some were streamlined. The bulky match safe compass was deleted, being replaced by a new match safe and a compass developed by the Waltham Watch Co. specifically for this kit, under their part number WTS-43. The new unit was submitted to the Military Intelligence Service – which now had additional authority regarding escape and evasion aids – for approval. It was approved and standardized, 1 February 1944, under Amendment-B as Specification 94-40441-B titled as Kit, Emergency Sustenance, (Aids Box) Type E-3B.

Being established by MIS-X, guidelines for this unit were more defined than its predecessors, requiring that it be free of any identifying markings as outlined by specification. Its components seemed to be free of this requirement, as the bulk have brand names and even locations on them. The exterior of the flask lists all components with basic instructions on their use.

The Type E-3B was principally assembled by the contractor, but MIS-X consigned empty flasks for shipment to the CBI, because it had been determined that the kit did not completely meet requirements for that theater. The E-3B was a step forward, providing better sustenance and medical supplements to what might already be available in a parachute kit, or carried on the individual. The empty flasks would be supplied at the discretion of the local intelligence sections and filled with escape and evasion aids such as maps, money and other barter items. The Type E-3A was also purported to have been dismantled and supplied with additional aids in the CBI.

(Above left and center) Type E-3B sustenance kit. The assembly includes: one flask, one match safe with 25 matches, one brass compass WTS-43, adhesive tape 1" x one yard, one vial each of halazone, aspirin and Benzedrine sulfate, one tube of antiseptic ointment, two 2 oz. D ration bars, two packages of Bouillon powder, four sticks of chewing gum and one box of dextrose tablets (six lemon, nine malted milk). Two contracts were awarded for the Type E-3B, with the first in March 1944, under contract 33083-AC-2284 for a quantity of 54,000 kits awarded to the Cappel MacDonald Co.

(Left) The hexagonal screw-top which enabled the empty case to be used as a water canteen.

(Above right) The unique match safe, specially designed for the E-3B.

(Right and far right) Following ration development and the discontinuation of dextrose tablets, the Type E-3B was revised under Amendment-2, dated 11 April 1944, to delete the dextrose and replace it with one 2 oz. box of caramels, and replace the old formula D type ration with sweet chocolate bars. Shown is the Type E-3B as per Amendment-2. The second contract awarded, again to the Cappel MacDonald Co., was 33083-AC-5264, in September 1944, for a quantity of 65,000 kits.

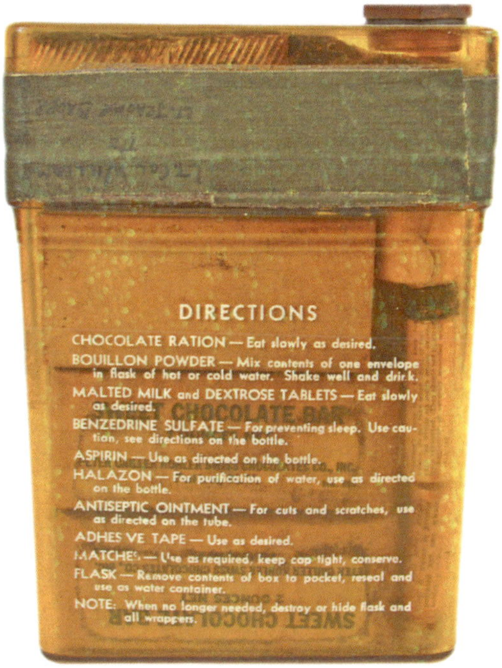

(Above) In October 1944, a new model flask was approved for the Type E-3 which had already been approved for the recently adopted Type E-17 sustenance kit, detailed under USAAF Dwg. No. 45B13490, molded by the Standard Molding Corp. under their Part No. SM-1080-D. The flask featured a new lid with a bail locking mechanism. Though approved, it was not integrated into the contract from September as it was determined that waiting for production of the new flask it would delay delivery, and the Type E-17 was priority for any production at the time. The Type E-17 would continue to use all stocks of the 45B13490 flasks into 1945. Due to the quantities of the Type E-3B procured in 1944 and the preference towards the Type E-17 for service use, approval for procurement of the new flask was not authorized until July 1945. The new Type E-3 unit would include the 45B13490 flask and outlined the use of the new match safe compass, USAAF Dwg. No. 45A9185. This unit would be adopted as standard as Specification No. 94-40441-C, dated 10 August 1945, titled Kit, Emergency Sustenance (Aids Box) Type E-3C. Production quantity is unknown. Illustrated is the Type E-3C sustenance kit. Most surviving examples still contain the older cylindrical match safe as detailed under Dwg. No. 45B13490, though the use of the 45A9185 was outlined in specifications.

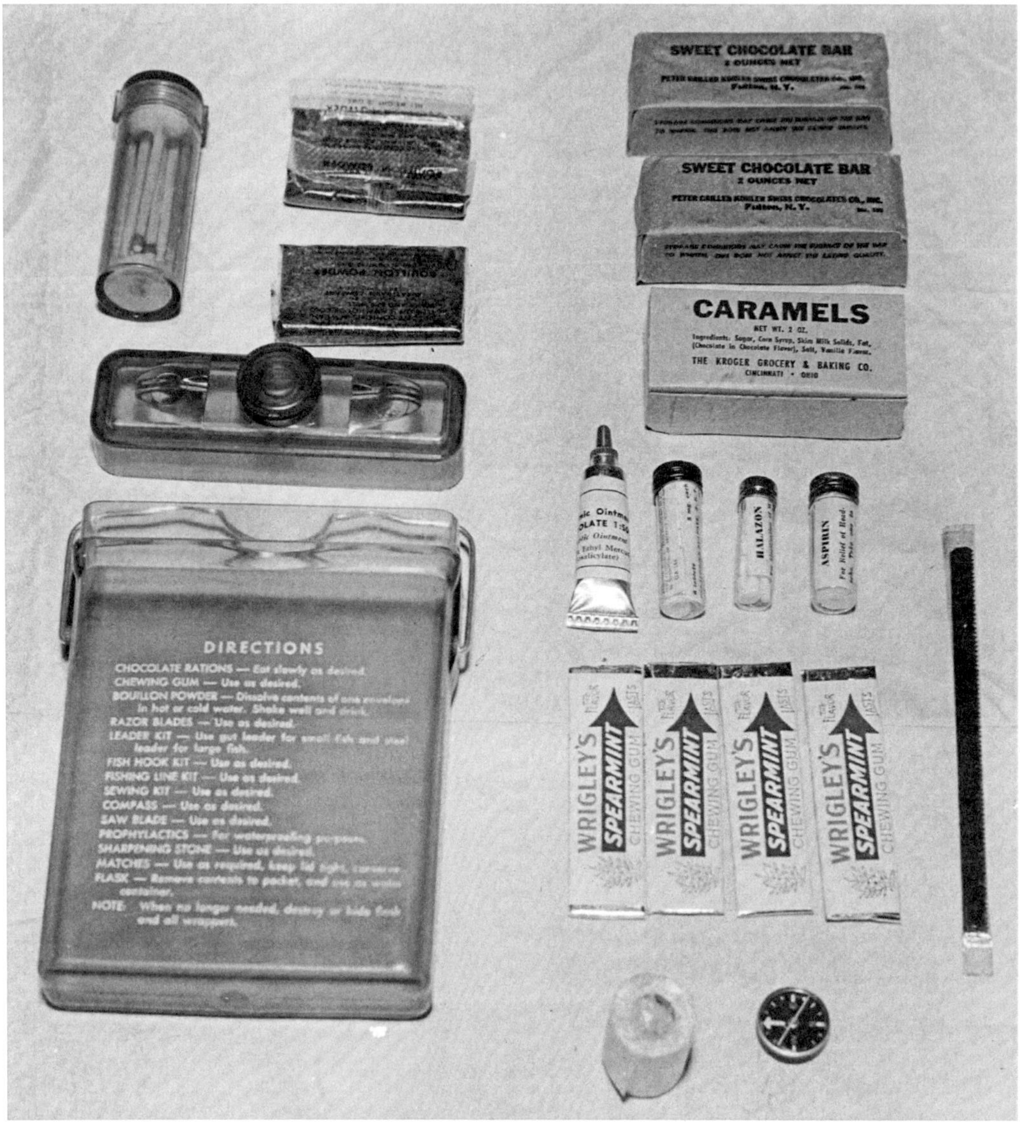

((Left) The components for both the amended E-3B and E-3C, the only differences being the match safe and flask. The first generation E-3B utilized D-rations with malted milk and dextrose tablets. All remaining articles are common to all three versions. This compilation was pictured prior to formalized production of the E-3C and utilizes the sustenance flask from a Type E-17 kit. The assembly illustrates the eventual new revision and is featured in the Manual for Personal Equipment Officers printed in June 1945. This has led to the erroneous designation of Type E-3A, since that is how it is titled in the manual. Formal adoption of this configuration would not occur for several more months after the printing of this manual. (USAF)

To meet service requirements in the PTO and CBI, flying units in both aeronautical services would often assemble compact emergency kits).

One reason for the improvisation was the universal dislike of the emergency parachute kits, due to their emphasis on compactness. The Type E-3 sustenance kit provided sufficient nutrition, but lacked medicinal aids, being unsuitable for use in tropical regions. There were multiple ailments peculiar to that climatic zone, and medical items were viewed as an essential aid for maintaining health in survival situations.

For convenience, the E-3 kits were often stored in the parachute kit, but while evaluating the addition of the Type E-3 as a new emergency ration for parachute kits, as early as 1942, it was determined that plastic containers were not optimal for long term storage. Parachute kits were often lost after release from the harness, being stuck in the trees or descending into water. Also, most flying personnel wore regular chino type uniforms, which had no facilities for carrying supplementary equipment.

Engineering divisions of both aeronautical services worked to streamline individual emergency kits, such as a combination seat pack for the BuAer and development of a sustenance type vest for the USAAF. In the interim, units operating in the combat zone fabricated kits to suit their needs. Each service had differing opinions but shared the same goal: that a survival kit should be compact, orientated towards jungle use, and needed to be attached to the individual.

By the close of 1943, a solution was presented to meet these requirements from the Far East Air Forces and MIS-X (SWPA). A personal aids kit was devised, comprising essential components and utilizing the Type E-3 sustenance kit, supplemented by additional items in a webbing container that could be attached to a belt. The new kit was titled as the E-3J (Jungle Kit) by MIS-X.

The exact date of origin for the concept of the E-3J is not known, but records indicate that fabrication began sometime around October-November 1943, being assembled by MIS-X HQ in Brisbane, Australia, and continued to be assembled and distributed from that location.

Commanding General FEAF submitted a report in January 1944 to the Materiel Command, recommending that the kit be approved for procurement in the continental United States. The report stated that 3,500 E-3 kits had finally been located at Port Moresby and were being shipped to Brisbane for conversion to the E-3J. In the following month, MIS-X reported that the quantity was 5,000.

The new jungle kit would retain the following from the E-3 kit: combination match safe compass supplied with blue vestas (wax coated matches), D ration, water bag, bouillon powder, chewing gum, hacksaw blade, dextrose malted milk tablets, Benzedrine sulfate, halazone and adhesive tape. Additional items were to be obtained from the Medical Department and QMC to include sulfadiazine tablets, six Band Aids, atabrine, sulfaguanidine, salt tablets, vitamin tablets, iodine, three rubber sheaths for protecting items after watertight seals were broken, toothbrush, sharpening stone, fish hooks, fishing line, signal mirror and razor blades. The components were to be stored in a hinged-lid tin and contained in a web container for attachment to a belt. For the Materiel Command ComGen, FEAF recommended the inclusion of Argyol to help in the removal of leaches. MIS-X also recommended the addition of a pocket compass and tweezers.

Being locally fabricated and assembled, components most likely varied in the distributed kits to service activities.

(Above and left) Surviving example of an E-3J Jungle Kit. The webbing pouch was fabricated locally in Australia and had provision for a single belt loop only. The contents, contained in the tin container, were controlled items, distributed through MIS-X to the BuAer, USAAF, and RAAF (Royal Australian Air Force). Due to the limited quantities available the basis of issue was one kit per fighter and two per bombardment aircraft.

(Below) The contents of the E-3J Jungle Kit (pictured on previous page) comprise a mix of US and RAF. RAF components are from a ration box.

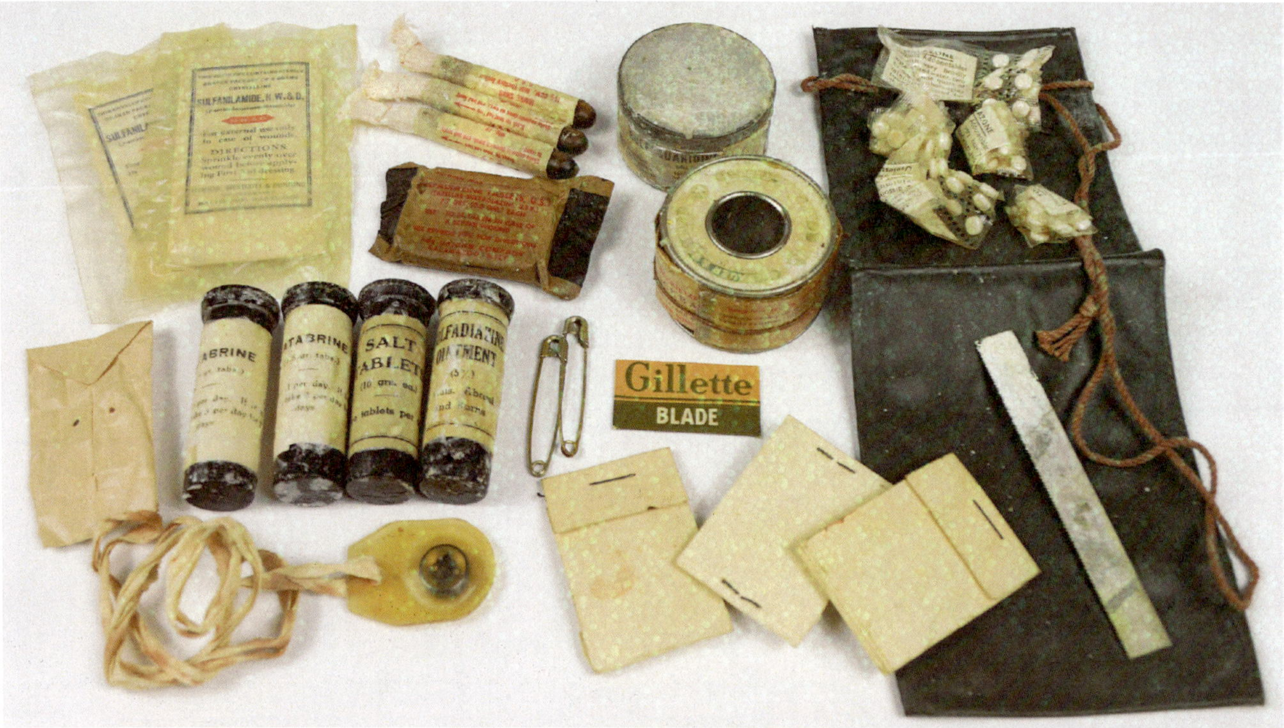

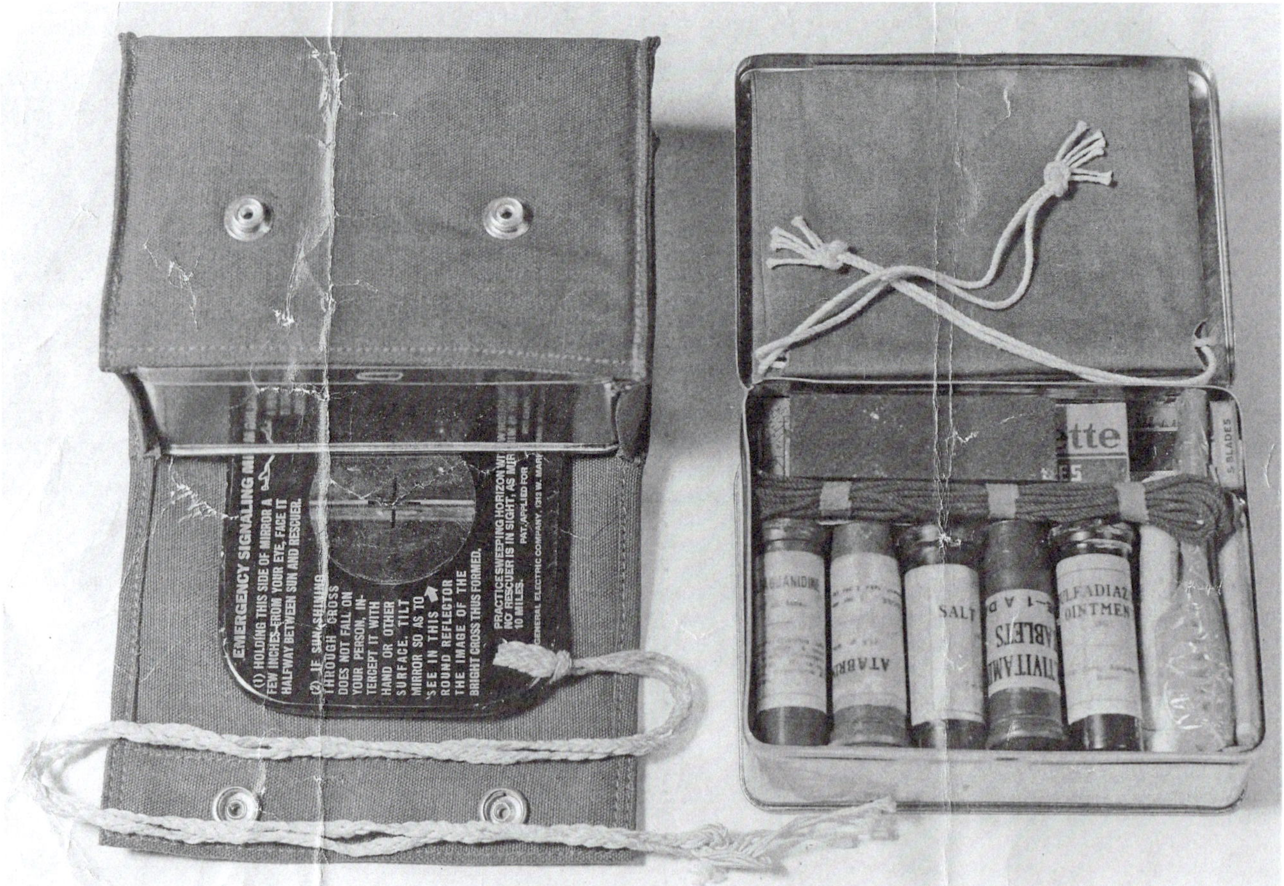

(Above and top of opposite page) Wartime images illustrating the E-3J as assembled by MIS-X. One discernible difference between the two is the signaling mirror in the image above. In the MIS-X report of February 1944, the option of the tempered glass mirror (ESM/1), or a metal mirror 5" x 3-1/2" with center sighting hole is listed. MIS-X had reported that 3,708 were available in theater and that another 2,700 being assembled would be available as of October 1944. (National Archives)

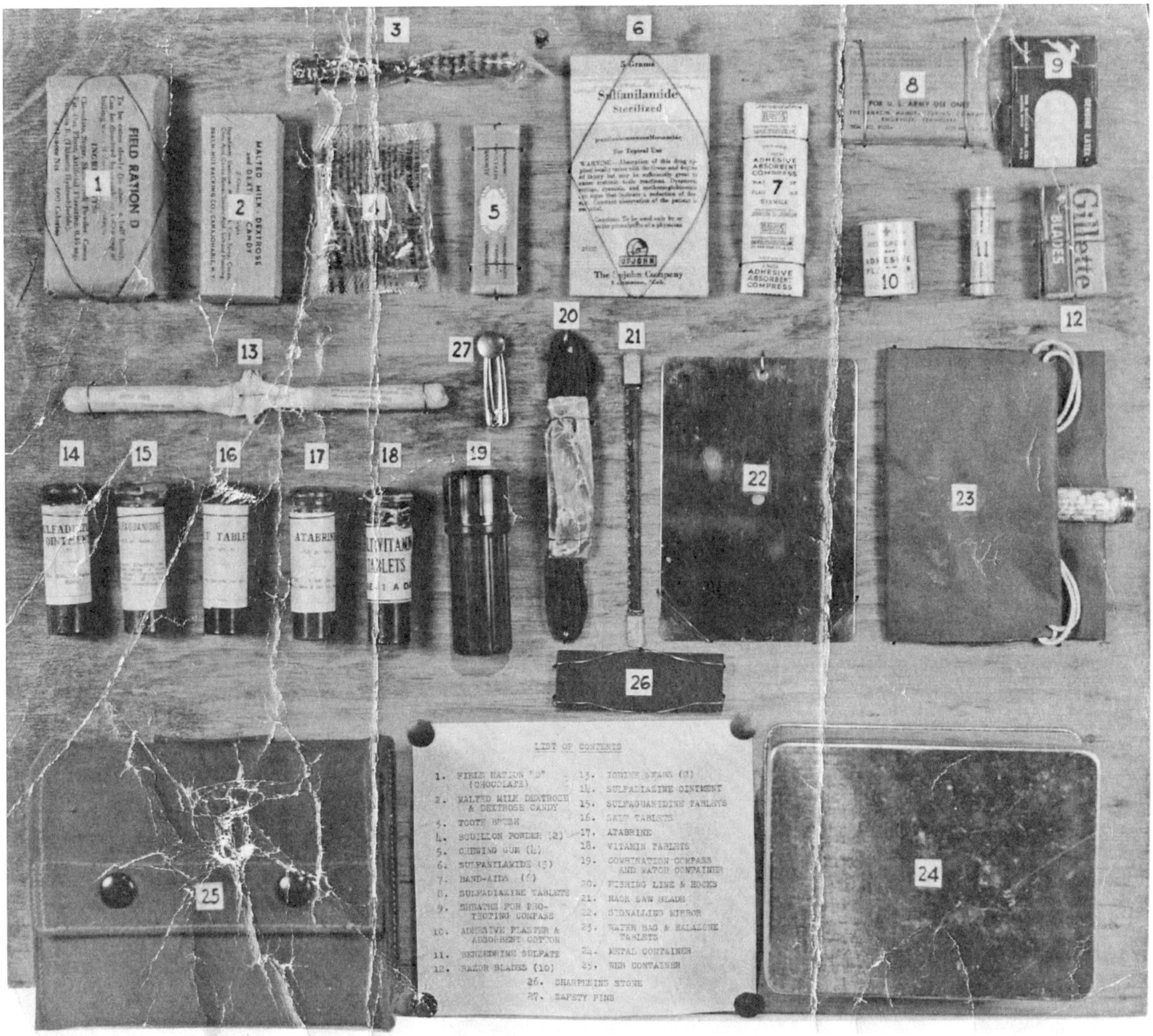

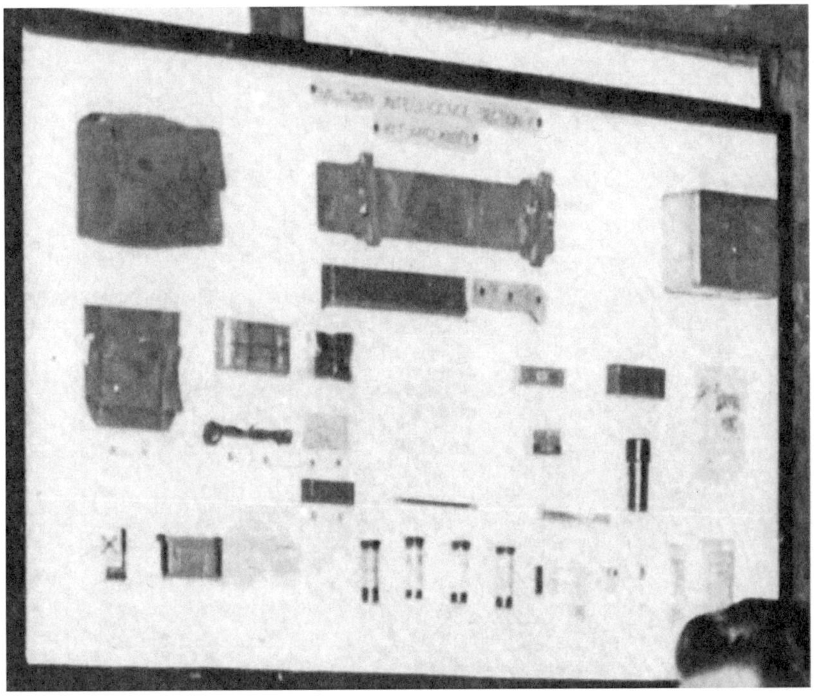

(Left) An E-3J kit is being displayed in an orientation lecture room of the 2nd Medical Establishment FEAF on New Guinea, c.1944. This is a segregated section, orientated towards items peculiar to MIS-X. The label on the top of the board says "MILITARY INTELLIGENCE SECTION". Also included is the Australian leg machete and the manual "Survival Hints". (National Archives)

(Left) VC-86 TBF/TBM pilot Lt. Dewyn Severy wears the E-3J in a typical position for comfort while flying, aboard the USS Bismarck Sea, January 1945. With the limited quantities available the kit was issued to the pilot's position in multi-seat aircraft. The gunner and radio operator, Robert Lustig ARM2/c and Robert Hollbech ACOM accompanying him would not be issued the unit. (National Archives)

(Below) A PBY pilot preparing for a mission at a base on New Amsterdam Is., Dutch New Guinea, September 1944. As part of his personal compliment is a side arm, canteen, sheath knife and E-3J kit. One end of the belt loop strap has the unique feature of being secure to the top of the container. (National Archives)

(Right) TBF Avenger pilot Lt. jg. Delbert Gerlach of VC-85 aboard the USS Lunga Point, March 1945. (National Archives)

(Below and below right) Jungle kit fabricated in Australia for use by RAAF personnel. It was assembled along the same general guidelines as the MIS-X E-3J and contained similar components by description. It comprises a web pouch with two belt loops. The flap also has eyelets for sealing the unit after assembly. (R. E. Baldwin Collection)

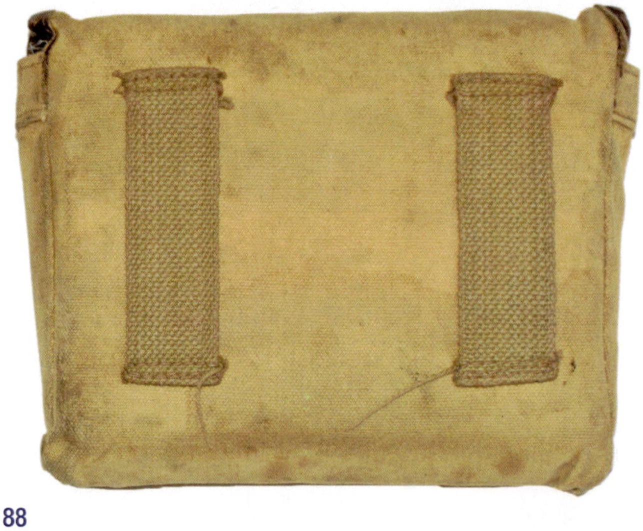

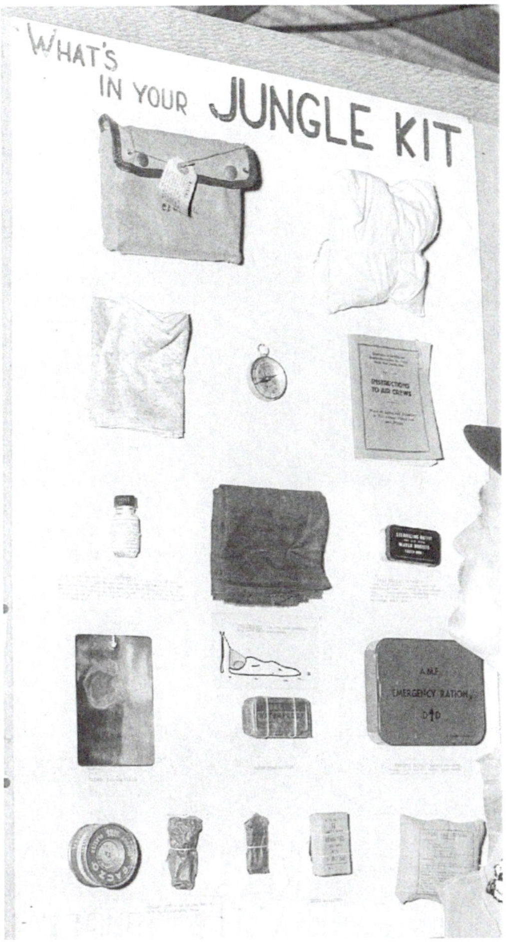

(Above) Contents of the The RAAF jungle kit on a display board for a RAAF Catalina squadron, January 1944. (Courtesy Australian War Museum)

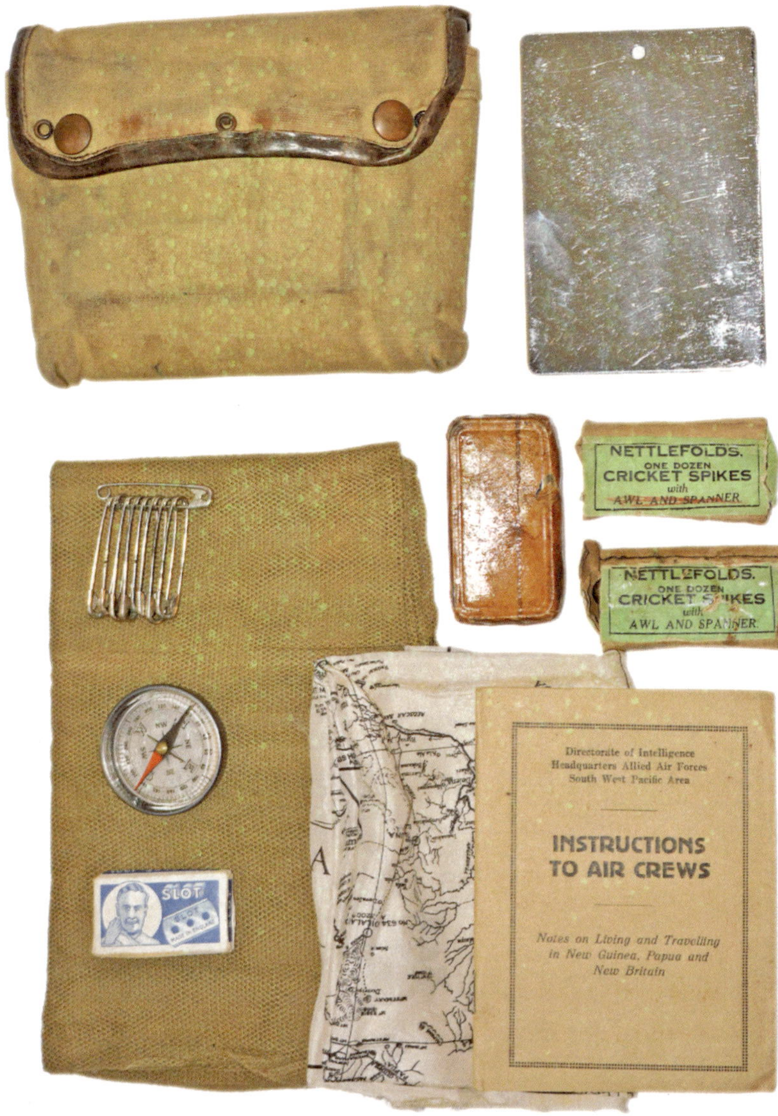

(Above) Typical contents of the The RAAF jungle kit: first aid dressing, adhesive tape, safety pins, 12 cartridges for a revolver, waterproof matches, emergency ration, signal mirror, mosquito netting, water sterilizing kit, atabrine, cloth map, compass, survival manual, razor blades and cricket spikes for service shoes. (R. E. Baldwin Collection)

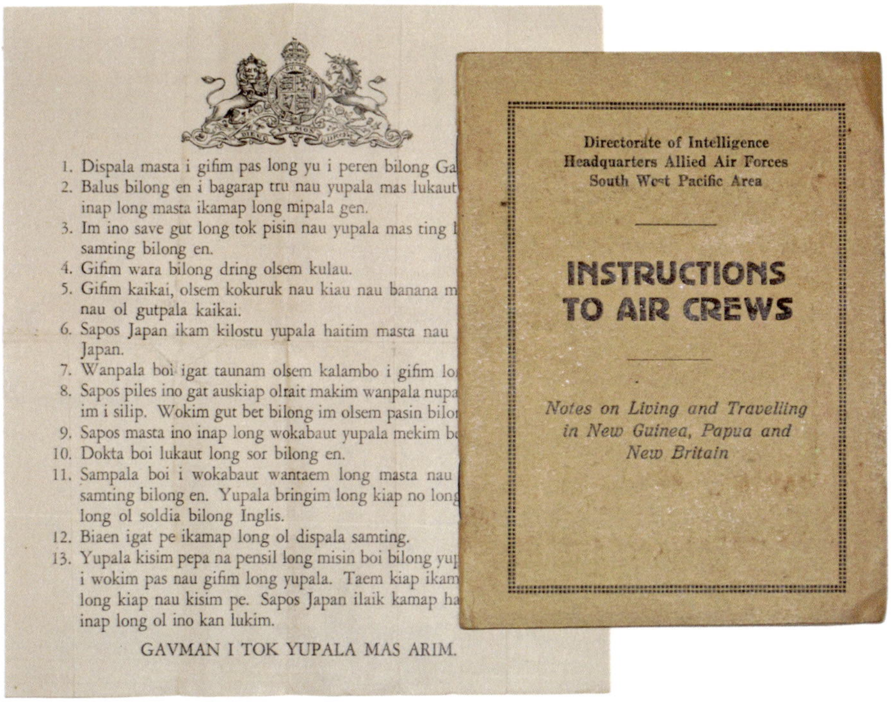

(Left) This example that had been slightly modified with the inclusion of a two-sided paper blood chit in Pidgin and English. (R. E. Baldwin Collection)

(Below) Australian manufactured waterproof wax vestas were preferred over the QMC matches for use in the tropics. The manual "Survival Hints," distributed by intelligence SWPA, recommended one box of blue headed vestas to be carried as a survival accessory.

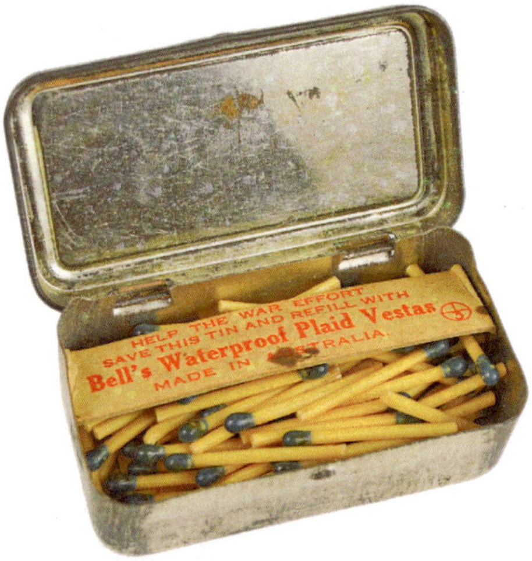

(Above) RAAF B-24 air crew. Note the front-most crewmember wears both the E-3J and RAAF kit. It is equally likely that the RAAF jungle kit was used by US airmen but no photographic evidence has surfaced.

(Below left) RAAF pilot wearing the jungle kit with the leg machete. (Courtesy Australian War Museum)

(Below) Fighter pilots of 78 Squadron, RAAF, carry a combination of RAAF jungle kit and the E-3J, Netherlands East Indies, c.1945. (Courtesy Australian War Museum)

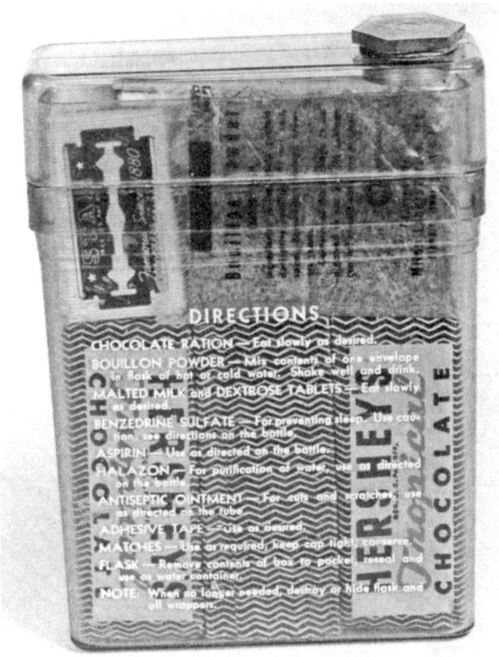

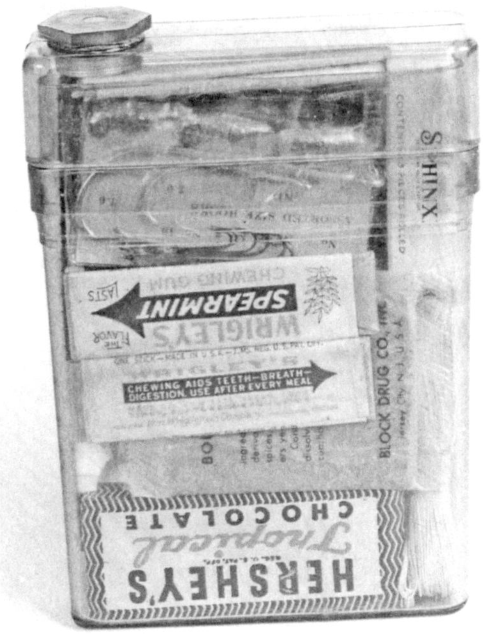

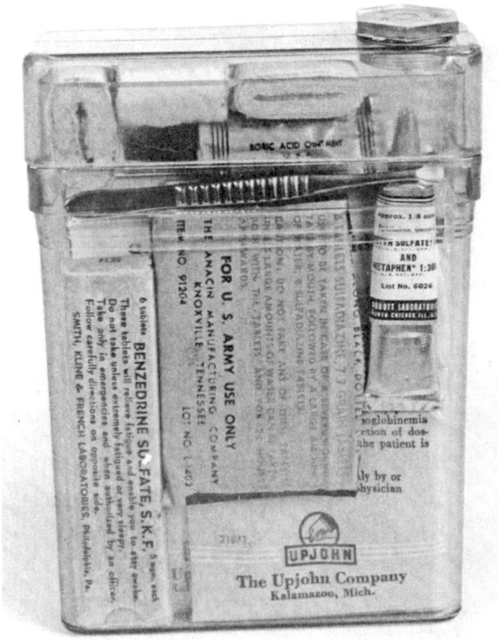

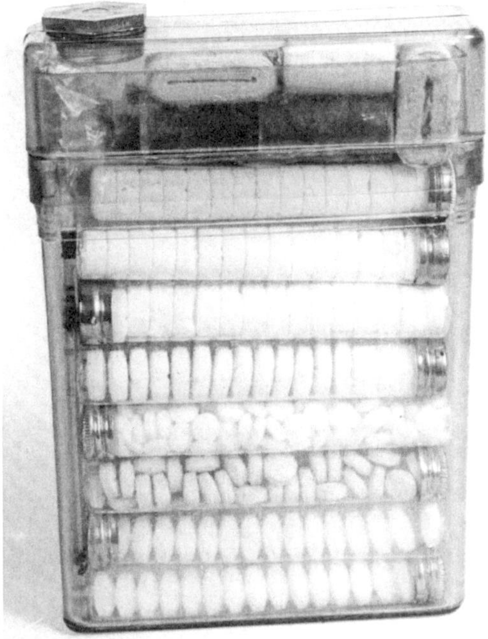

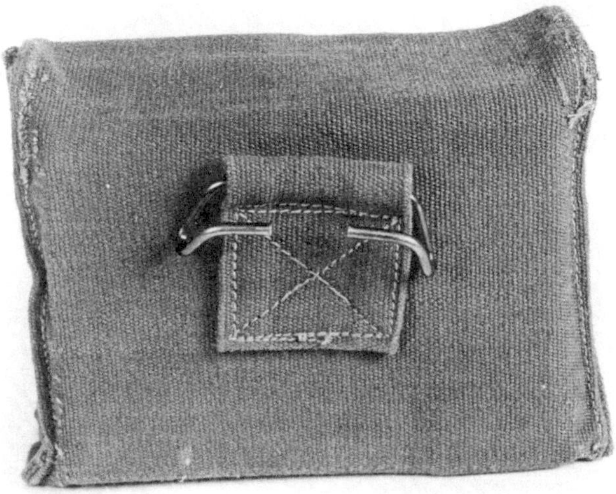

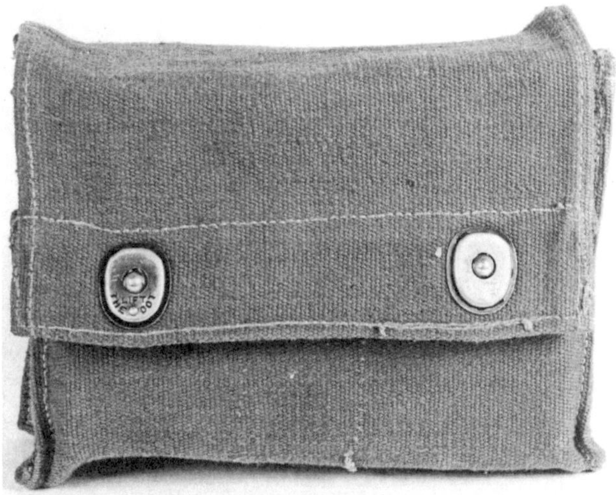

(This page) The concept model for a new type of personal aids kit to overcome the deficiencies of the E-3 models found with jungle use. Two E-3B flasks were used, one as a sustenance flask and the other a medical flask. The components were assembled based on suggestions and recommendations from multiple sources. Two flasks allowed for larger quantities of rations, and medical items, as well as other articles. The container for the flasks was based on the USMD M2 jungle medical kit, this container was used for the preliminary mock up of the unit and later revised to the type illustrated below. (USAF)

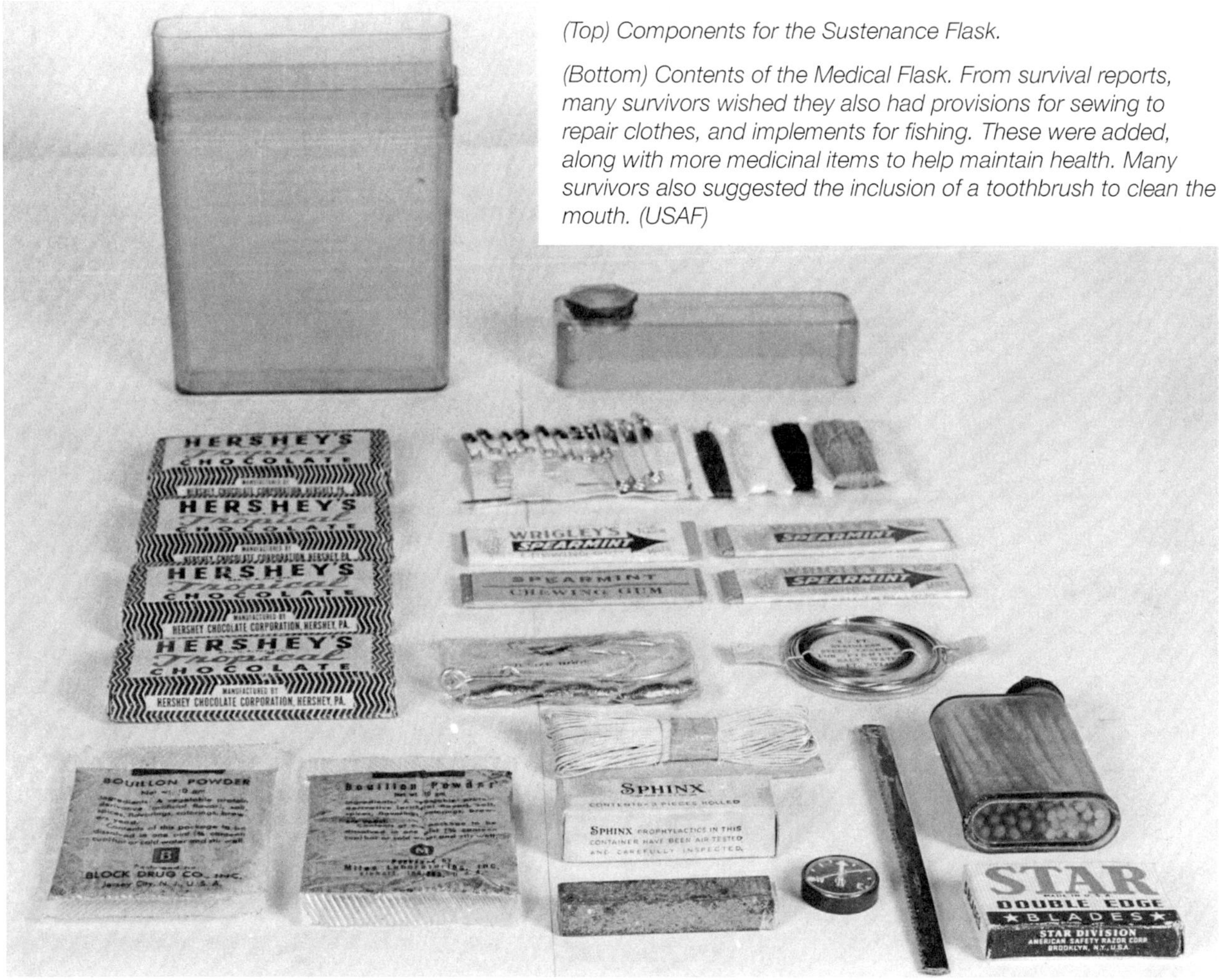

(Top) Components for the Sustenance Flask.

(Bottom) Contents of the Medical Flask. From survival reports, many survivors wished they also had provisions for sewing to repair clothes, and implements for fishing. These were added, along with more medicinal items to help maintain health. Many survivors also suggested the inclusion of a toothbrush to clean the mouth. (USAF)

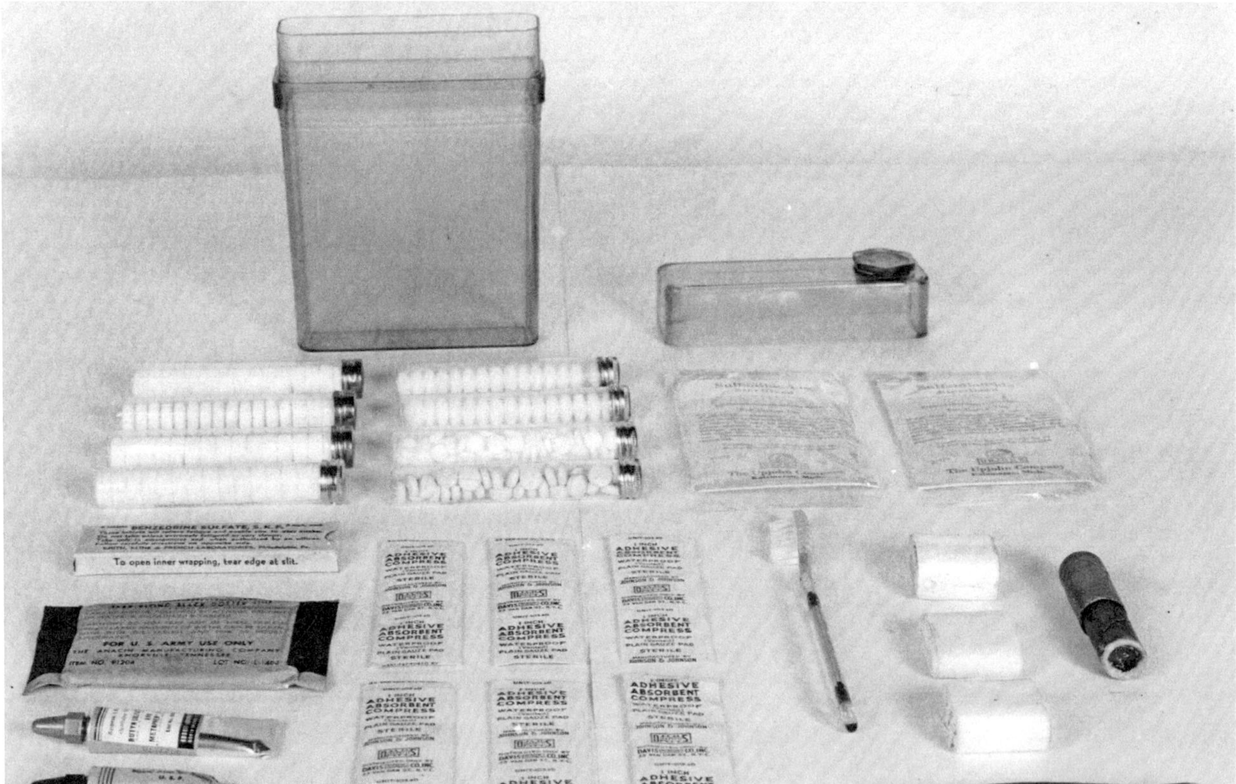

Immediate interest was expressed by ACAS, and for correspondence purposes they approved of the designation E-3J. Discussions commenced about the possibilities and configurations of suitable designs for procurement. One opinion, as early as February 1944, suggested the items be separated into two packages with a possibility of incorporating them into the Type C-1 sustenance vest. A sample kit was submitted to the Engineering Division, USAAF, at Wright Field in March - brought back from the combat theater by Maj. V.H. Mersen of the 5th AF.

With the unit in hand, the expedited engineering process began. A major stipulation was that the unit needed to be waterproof. The plastic container of the Type E-3A was reviewed before moving onto the flask type of the Type E-3B. Building upon the principle of two containers, more essential items could be included in the unit.

Utilizing the USMD first aid pouch, stock number 97109, in conjunction with two Type E-3B flasks, the base model was established.

In June 1944, the Engineering Division coordinated with the Surgeon General's Office on proposed components with regard to nutrition and medical items. The unit was assembled based on service requirements and suggestions, with the emphasis on survival. It was decided not to incorporate it into the Type C-1 vest, but as a separate unit, issued as a supplementary item. Development progressed quite rapidly with specifications drafted the following month, being approved and standardized by ACAS on 31 July 1944, under USAAF Specification No. 94-40840, titled Kit, Emergency Sustenance, Type E-17 (Personal Aids).

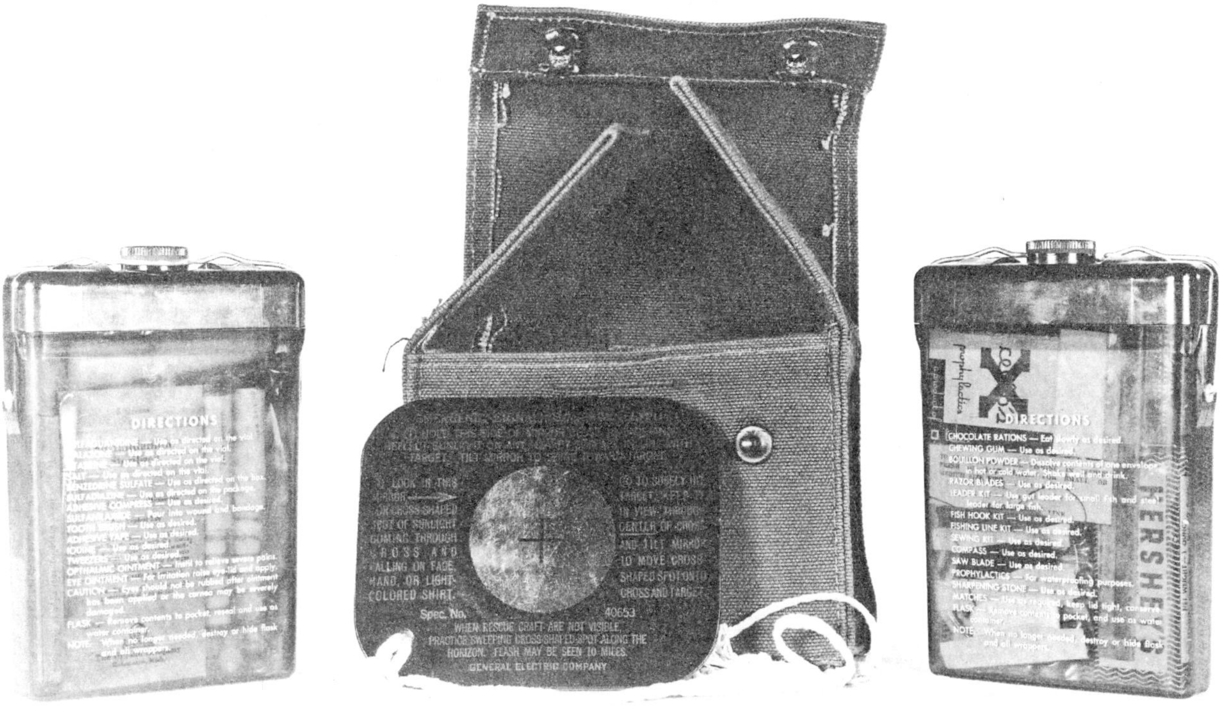

(Above) Wartime image of the standardized Type E-17 sustenance kit. The complete unit consisted of a cotton duck No.4 container, Dwg. No. 45G9001; two flasks, SM-1080-D; and an ESM/1 (40653) signaling mirror. As per guidelines established by MIS-X, the kit was to be void of any identifying markings.

Upon standardization, the first contract was awarded, in August 1944, to the Sears & Roebuck Co. for 20,000 kits under contract 33083-AC-4855.

Initial production was slow due to the multiple suppliers required for the complete assembly. Limited quantities did begin to arrive in the combat theater in October and November 1944, with priority for the PTO and CBI. Like their predecessors, distribution was controlled through MIS-X and S-2 intelligence sections but not accountable. These could be requisitioned by Personal Equipment Officers.

Due to the high volume of demand, the Materiel Command awarded an additional contract, 33038-AC-5617, in October 1944, this time to Clark WLM INC., for a quantity of 75,000 more E-17s.

They utilized the same style flask as supplied to Sears for their production, but the duck containers as supplied were modified with belt loops, approved 9 November 1944.

(Top row) This flask is identified by specification as the Sustenance Flask. A first-generation flask as supplied by Sears & Roebuck, its distinguishing features were the use of upper and lower case letters on the printed list of contents, and the clear cellophane packages for bouillon powder, which were exclusive to this type. It contained chocolate rations, chewing gum, bouillon powder, razor blades, fishing leader, fish hooks, fishing line, sewing kit, compass, hacksaw blade, rubber sheaths, sharpening stone and matches in a match safe.

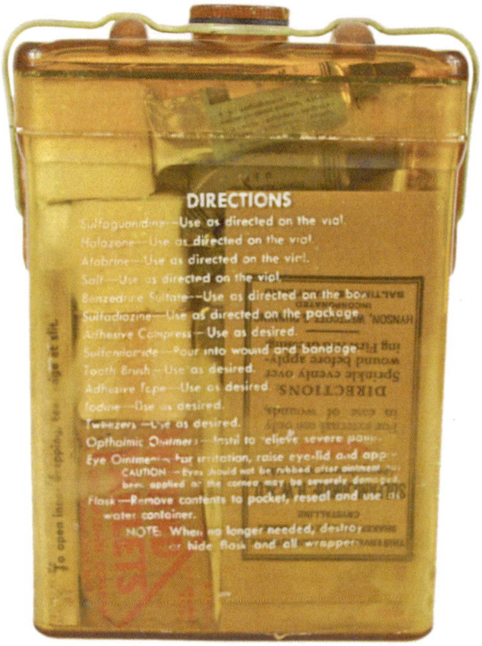

(Middle row) The Medical Flask which contained items for treatment of tropical ailments. Another first-generation flask with upper and lower case lettering, its components included: sulfaguanidine, halazone, atabrine, salt, Benzedrine sulfate, sulfadiazine, Band Aids, sulfanilamide, toothbrush, adhesive tape, iodine, tweezers, and eye ointment..

(Below) The duck container, Dwg. No. 45G9001, as supplied by Sears & Roebuck, included a tab and M-1910 belt hook only.

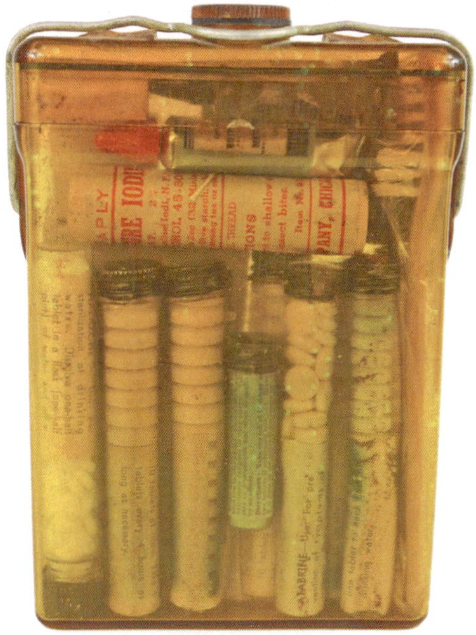

(Top row and middle row) To meet further projected service requirements, procurement for 300,000 Type E-17 kits was approved in February 1945. The order was split, and in March, contracts were awarded for 150,000 each to the Cappel MacDonald Co. under contract 330383-AC-9185, and to Bogardus Bros. under contract 33038-AC-9185. Deliveries for these units were projected between June and September. The flasks differed slightly from those provided by Sears and Roebuck and Clark in that they used upper case letters for the printed contents list and included the new foil lined packaging for the bouillon powder.

(Bottom row) This sustenance flask could be considered a transitional packaging, since it includes the older, clear cellophane wrapped bouillon powder.

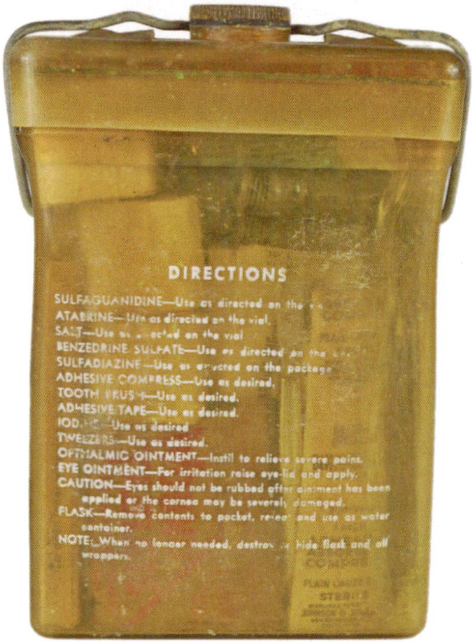

(Top and middle rows) War Department Circular No.77, dated 10 March 1945, directed that the use and self-administrating of sulfanilamide be discontinued, and that it was to be removed from first aid kits. Following the directive, Materiel Command amended Specification 94-40840, 24 May 1945, to delete the sulfanilamide and replace it with an additional six Band Aids. Contractors and suppliers were notified and new flasks were fabricated to reflect the change. Halazone was transferred to the sustenance flask and sulfanilamide was deleted from the medical flask. As part of the amendment, the inclusion of the new match safe compass, Dwg. No. 45A9185, was added. The new packaging would have occurred late in the assembly cycle for Bogardus Bros. and Cappel MacDonald.

(Bottom row, left) The Type E-17 was to include the flask and match safe as detailed in Dwg. No. 45B13490. The match safe is the same as that of the Dwg. No. 42A6930, but the cap has been modified by removing the knurled edges and instead using two extending tabs for leverage. Also included in original specifications were the matches, which were to be in halves for a count of 56. An incomplete sample is shown to illustrate the layering method.

(Bottom row, center) Some examples of kits utilize full length matches as pictured.

(Bottom row, far right) Match safe compass, Dwg. No. 45A9185, which was included in late assemblies of personal aids kits.

(Left) The ESM/1 signaling mirror, USAAF Specification No. 94-40653, was a part of the standard compliment for the E-17. Initially the mirror was provided with a cardboard sleeve.

(Above right) After Amendment-A of December 1944, mirrors were provided in a Kraft envelope and included in later assemblies as they became available.

(Right) The Type E-17 was provided with an OD duck No.4 container of basic rectangular construction referenced under Dwg. No. 45G9001. Modeled after the M-2 jungle first aid kit, some of its characteristics were carried over, for example: having two inner flaps, a top flap secured by two fasteners, and an M-1910 belt hook. Two belt loops were added as approved in November 1944. The only known contractor for the containers was the National Carbon Co.

(Below) Variations of the belt loop modifications.

(Below right) In the process of negotiation to obtain supplies for the 300,000 E-17 kits, the Jefferson QMD Hardware Unit informed Materiel Command, Procurement Division that all supplies for M-1910 belt hooks were tied up by Army Ordnance contracts. They in turn forwarded the names of other contractors for the supply. The contractor was not disclosed, but the hooks supplied have the unique distinction of being associated with Marine Corps equipment. An existing example of the container with the USMC style hook is shown.

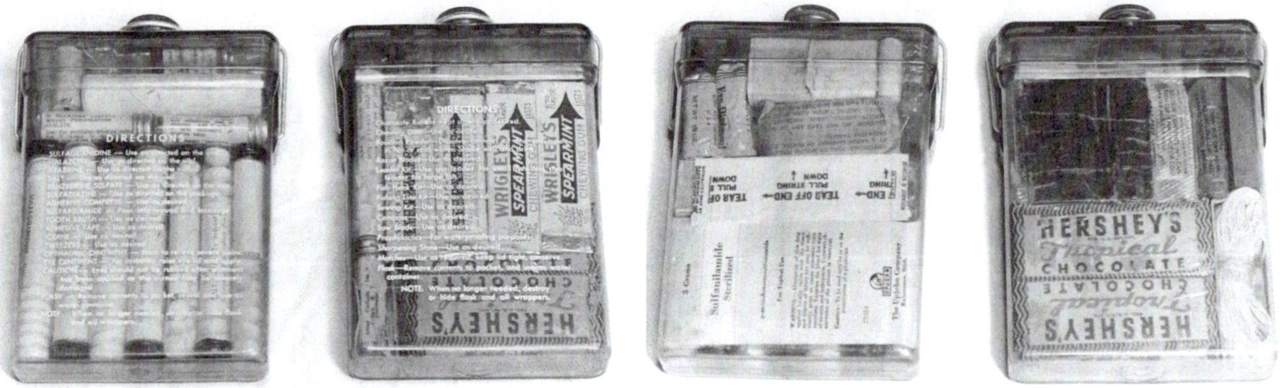

(Above) The standardized Type E-17 as issued.

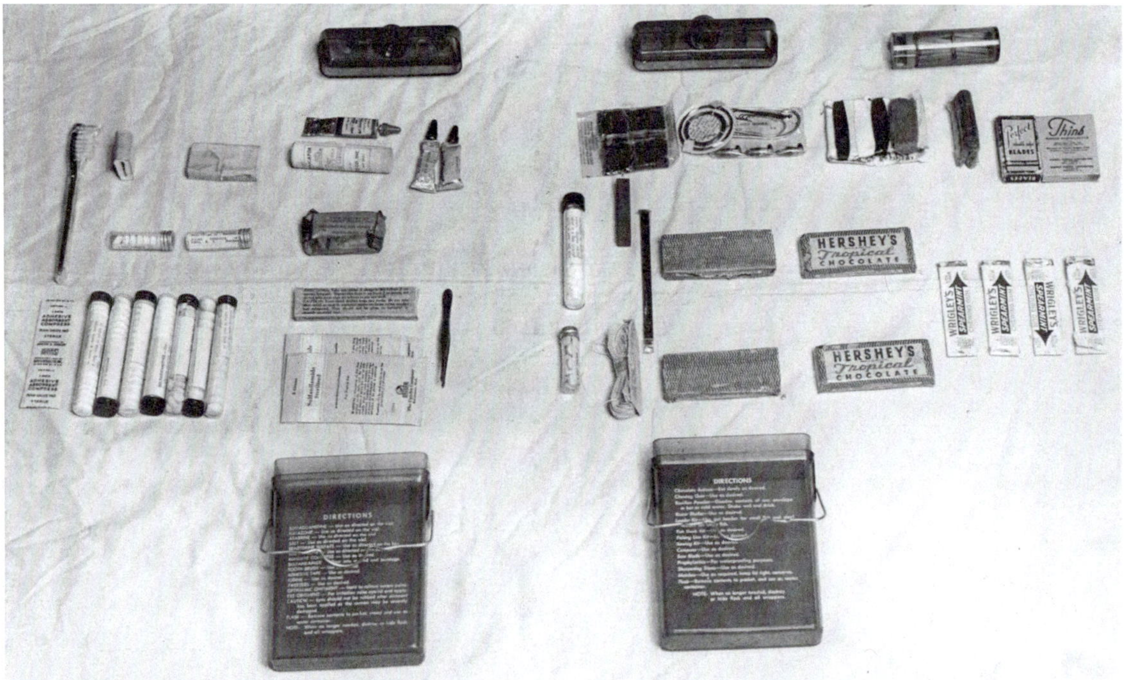

(Right) As can be seen from this layout of contents, very little changed from the initial concept model. (USAF)

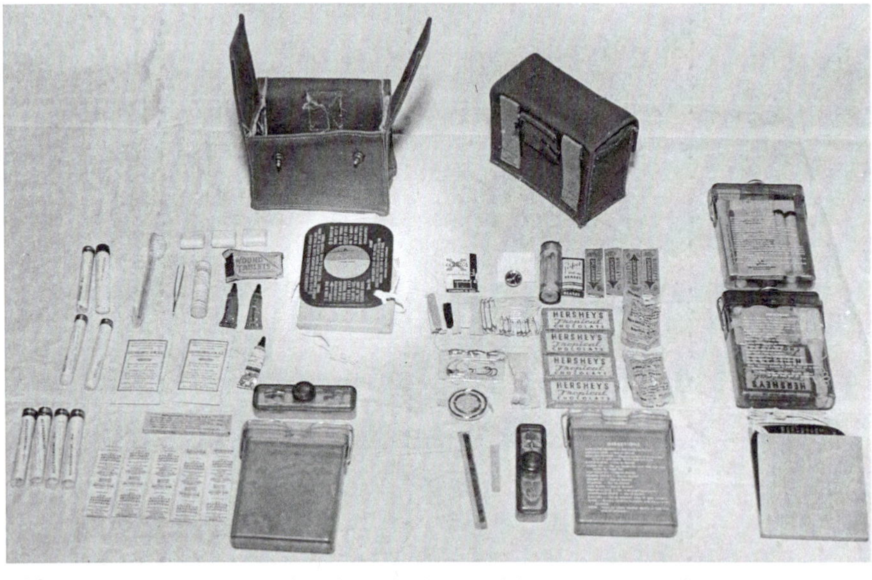

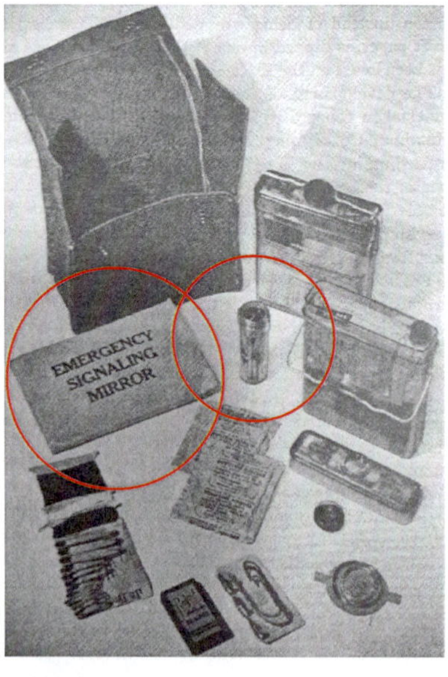

(Above left) Wartime image showing a full layout of the components for the E-17 sustenance kit. (USAF)

(Above right) Late war packaging of the E-17 by Bogardus Bros. The two features to note are the match safe compass, 45A9185, and the new packing method of the 40653 signaling mirror in a Kraft envelope as detailed per Amendment-A. (USAF)

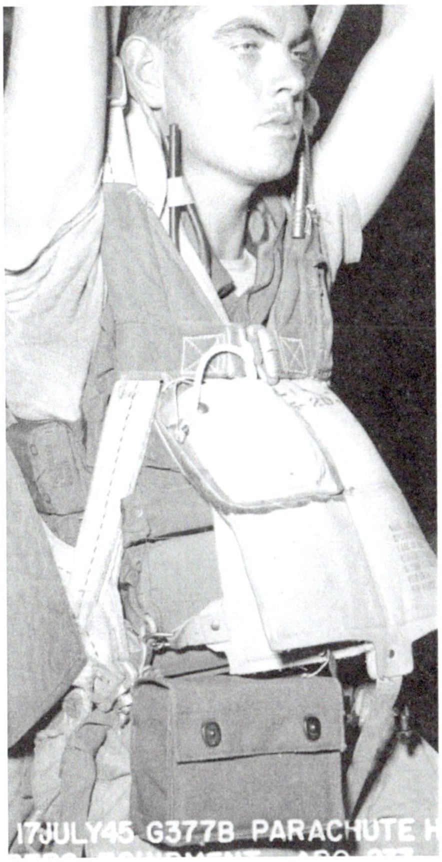

(Above) A USAAF aircrew member of the 500th BG demonstrates his safety and emergency equipment. He wears a C-1 vest and a B-4 life preserver modified by adding a signaling mirror, and carries an E-17. Saipan, July 1945.

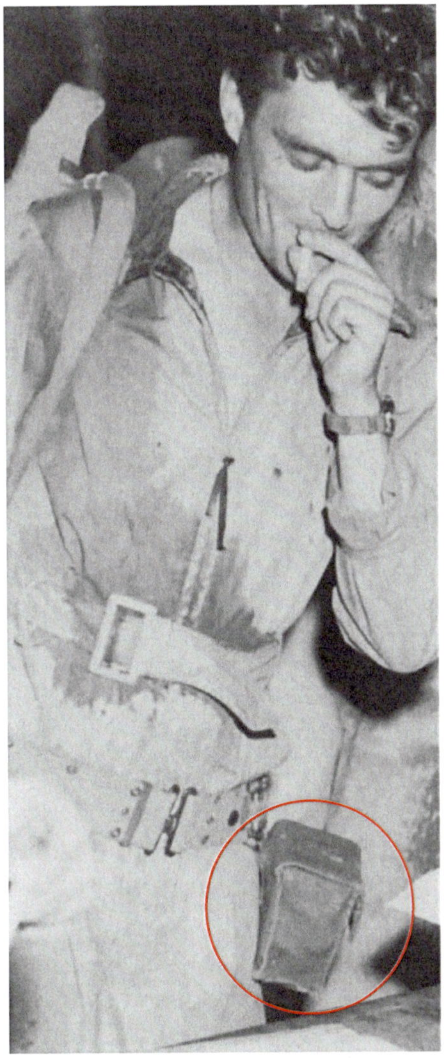

(Above) Two members of VMF-312 wear E-17 sustenance kits on their pistol belts. These would have typically been obtained through ACI sources. (USMC)

(Left) The US Navy Air Combat Intelligence (ACI) section worked in coordination with ACAS and MIS-X. ACI would requisition escape and evasion aids, such as the E-17, through local channels directly from MIS-X or USAAF stations. A PV-1 pilot is undergoing a briefing at an airfield in the Philippines prior to air strikes over Borneo, May 1945. Attached to his M-1936 pistol belt is a USAAF Type E-17 sustenance kit. (National Archives)

In August 1944, Commander, Air Force, Pacific Fleet, submitted a report detailing the considerable interest in the E-3J by US Navy squadrons. Such a kit was particularly desirable for rear seat personnel in TBF/TBM Avengers and other carrier based aircraft, in which space restrictions did not allow use of the back-pad kit.

ComAirPac recommended the development of an equivalent kit, intended for personal issue, supplementary to the pararaft, when the back-pad kit could not be worn. In the interim, E-3J kits were requisitioned for distribution whenever possible.

In a conference held in October between ComAirPac and the BuAer, it was stated that in view of the large-scale operations over malarial areas, it was imperative that certain basic components for jungle survival be provided for all flying personnel not equipped with the emergency back pad kit. It was also considered desirable that items provided for this purpose be packaged in small, compact kits, designed for availability in case of emergency over jungle areas, without interfering with the movements of the user in the aircraft. The USAAF Type E-17 was evaluated, but found unsuitable due to bulk and added weight. Emphasis was placed on the essentials, in particular to need to avoid diseases prevalent in the Asiatic area.

The development of the PK-1 pararaft combination seat pack was intended to eliminate the use of the back-pad kit and, until such times as they became available, it was decided to procure a personal aids kit as an expedient, to satisfy fleet requirements for rear seat personnel.

A unit was designed utilizing a stripped-down E-17 kit, reducing the components to one flask, and designated as Kit, Flask, Emergency, Mark-I. To meet service requirements, 15,000 kits were procured from the Sears & Roebuck Co. under contract NOAs-4839, awarded in October 1944, with expected deliveries by November. However, due to production delays, actual deliveries did not commence until late January 1945. Upon availability, Air Force, Pacific Fleet Technical Bulletin No. 13TR-45, dated 31 January 1945, detailed the new kit. It stated that due to the limited quantities available, they were only to be issued to personnel not equipped with the back-pad kit, and that it only be distributed to flying personnel operating over jungle areas.

The Mark-I utilized the SM-1080-D flask containing a match safe, magnetic compass, one vial each of sulfaguanidine, atabrine, halazone, aspirin and Benzedrine sulfate, one tube of antiseptic ointment, four sticks of chewing gum, three tablet rations M-539B each containing five sucrose tablets and seven malted milk tablets, one yard of adhesive tape, hacksaw blade, and two packages of bouillon powder.

Initially the Mark-I flask kit was procured as a temporary measure, but proved viable as determined by fleet operations. It transitioned into a supplementary role to the PK-1, and complimented the standard list of Pilot's Personal Equipment.

It was reported that carrier pilots were discarding the back-pad kit and exchanging it for the flask, which was becoming a new standard.

Two additional procurements of the Mark-I were executed to meet these new service requirements, but they were urgently needed. On 23 April 1945, ComAirPac released Technical Bulletin 71TB-45, for the local manufacture of supplemental kits to compensate for the undelivered Mark-I kits. In accordance with the bulletin, the kits would contain special Asiatic survival items consisting of two parts. The original concept consisted of a plastic soap dish with items similar to that of the Mark-I, and a 2 oz. can of insecticide powder, stock number S13-451, with both items comprising what was titled the Asiatic Kit.

These were specifically intended for carrying in pararafts by units engaged in operations over or near the Asiatic mainland, and were directed not to be distributed on a personal issue basis.

The Mark-I was a personal issue item and would often be placed in the pararaft. With limited quantities of the PK-1 in service at that time, one desalting kit would be removed and replaced with the Asiatic Kit.

(Top row) The US Navy Mark-I emergency flask kit. Like its USAAF counterpart, each component and its basic use is listed. This unit is of the initial procurement and does not list aspirin, even though it was included. For requisitioning, the nomenclature was used in lieu of a stock number, obtained through normal ASO supply channels.

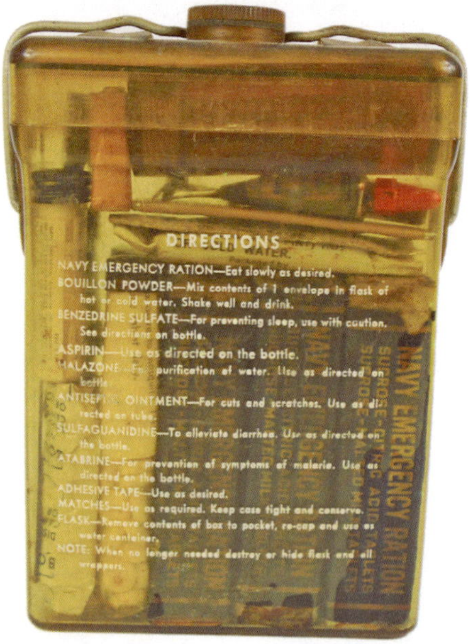

(Bottom row) Contracted flasks produced after amendments 1 and 2. They are identical except these flasks list aspirin in the contents. In February 1945, Amendment-1 to contract NOA-4839 was approved for the procurement of an additional 10,000 kits. 30,000 more kits were procured in May under amendment-2. ComAirPac found it desirable to provide special survival equipment for operations over Asiatic mainland. The Mark-I, in conjunction with standard equipment, further assisted in survival on land and avoidance of diseases prevalent in the tropical islands of the Pacific.

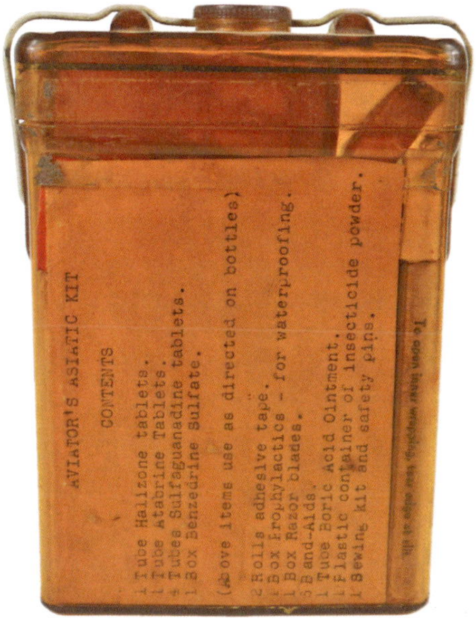

(Left) Difficulties were encountered assembling the units in the soap dish container as originally outlined. Type E-17 kits were requisitioned, disassembled, and repacked according to the requirements of the bulletin. One flask would be utilized and the components list scraped off. A new component listing was printed on paper and inserted inside the kit. As outlined in 71TB-45, the flask was then sealed with waterproof tape. The insecticide was integrated into the match safe and contained in the flask. These Asiatic Kits were assembled by the Fleet Parachute Loft, SOSU-1 at NAS Pearl Harbor. The example shown primarily comprises components from the USAAF Type E-17 sustenance kit.

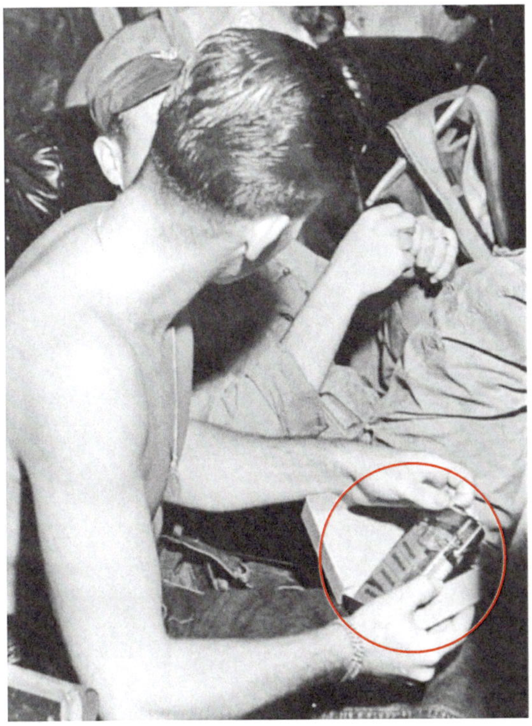

(Above) An aircrew member of VB-87 aboard the USS Ticonderoga relaxes in the ready room, July 1945. He is examining a Mark-I emergency flask kit. Being available from normal supply channels, these would be distributed through either the AESO or ACI officers. (National Archives)

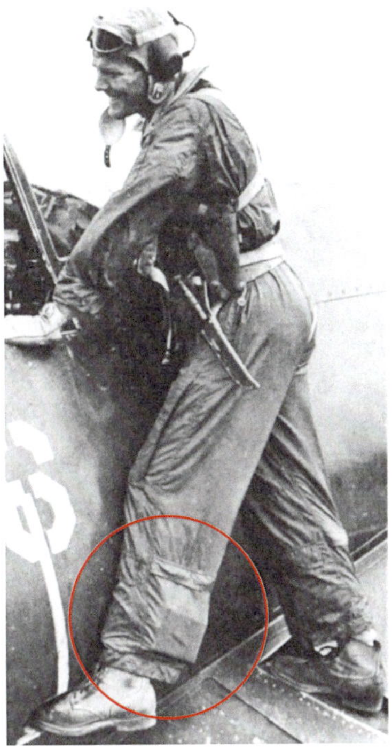

(Above) USMC 1st Lt. Fred Borwell of VMF-312 carries the emergency flask kit in a leg pocket on his flight suit. Borwell stands on the wing of a squadron mates Corsair, April 1945. (National Archives)

(Above) Lt. jg. W. Patterson of VT-9 carries his emergency flask kit in a rigger applied thigh pocket, USS Yorktown May 1945. (National Archives)

To help meet growing service requirements, the USAAF developed two supplemental emergency sustenance kits, with the design emphasis on supplementing the parachute emergency kit. These were intended for pilots of pursuit type aircraft, so that in the event of a bailout or forced landing, the individual would have additional water and ration provisions.

Samples were submitted for evaluation in October 1942, and approved for standardization 27 November, titled Kit, Emergency Sustenance, Type E-6 (Individual Bail Out Rations), USAAF Specification No. 94-40488; and Kit, Emergency Sustenance, Type E-7 (Individual Bail Out Water) USAAF Specification No. 94-40489.

Specifications were drafted and approved 14 January 1943. Each kit consisted of a canvas container with snap hook for fastening to the parachute harness, detailed under Dwg. No. 43G9962. The Type E-6 contained two Type K rations, one breakfast and one dinner unit. The Type E-7 contained two water cans, AN-W-5A (Flat). Both units utilized the same type container detailed under Dwg. No. 43G9962.

Upon release of specifications, service test quantities were approved for the procurement of 2,500 of each kit. The kits would be available through standard Class-13 supply channels under stock number 8300-55970 for the Type E-6 and 8300-559174 for the Type E-7.

Issuance of the kits was as follows: when operating over areas where water was plentiful the E-6 was to be carried. When operating over deserts and ocean, the E-7 was to be carried.

(Below) Kits were intended to be stored in a convenient location, such as attached to the airplane fuselage. (USAF)

(Right) During an emergency, the kits were clipped to the parachute harness. (USAF)

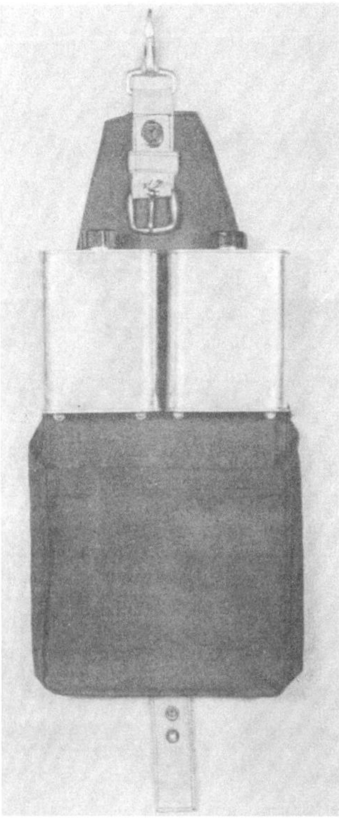

Images from a training aids poster show both units open with their components. The rations and/or water were stowed inside the pocket, then folded up and secured by a strap and buckle. A snap hook was provided for securing the kit to a convenient location in the aircraft or on the parachute harness. (USAF)

(Far left) The E-7 with flat water cans. (USAF)

(Left) The E-6 with K Rations. (USAF)

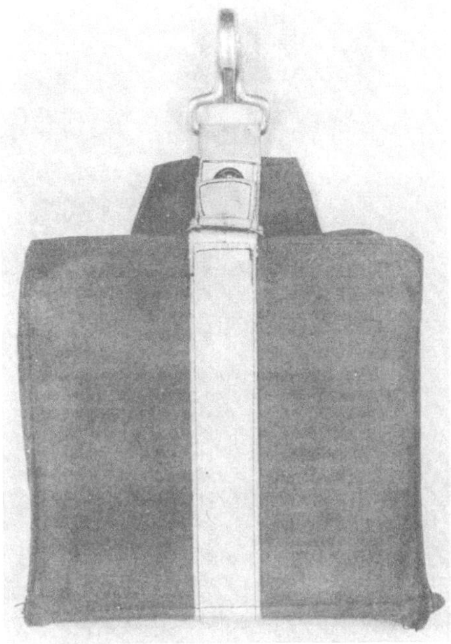

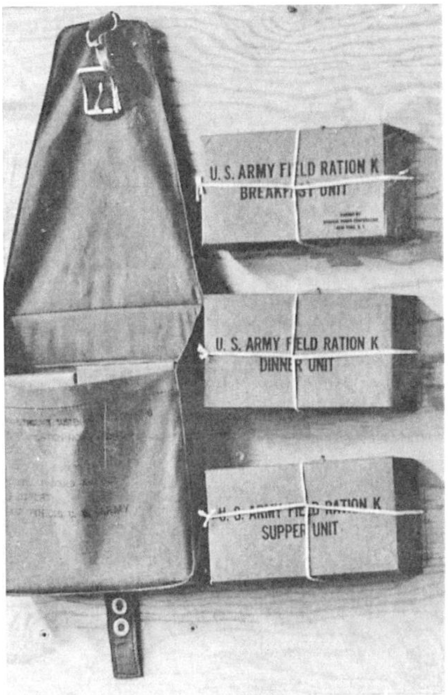

(Far left) Type E-6 / E-7 container from the same training aids poster.

(Left) Type E-6 on a display board showing contents. This was one of the initial 2,500 service test kits contracted from the Cappel MacDonald Co.

Further procurement of the Types E-6 and E-7 was withheld, pending the development of the Type E-3 sustenance kit. In the interim, a signal kit was submitted to the Material Command by the 26th Antisubmarine Command that they were recommending for standardization in April 1943. This unit was specifically intended for use by personnel flying over water, and to assist in the event of a forced landing.

The container strapped to the user's leg and secured to the pistol belt. Specifications were drafted and recommended for standardization in May, titled Kit, Emergency Sustenance, Type E-13 (Individual Signal), under USAAF Specification No. 94-40654.

The kit included a canvas container, whistle, signaling mirror, yellow cap, A-7 floating light, one can of water, and one can of sea marker.

After further evaluation, it was decided not to adopt the new unit as it was determined signal items were already contained in life rafts and other sustenance kits. However, the canvas bag was repurposed for use with the Types E-6 and E-7 under Dwg. No. 43D7591. Specifications 94-40488 and 94-40489 were revised under Amendment-1 dated 8 July 1943, deleting the 43G9962 and replacing it with the 43D7591.

In August, new procurements were approved to meet Fighter Command requirements, awarding a purchase order for 5,500 of each type to Breslee MFG. Co., under P.O. (11-107) 44-10982.

(Left and far left) The new container, Dwg. No. 43D7591. The OD canvas pouch featured a zippered closure, two horizontal leg straps and a vertical belt strap, all with spring clips. These wartime images show the container with amended Type E-6 and E-7 sustenance kits. The Type K rations were stored face to face.

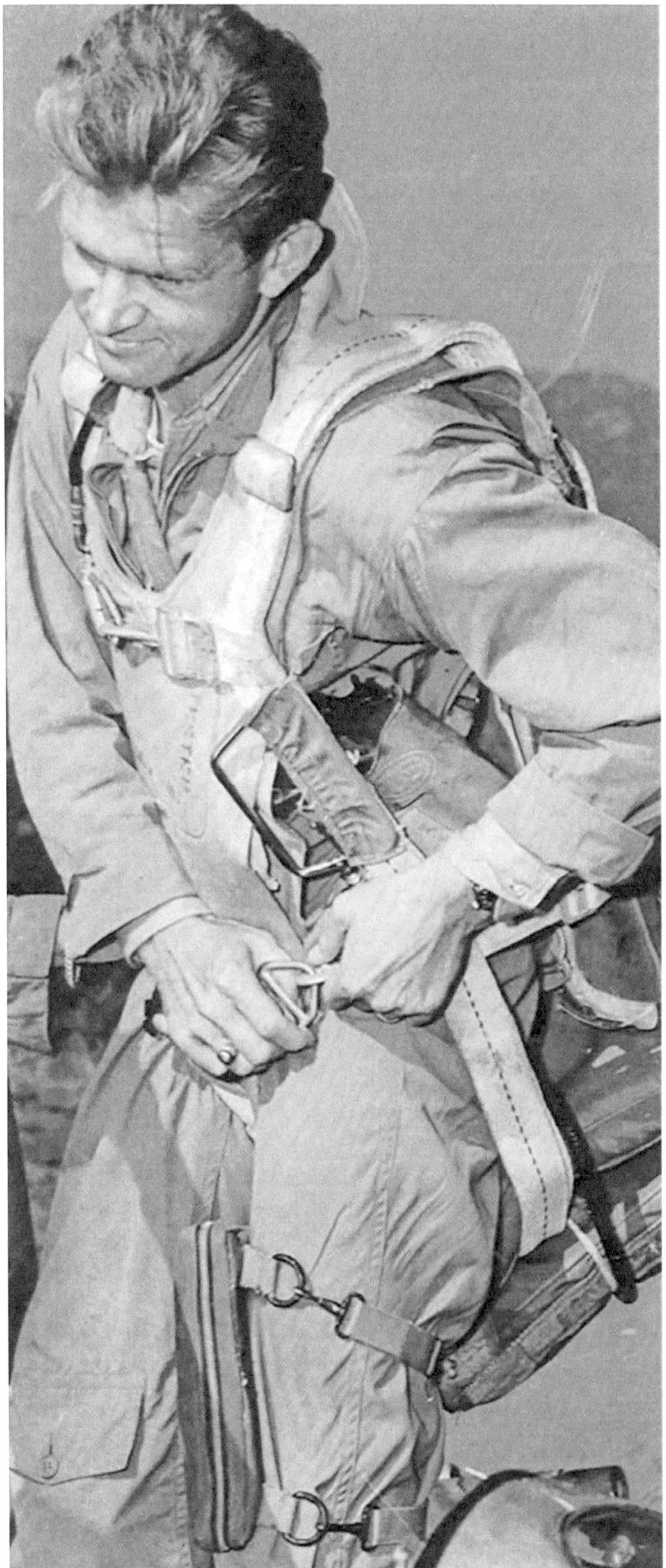

(Above) P-38 pilot of the 94th FS wears a Type E-6 sustenance kit on his left leg. Italy c.1944. As per guidelines the E-6 would be the issued unit in that region. (National Archives)

(Above) Another pilot of the 94th FS has a Type E-6 kit under is right arm. (National Archives)

(Right) 1944 image showing a Type E-6 kit manufactured by the Breslee MFG. Co. (National Archives)

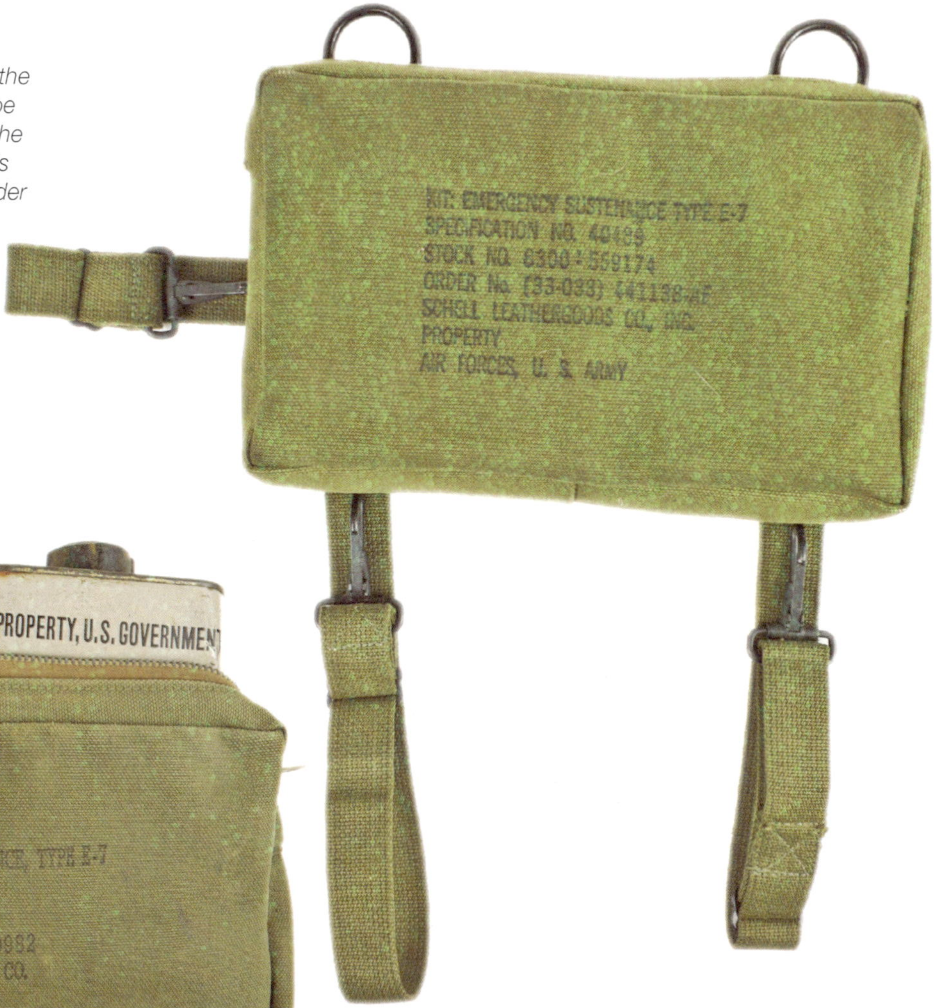

(Right) Due to extensive operations in the PTO, an additional order for 5,000 Type E-7 sustenance kits was placed with the Schell Leathergoods Co. Inc. Shown is a surviving example from purchase order (33-038) 44-1138.

(Above) Breslee contracted E-7 sustenance kit.

(Right) Breslee contracted E-6 kit. In December 1943, it was recommended to discontinue the use of the Type K rations and replace them with the Parachute Ration tin. By spring of 1944, shipments were allocated to Breslee to supply the Type E-6.

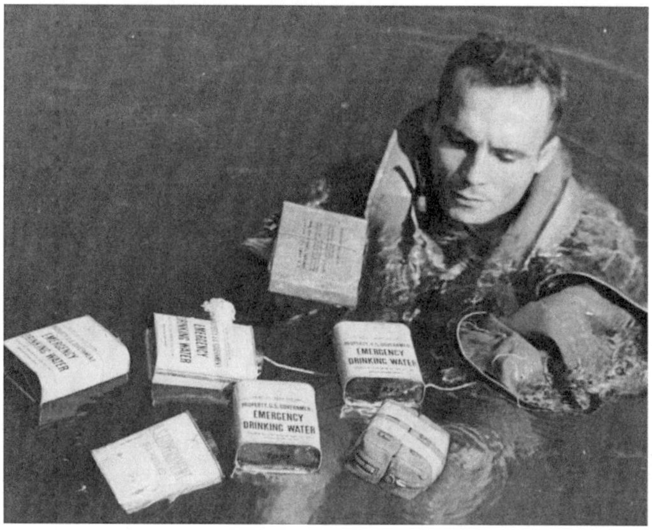

(Above) Locally fabricated USAAF sustenance kit. Its design basically follows along the same lines as the proposed Type E-13 signal kit, with the addition of rations and first aid. Proposed units such as these, and the eventual discontinuation of the Types E-6 and E-7, gave way to the personal aids kits of the E-3 series and the Type E-17. The flask kits were much more compact, providing sustenance and provisions for containing potable water. (National Archives)

In response to conversations with personnel returning from the combat theater, and service requests, the Materiel Command, in coordination with ACAS, initiated a program to develop a trading packet for flyers. The purpose was to furnish trading material to personnel operating in areas where the population was predominately primitive, with an assortment of trinkets, jewelry and other items that would appeal to primitive people.

The development process occurred in the spring of 1943, by compiling components in a container similar in size to the Type E-3A, the intention possibly being of making a dual component escape and evasion kit. It was later decided not to incorporate the two, and that the trading packet needed to be limited in size.

Specification No. 94-40685 was assigned to the new kit, titled Packet, Trading. From further conversations, it was determined that a suitable size for the container would be 3" x 3" x 1," but that to adequately contain all the items proposed, a second container of the same size was added. Formal specifications were submitted to ACAS in September 1943, for a trading kit that consisted of a disposable cardboard carton with two cellulose containers, 3" x 3" x 1".

One container held safety pins, fish hooks, needles, red ribbon, lipstick, tobacco and thread. The other unit contained a necklace, mirror, buttons, pendant (junk jewelry), rings and bats (ornaments).

The kit was approved as Kit, Flyer's, Trading, under USAAF Specification No. 94-40685, 1 October 1943, available under standard Class-13 supply channels as stock number 8300-559285.

Approval for a service test quantity of 5,000 kits was awarded in October 1943, to the Cappel MacDonald Co. under P.O. (14-055) 44-10111. The basis of issue was intended to be one per operational crew member, in the 5th and 13th Air Forces only, with further procurement pending evaluation in the combat zone.

By the time the packet became available, the requirement for such an item had declined, as combat operations in the region abated, and no further purchases were made.

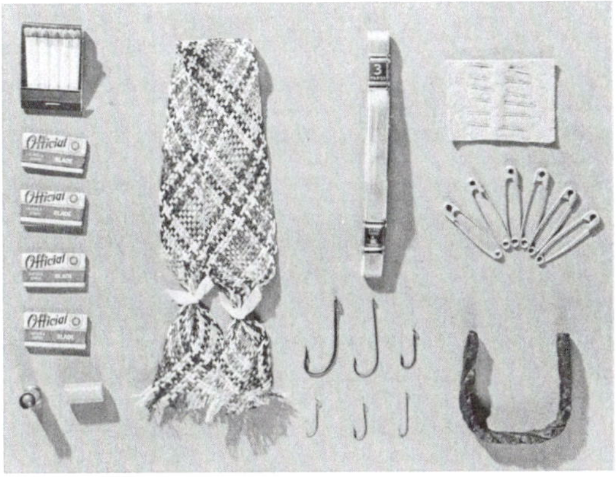

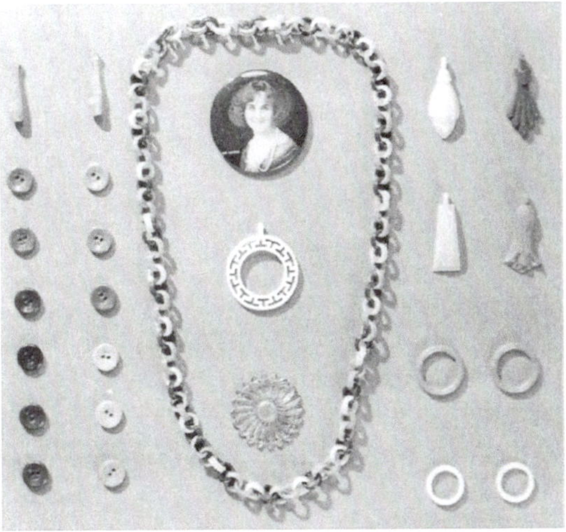

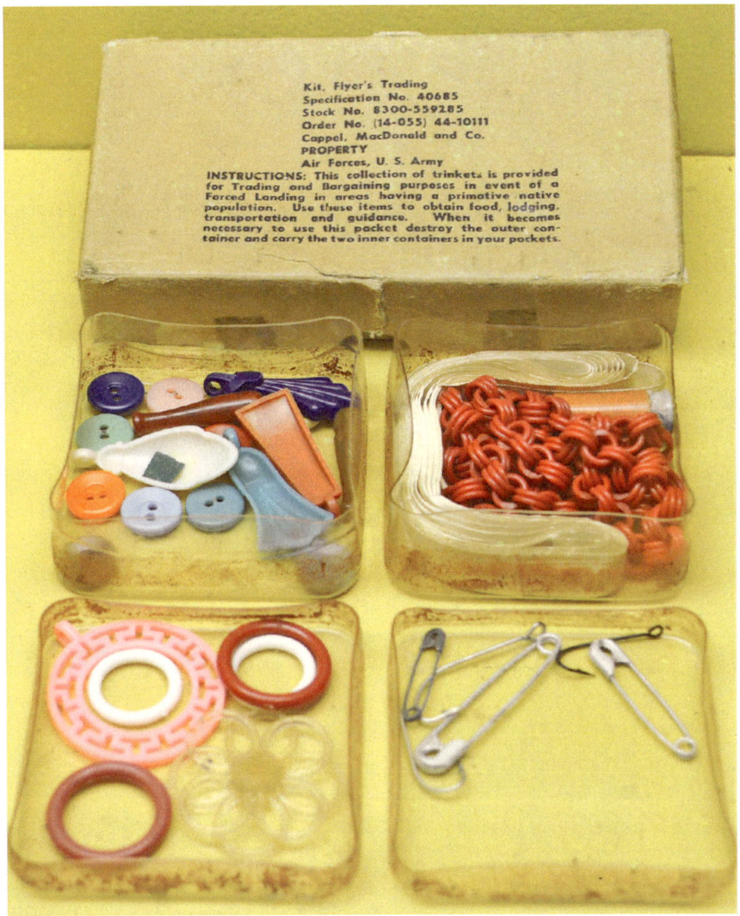

(Left, top and bottom) Wartime images showing the components for the trading packet.

(Right) A surviving example of the containers illustrating the barter items. These were cheap, colorful, cost-effective trinkets thought to appeal to primitive peoples. (USAF)

PROTECTIVE CLOTHING

In the early stages of the war, development progressed towards life-saving measures for shipwrecked (or ditched) individuals, with no consideration towards actual protective clothing.

While standard flight clothing could be considered protective, it was orientated towards flight operations rather than for emergency purposes.

The first article of clothing to be incorporated as standard was a mosquito headnet. By the summer of 1942, the BuAer was working on the development of emergency parachute kits for flight over jungle terrain, and utilized the newly standard headnet of the USAAF and commercially available types. It was also established at this time that some form of protective headgear be provided for life rafts. As an expedient measure, specifications were drafted, based on a commercially available headnet, and standardized as the Mosquito Headnet under NavAer Specification M-565.

(Top left) A commercial headnet utilized by both the USAAF and BuAer, initially stocked as the Headnet & Fish Seine ASO stock number R27-N-261. (National Archives)

(Above and left) The headnet consisted of green cotton netting, a cloth top piece, and a wire face frame. A mouthpiece was provided inside the frame. This headnet was called out for use with the USAAF Type B-1 Alaskan parachute kit in its specifications from 1941, and later by the BuAer, where it became the pattern for the formal M-565.

(Above) The mosquito headnet Specification M-565 was standardized 26 January 1943, with quantity contracts awarded the same month to Dadant & Sons, the only known contractor.

(Above right) The contracted headnet copied the commercial type in all regards except the top piece, which was changed from a rectangular khaki top to a 12" diameter green cotton fabric.

(Right) The headnets were packed in a Kraft envelope and stamped with its nomenclature and specification number. These nets were supplied as standard, one per person for multi-person life rafts, and for the M-592 back pad kit available under ASO stock number R27-N-261 with the new nomenclature Headnet, Mosquito.

(Below, left and right) The M-565 provided protection from insects, but was found inefficient in providing any protection from weather elements and unsatisfactory for use in life rafts. In July 1943 the Naval Medical Research Institute conducted sea trials on methods of preventing sunburn, and within these tests evaluated an anti-sun hat developed and standardized for use in life rafts by the Transcontinental & Western Airlines. This cap consisted of a neck shield and acetate tinted sun shield for the face. The overall purpose of this study was to determine the best methods for preventing minor injuries, such as sunburn, by protecting a survivor from the glare of the sun off the water and additionally to help prevent dehydration- with both aspects leading to fatigue and shortened survival duration. (National Archives)

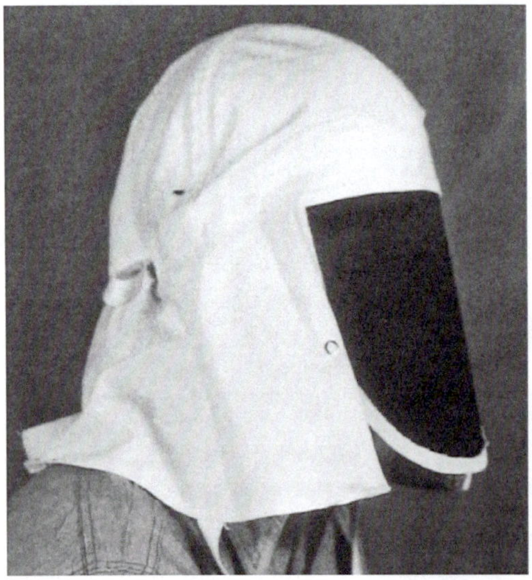

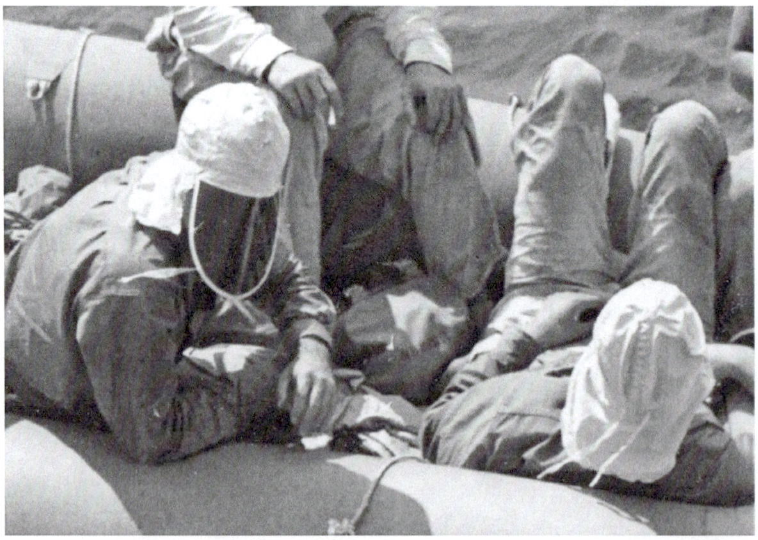

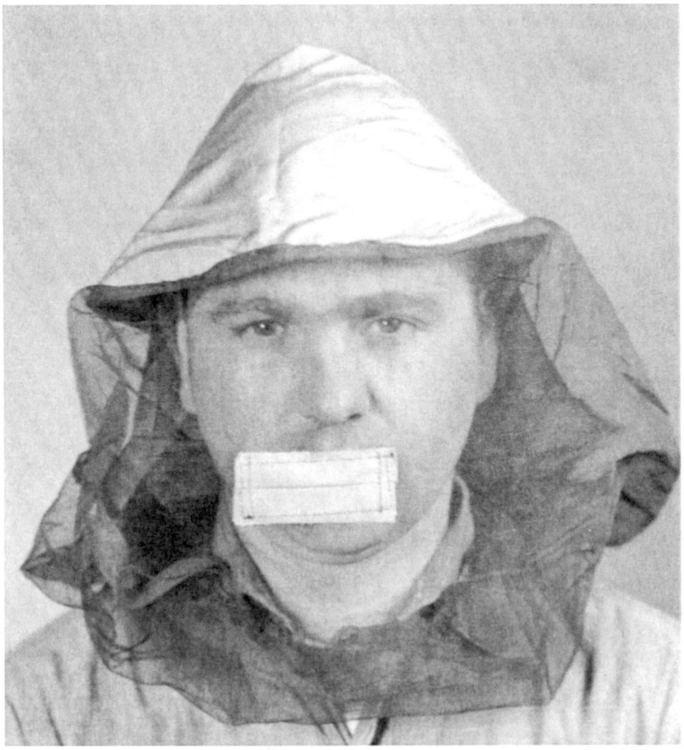

(Above) In September 1943, a sample headnet water bottle combination was submitted by the Strato Equipment Co., designed with a mouth piece feature like that on the M-565 and similar features to the TWA anti-sun hat.

(Top right) The design incorporated a neck shield with cotton netting.

(Below) Period drawing illustrating how it was used.

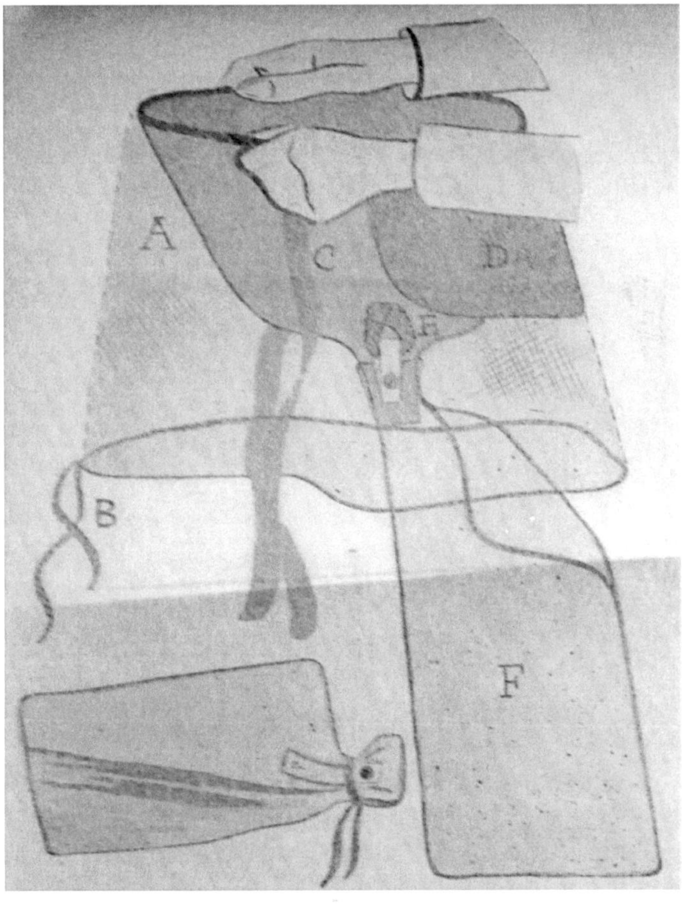

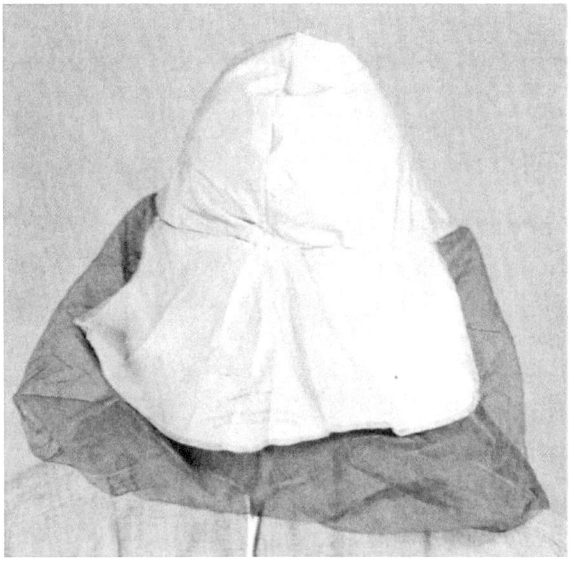

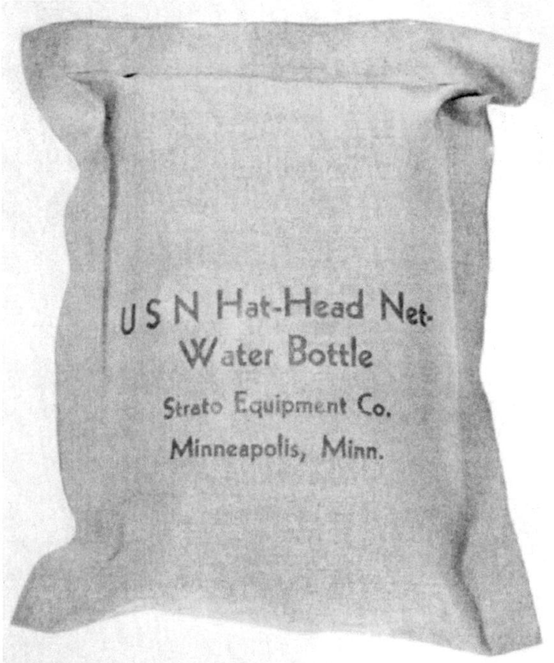

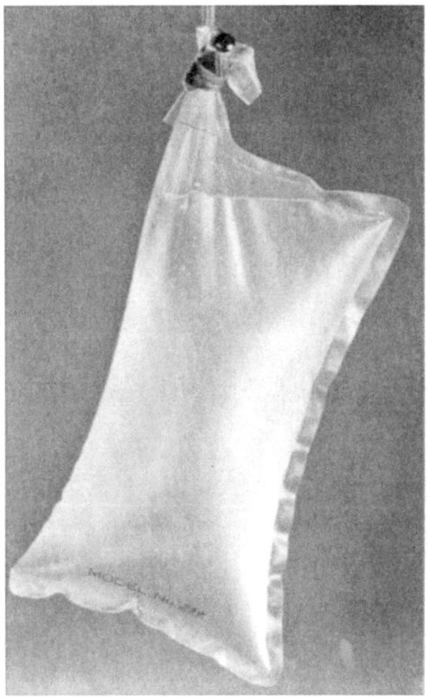

(Left) The hat also functioned as a water catcher, with an integral water bottle for collection.

(Above) The complete unit was compactly packaged. (National Archives)

(Right) Based on the positive field test results of the TWA and Strato head gear, the NMRI drafted tentative specifications to adopt a suitable hat to protect survivors in life rafts from UV radiation. The recommendation was to adopt the design submitted by the Strato Equipment Co. but eliminate the water bottle feature (because a five quart bladder had been separately approved), and to add a face shield like that of the TWA cap. Strato then redesigned their protective hat, submitting samples in February of 1944 for field testing. Meeting approval of the BuAer, procurement was authorized to supply multi-person life rafts. A quantity contract, NOAs-3528, was awarded to the Strato Equipment Co. in April of 1944 enough to supply all life rafts in service and on procurement referenced under NavAer Specification M-565A. Limited quantities began arriving at supply points and to life raft contractors as CFE in July 1944.

(Far right) The hats were packaged in a tarred Kraft paper envelope and available from the ASO under stock number R83-H-200000 as Hat, Headnet, Protective. The M-565A was adopted as standard 2 December 1944.

(Left) Initial production of the hats by Strato included a white cotton face shield and a company tag. Later versions transitioned to rayon material for the top and neck shield and eliminated the company tag.

(Right) The bulk of the production by Strato was consumed by life raft contractors and service issue, leaving none for stock purposes. In January 1945 two additional contracts were awarded, but Strato was not chosen as one of the two contractors and instead the awards went to the Winget Kickernick Co. ASO contract 288s-27763 and the Blatter Titus Co. ASO contract 288s-28485. Their production hats would include the khaki rayon face shield. This headnet was not included in the 1944 contracts of the M-592, as they were not available for their assembly but were available for replenishment purposes. As supply permitted they were to replace the M-565 in multi-person life rafts. By the close of 1944, the PK-1 pararaft was adopted, and within the design emphasis the headnet was placed under the authorized list of Pilot's Personal Equipment to be issued to the individual.

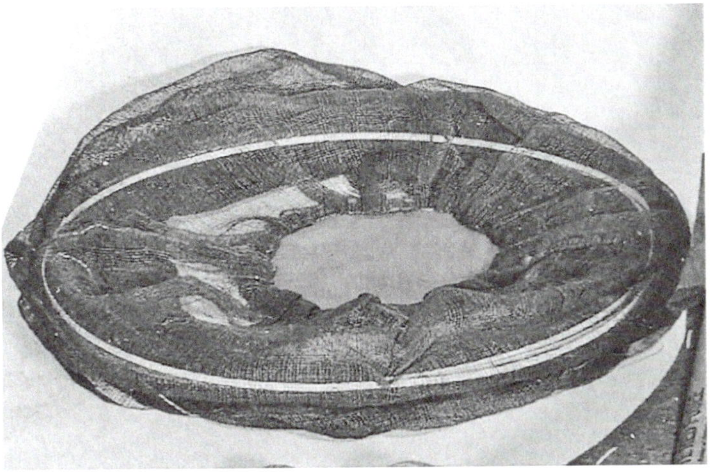

(Above) In the inter-war years the USAAF utilized commercial mosquito headnets dating back to the mid-1930s. Shown is an example of a commercial headnet incorporated into USAAF parachute kits. (National Archives)

(Above right) In an effort to formalize, the USAAF adopted a simply constructed mosquito headnet with a cotton mesh and round solid top authorized for procurement in July 1942 under Dwg. No.43G3868, available under stock number 8300-384000 titled Headnet, Mosquito. At an unknown date it was revised to stock number 8300-601200 with the new nomenclature of Net, Parachute, Emergency Kit, Mosquito Head. To complicate (and further confuse) the issue, with the adoption of new nylon headnet the nomenclature and stock numbers were changed. Nomenclature for the headnet under stock number 8300-601200 was revised to Net-Head, Mosquito, Cotton, Olive Drab, Emergency Sustenance Kit.

(Above left) The new green nylon type was tilted Net-Head, Mosquito, Nylon, Green and listed under stock number 8300-601250. These headnets would be standard for parachute and sustenance kits for the duration of the war. Shown is the new green nylon headnet, adoption date unknown. The stock listings for this headnet began to appear in late war publications, which would suggest its adoption sometime in 1944.

(Right) Wartime photograph of the headnet.

(Left) USAAF liaisons were also a part of the sea trials conducted by the NMRI, relating to many equipment aspects, but as part of the field tests on sunburn prevention, US Army fatigue hats were used. The USAAF came to the same conclusion as the Navy representatives in regards to UV protection but approached the solution in a different way. Based on the standard HBT fatigue hat, the Material Command developed a hat suited to their requirements to aid in protection. The result was a one size fits all fatigue hat, drawing No.44G5062.

(Left below) The hat was OD one side and orange yellow on the other, for recognition purposes. The original intention was to supply them as standard for multi-person life rafts, but with the adoption of the C-1 vest and under the premise that all flight personnel would be so equipped with the vest they were not added to the standard compliment. However, the hats were included in multi-person sustenance kits and were available under stock No. 8300-383860 as Hat, Reversible, Sun.

(Bottom, left and right) In 1944, the XX Bomber Command found the standard mosquito headnet, 43G3868, highly unsuitable for use in the CBI theater. In spring of 1944 they submitted samples of a cap assembly headnet combination to the Material Command. Building off this and from further requests, the Material Command devised a new concept of flight clothing to be utilized in jungle terrain. The new assembly would consist of what would be eventually standardized as the Type K-1 flight suit, Type K-1 Mosquito Resistant flight gloves, and an experimental mosquito cap assembly. Efforts were made to incorporate the contents of the C-1 vest into the K-1 flight suit, but were dropped due to practicality issues. Sample bundles were sent to the XX Bomber Command for service testing in October 1944 that included an experimental Cap Assembly, Jungle, Mosquito Netting under drawing No.45J9037 colored OD one side and Universal Orange on the other. A revised version of this cap would be adopted and standardized as the Cap, Jungle, Very Light, Type K-2 in 1946.

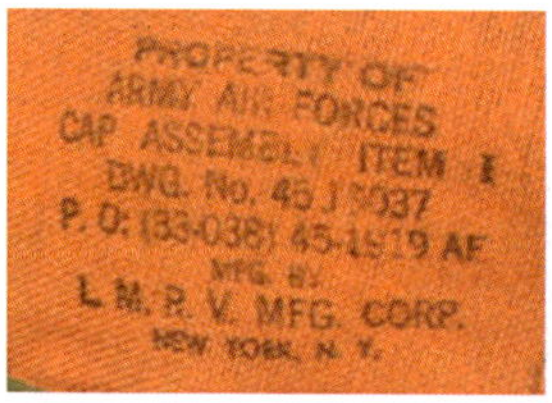

(Above and right) The USAAF implemented other pieces of protective headgear in the form of a knitted yellow skull cap unique to the Type E-11 sustenance kit. Specifications detail it simply as a water-resistant cap to be supplied as CFE. (National Archives)

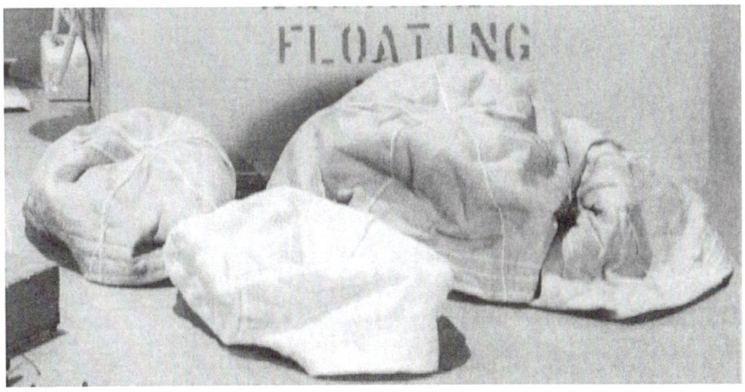

(Right) Another skull cap utilized by the USAAF. This example is made from a cotton fabric orange yellow on one side and white on the other.

(Below, left and right) Surviving examples typically have the USAAF distinctive insignia stamped on the white side.

(Below) A similar cap is illustrated in this wartime image of an emergency rations and equipment kit of the 18th AAF Base Unit. No information has surfaced discussing this model. (National Archives)

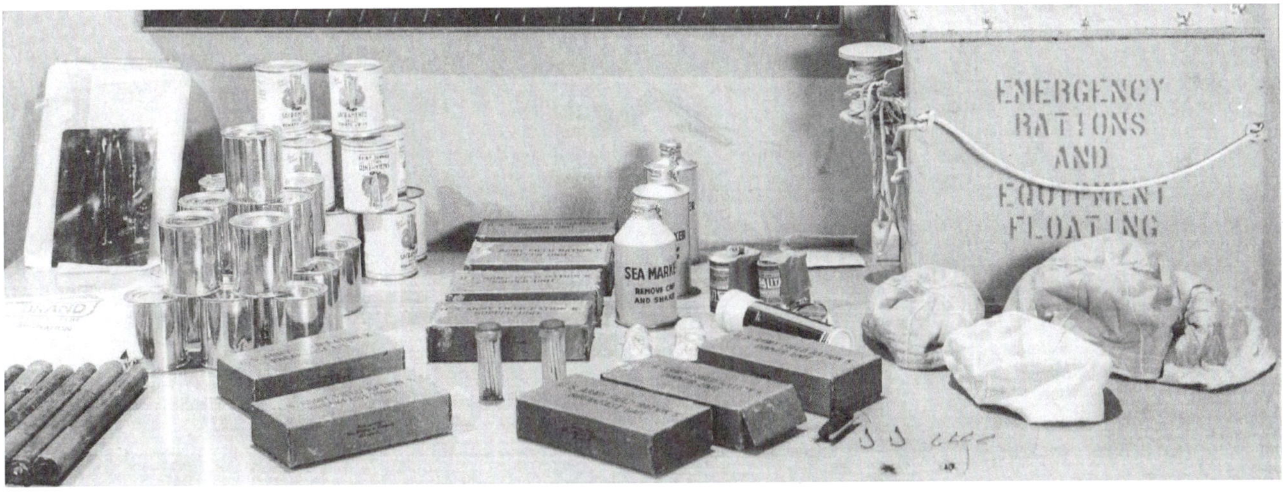

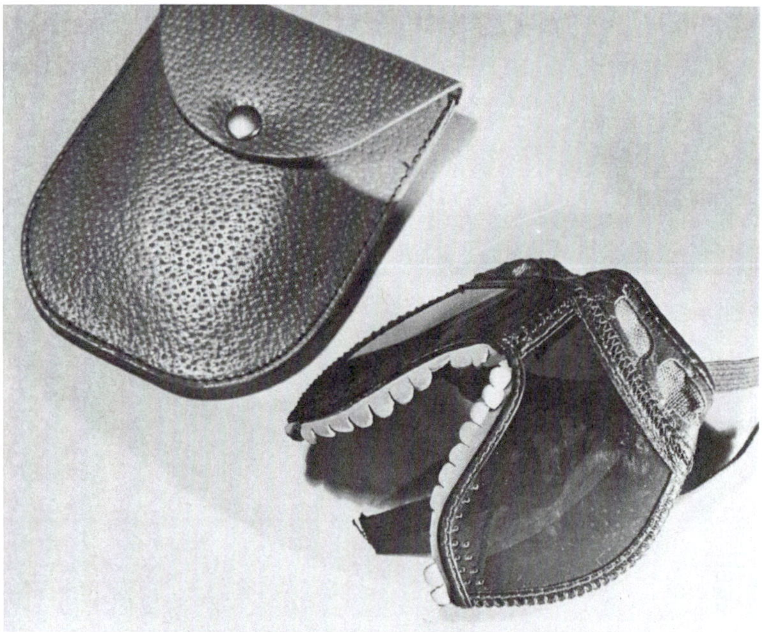

(Left) In January of 1943 the USAAF standardized the dark adaptation goggle under Specification No. 94-3142, later revised in March to include a dark green lens. The purpose of the revision was to provide a sun goggle for inclusion in the Type B-4 emergency parachute kit. Pictured is the original developed Goggle, Dark, Adaptation, Type E-1 and its case. (National Archives)

(Left and below left) Upon the revision in March the large oval vents along the top of the goggle were omitted, being manufactured with red and green lenses. Pictured left is the Goggles Assembly, Dark Green Lens USAAF Specification 94-3142 as included in the B-4 emergency parachute kit. As part of the original intended assembly for sun protection on life rafts a pair of these goggles was to be used in conjunction with the sun hat Dwg. No. 44G5062, but like the hat, the goggles were transferred to the C-1 vest.

(Below) The case for the goggles.

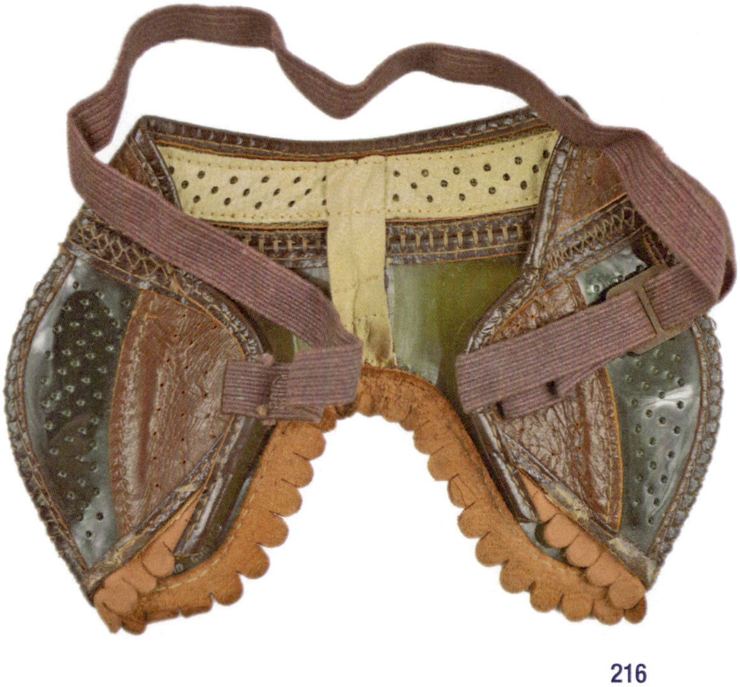

(Above) Two USAAF aircrew members wearing the goggles in simulated desert survival training, 1944. (National Archives)

(Left) These goggles were eventually adopted by the QMC and re-designated as the Goggle M-1943, available in three lens variations: clear, red and green. There were two discernible differences between the Specification 94-3124 and M-1943: the size and the addition of a felt application along the bottom frame of the goggle on the M-1943. The folded goggles clearly show the size comparison, with the 94-3124 on the left.

(Below) The back side of the M-1943, showing the felt covered face frame. By the time of standardization of the C-1 vest, the M-1943 was included under a new designation of Type H-1, standardized under USAAF Specification No. 94-3227, 31 August 1944.

Standard flight clothing was found satisfactory for protection in warm and intermediate climate zones, but the exposure to frigid waters was an entirely different matter. The idea of a protective suit for exposure to protect a survivor in cold water dates back to WWI and remained a consideration through the decades leading to WWII, with the Merchant Marine adopting lifesaving suits for use in the event of abandoning ship.

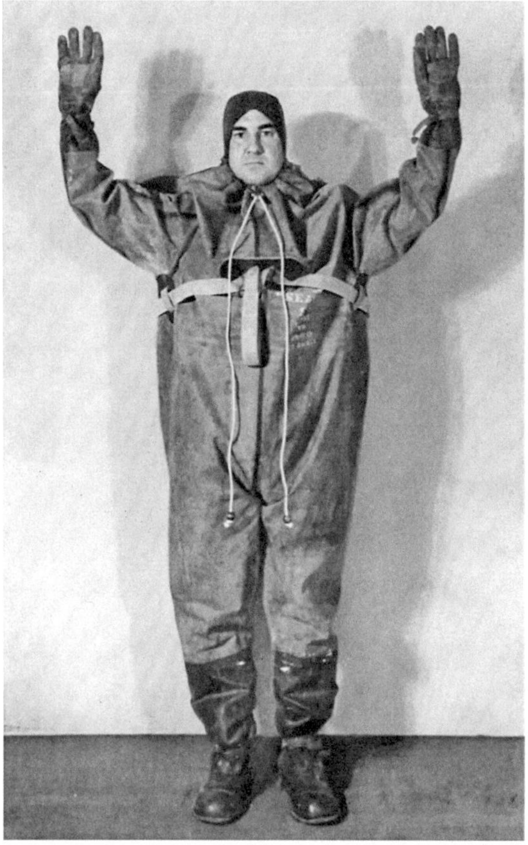

(Left) An example of a life saving suit, which were often referred to as "Zoot Suits," utilized by the Merchant Marine in the inter and war years.

(Above) Three Merchant Marine Seaman demonstrate emergency clothing for abandoning ship in cold water. The suit is designed to accommodate the bulk of clothing and a kapok life preserver. (National Archives)

(Left and below) The Civil Air Patrol was delegated responsibilities for patrolling the coast lines of the continental United States, relieving the burden on military assets that could be better applied towards the war effort abroad. Exposure suits became standard wear or those operating of the Atlantic coast. (National Archives)

(Left) Detail of the hood fastening on an exposure suit worn by a CAP pilot.

(Above) Suits developed for field testing conducted in November and December 1943. The purpose of the exposure suit was to create a watertight seal, keeping the wearer's clothes from getting wet and helping the body maintain temperature to prevent hypothermia. By a joint committee of a multitude of services, it was decided to develop suits for aviation personnel in October 1943. Development became a medical study of exposure and materials suitable for the task. The older type suits used by the Merchant Marine were not considered optimal for use in fighters and bombardment aircraft, where the emphasis was toward compact and constant wear. (National Archives)

(Left) A lightweight exposure suit was field-tested where the suit doubles as a life preserver. The white fur lined garment was worn underneath, replacing the standard fleece lined leather type flying suits June 1943.

(Below) New aviation exposure suits being tested under a joint Canadian and US field test off the eastern coast of Canada. These three RCAF airmen spent 16 hours in 40-degree water. (National Archives)

(Right and below) Testing continued through the Liaison Committee on Emergency Rescue Equipment, then the Material Command conducted its own tests on suits in 1944. The USAAF eventually adopted this yellow, lightweight rip-stop nylon suit, 19 January 1945, as the Type R-1, available under stock number 8300-783900. A large opening was provided at the shoulders for quick donning, then secured closed by a drawstring. The suits were designed as one-size-fits-all to accommodate heavy flying clothing.

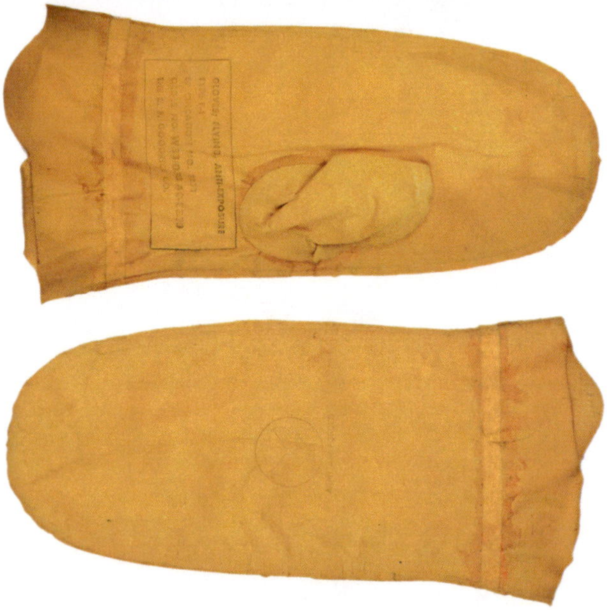

(Left) A pair of gloves (Type F-1) were provided for a complete assembly, stowed one each in the thigh pockets of the suit.

(Right) Detail of the nomenclature.

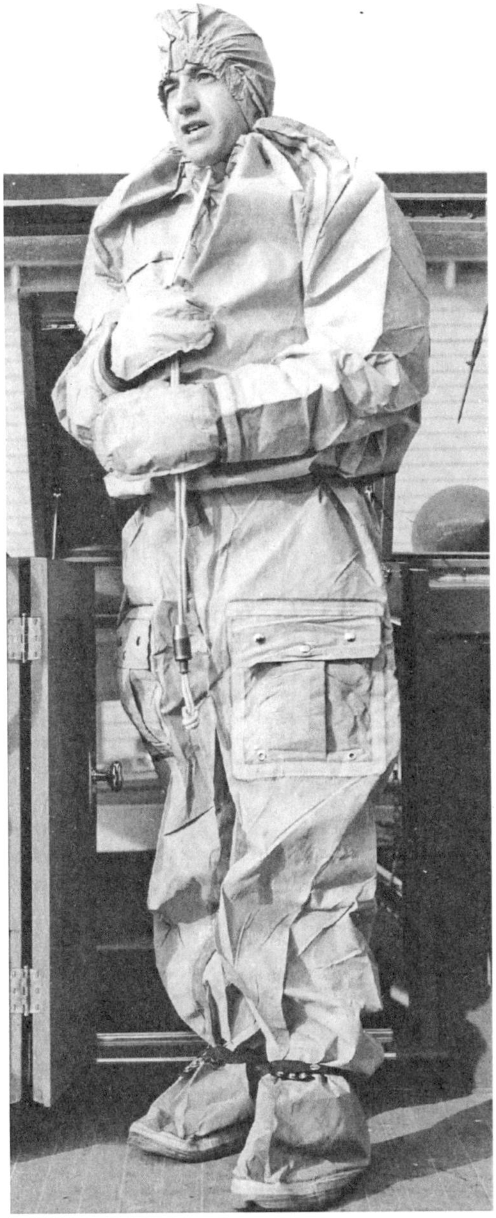

(Left) The R-1 suit being demonstrated.

(Below) The suit was stored in a cylindrical fabric container 12" long and 6" in diameter. Four grommets were provided for securing it to the aircraft fuselage, somewhere for immediate access in an emergency. The basis of issue would be one per crew member. If stored with the life raft, other items deemed not as critical would need to be removed. (National Archives)

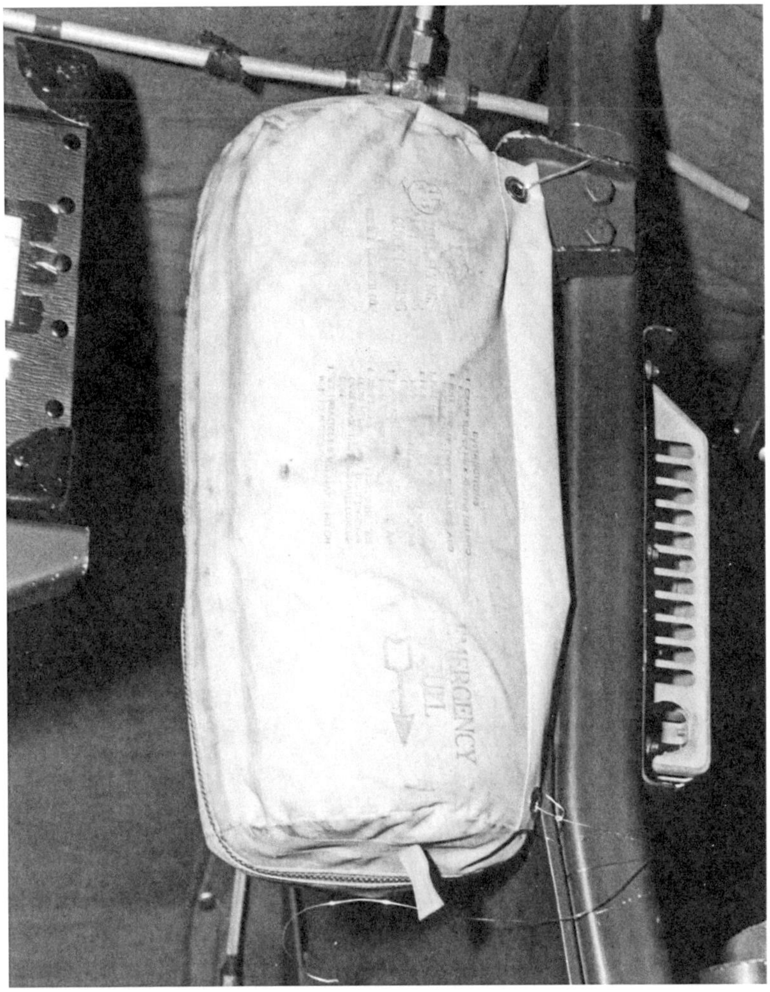

(Below) When worn correctly, the R-1 suit trapped air inside, assisting in buoyancy as well as keeping the wearer dry.

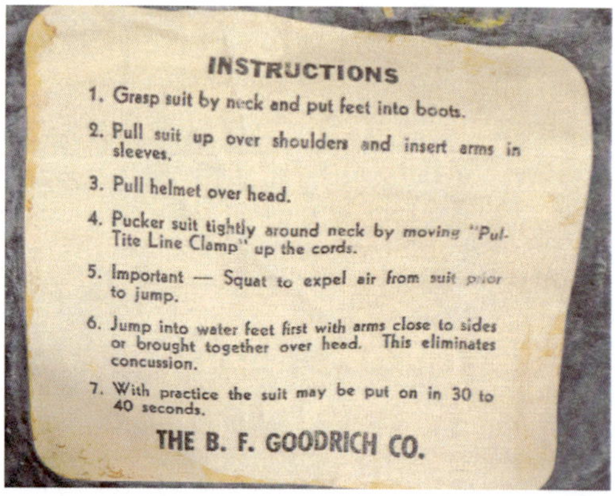

INSTRUCTIONS

1. Grasp suit by neck and put feet into boots.
2. Pull suit up over shoulders and insert arms in sleeves.
3. Pull helmet over head.
4. Pucker suit tightly around neck by moving "Pul-Tite Line Clamp" up the cords.
5. Important — Squat to expel air from suit prior to jump.
6. Jump into water feet first with arms close to sides or brought together over head. This eliminates concussion.
7. With practice the suit may be put on in 30 to 40 seconds.

THE B. F. GOODRICH CO.

(Above) The BuAer exposure suit adopted 28 February 1945 as NavAer Specification M-680. It was fabricated from black lightweight waterproof material and titled as Suit, Exposure, Mark-I, Quick Donning. The suits were to be distributed to all pilots and aircrew operating in the North and South Atlantic and Pacific Oceans available under stock number R37-S-5332. Each suit was equipped with two sea marker packets, M-566. The overall construction was like that of the R-1, except the gloves were integral to the suit and no pockets were provided. (National Archives)

(Top right two images) Fabric stowage bag for the M-680.

(Middle right) Printed instruction label inside the suit.

(Right) Period photograph of the suit in use.

(Left and far left) 5,000 Mark-1 exposure suits were placed on procurement in January 1945. (National Archives)

Coinciding with the tests for protective clothing, ointments were additionally tested as an optional alternative.

Clothing proved the more favorable option, as the ointments rubbed off easily and made the hands slippery. The principle focus was protection from the direct rays of the sun on the skin, which accelerated dehydration, and protection for the eyes to prevent infections and irritation.

With clothing, it was found most urgent to provide awnings as a primary means of sun protection, using sails and paulins. In conclusion, it was also recommended to include, at minimum, 2 oz of ointment to alleviate sunburn and to help maintain morale.

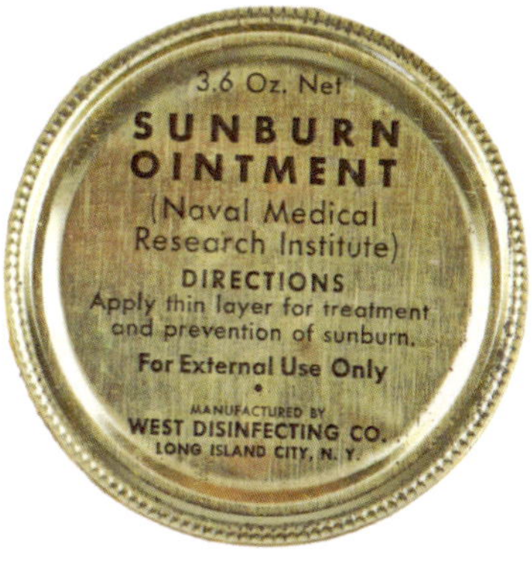

(Above) The Naval Medical Research Institute adopted a formula by the Eli Lilly Co. referenced under their Part No. P-40469. The containers were detailed for 4 ounces of ointment, but both types standardized were either slightly under or over that amount. For use in emergency kits, a tin was utilized available under ASO stock number R57-O-100. It was directed that one tin be included in the MK-II, two tins for the MK-IV and MK-VII life rafts, two tins in the shipwreck kit, and one in the M-592 back pad kit. For waterproofing, a foil lined Kraft paper was used, later transitioning to a scrim cover foiled wrapper.

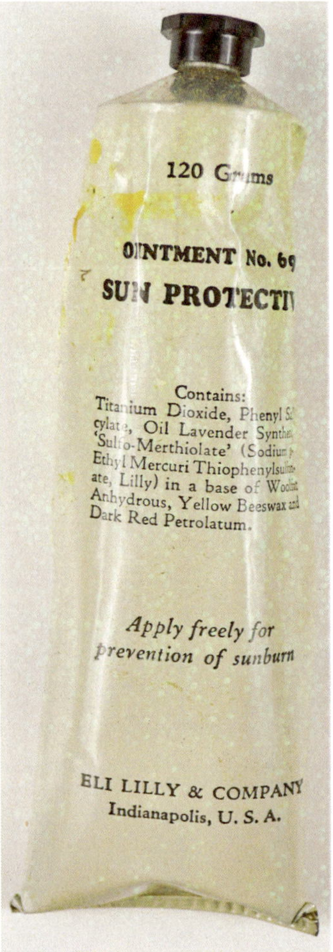

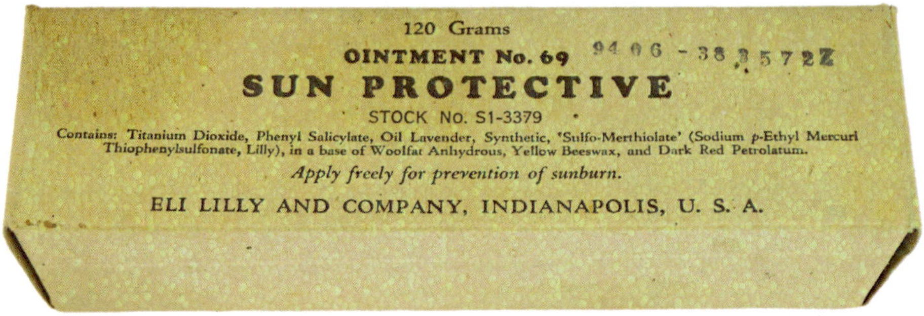

(Left and above) Ointment was also packaged in a tube with a protective cardboard carton. Sunburn ointment was integrated into BuShips first aid boxes, which utilized these tubes. For BuAer life rafts the tins were detailed to be included.

(Below) When the USAAF adopted the ointment for inclusion in their multi-person life rafts, they directed that four tubes be included, and one tube for the Type E-10 sustenance kit. They would be available for requisition under stock No. 8300-605500. The example illustrated here has a Bureau of Medicine & Surgery stock number, the USAAF requisitioned supply from them making no direct procurements.

SURVIVAL TRAINING

With the abrupt start of the United States' involvement in World War II, the military often found itself wholly unprepared for the massive task that lay ahead. Transitioning from a small peacetime military to the enormous force that would be needed to win the war would quickly stress training techniques and personnel to the limit. This would prove especially true for the men placed in charge of training flying personnel in the basics of land and sea survival.

Both aeronautical services experienced an influx in pilots and aircrew after the start of the war, with new schools opening across the country to handle the training of new recruits. Viewed as an essential in the instruction of pilots and aircrew, survival and air sea rescue training took on a new level of importance. The first few months of the war found survival and air sea rescue often unorganized, under appreciated, and lacking standardization between the different commands. For all intents and practical purposes, survival training was integrated into the already established Naval Aviation Physical Training Program. From there it was taken to overseas operational commands, out to the fleets, and even into the jungles of the South Pacific.

Before the war, US Navy pilot training consisted of elimination training courses to weed out those unfit, followed by six months of flight training at Pensacola. Just after Pearl Harbor the program was extended into a 23-month program. After an initial screening for the best mental and physical specimens, volunteers were sent to flight preparatory schools for 15 weeks, receiving instruction in physics, mathematics and other classes necessary for a Naval aviator orientated towards navigation. Next, they spent about 12 weeks at a Civil Aeronautics Administration War Training Service School. There, they would be indoctrinated in the theories and principles of flight and receive about 30-50 hours of flight training by CAA contractors. Those who qualified then went onto Pre-Flight Schools for 11 weeks, where the emphasis was geared toward physical conditioning. The course included boxing, wrestling, swimming, hand-to-hand combat, track, and other assorted obstacle courses. Those that continued to qualify would move onto primary flight training at one of 16 schools where they studied aerology, navigation, communications, gunnery and advanced flight training at Naval Reserve air bases for about 12 weeks. Afterwards, primary cadets were sent to either Pensacola or Corpus Christi for intermediate training operating more advanced aircraft and began the process of specializing as VF, VB, VT or VPB pilots for about 14-18 weeks. Only after this training they would earn their wings.

Another 8-10 weeks would be devoted to operational training, flying their respective combat aircraft, qualifying for carrier landings and preparation for fleet service.

From operational training, they were sent to the fleet, usually after additional practice while forming their squadrons at various Naval Air Stations on either coast. Those who did not go into operational training were assigned to patrol squadrons, Naval Air Transport Service or as instructors.

Marine Corps pilots would move on to Cherry

(Right) By 1943, the BuAer began to aggressively pursue an official organization and standardization of a survival training program. Survival experts John and Frank Craighead were recruited and commissioned by the Navy to help develop a syllabus and survival courses based on real world situations. An aviator would need to be physically tough to survive if they were forced down, possibly having to spend weeks in jungle or on islands, dependent on their own strength and resourcefulness. It was the logical choice to integrate survival techniques into the pre-flight school curriculum, which was designed to bring pilots to peak physical condition for combat, developing ruggedness and endurance essential for survival on land or sea. Courses and lectures were established at the five pre-flight schools located at the University of North Carolina, Chapel Hill, St. Mary's College, St Mary's California, University of Iowa, Iowa City, University of Georgia, Athens, and Monterrey Airfield in Del Monte California. The Craighead brothers built the survival training from scratch, and, as commissioned lieutenants, headed up the first courses at Chapel Hill. They taught trainees and groomed future survival instructors for other pre-flight schools and officers for overseas stations. For the program, they drafted a manual designed for use by both instructors and trainees detailing methods for surviving on both land and sea, titled "How to Survive on Land and Sea." This was the standard textbook issued to aviation cadets.

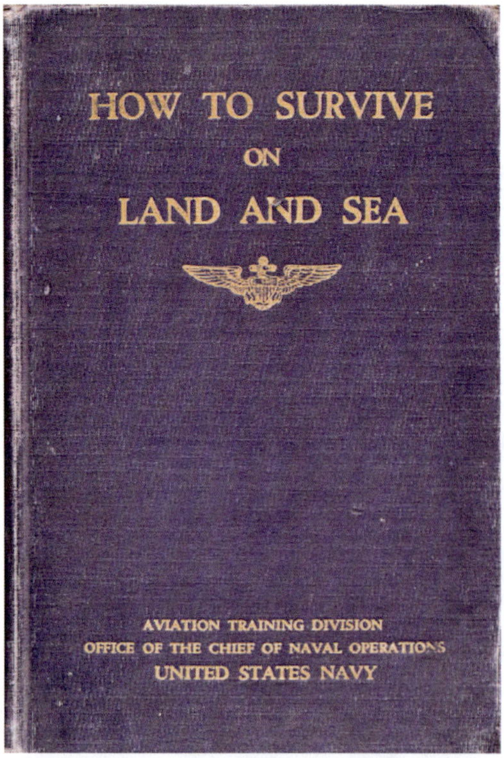

Point for operational training after the intermediate course. Lighter-than-air pilots went direct from pre-flight to lighter-than-air training centers at NAS Lakehurst or Moffet Field for a four month course to qualify as Naval Aviators (Airship).

In early 1944 the preparatory and CAA training schools were dropped, expanding the pre-flight, primary and intermediate courses to a 26-month training period.

(This page) Images are from the Iowa City pre-flight school. The pre-flight survival training curriculum was designed to be an elementary and basic indoctrination. A typical class consisted of 200-300 cadets who were taught the fundamentals of land and sea survival consisting of repelling, raft construction, gathering of food, cooking and fabrication of shelters. The course comprised 12 classroom lectures, averaging about 45 minutes, supplemented with color slides and survival training films. Demonstration rooms were also available, familiarizing cadets with current emergency equipment. After lectures, cadets would be outfitted with a 45-pound rucksack, supplied with only the essentials for survival off the land, which included a sleeping bag, canteen, compass, mess kit, headnet, signal mirror and machete. Three to five half-day field trips or hikes would be conducted, emphasizing navigation and how to travel on land and on the water. One overnight field trip would be conducted, supervised by four to five instructors. At various points during the field trips the instructors would conduct demonstrations. (National Archives)

(This page) Images from NAS Pensacola. Survival training was extended when the cadet entered intermediate flight training. This course was more advanced, giving more specific information and specialized instruction about specific climatic zones. As in pre-flight, the course included four lectures - one each on the subjects of survival at sea, survival along the sea shore, survival in the arctic and survival in the tropics, emphasizing plant life. A one-and-half-day field trip was conducted, giving the pilots practical experience in survival at sea and along the sea shore. Pilots would be provided with minimal equipment, principally only what they would be carrying in the event they were forced down. The course started one mile offshore where life preservers and life rafts were inflated, then they would navigate towards shore. Once ashore they were shown methods of obtaining fresh water, food, and erecting shelters. (National Archives)

(Above) In 1944, a survival exhibit was created for the indoctrination of trainees and equipment officers at NAS Pensacola. Prior to the field trip, groups of pilots would be taken on a two-hour tour of the facility designed to illustrate visually survival procedures, emergency equipment and technical information. The building consisted of various themed rooms and an outside field exhibit.

(Below) When visitors first entered the lobby, they were welcomed by a large map illustrating the combat theaters, native peoples and vegetative regions of the world.

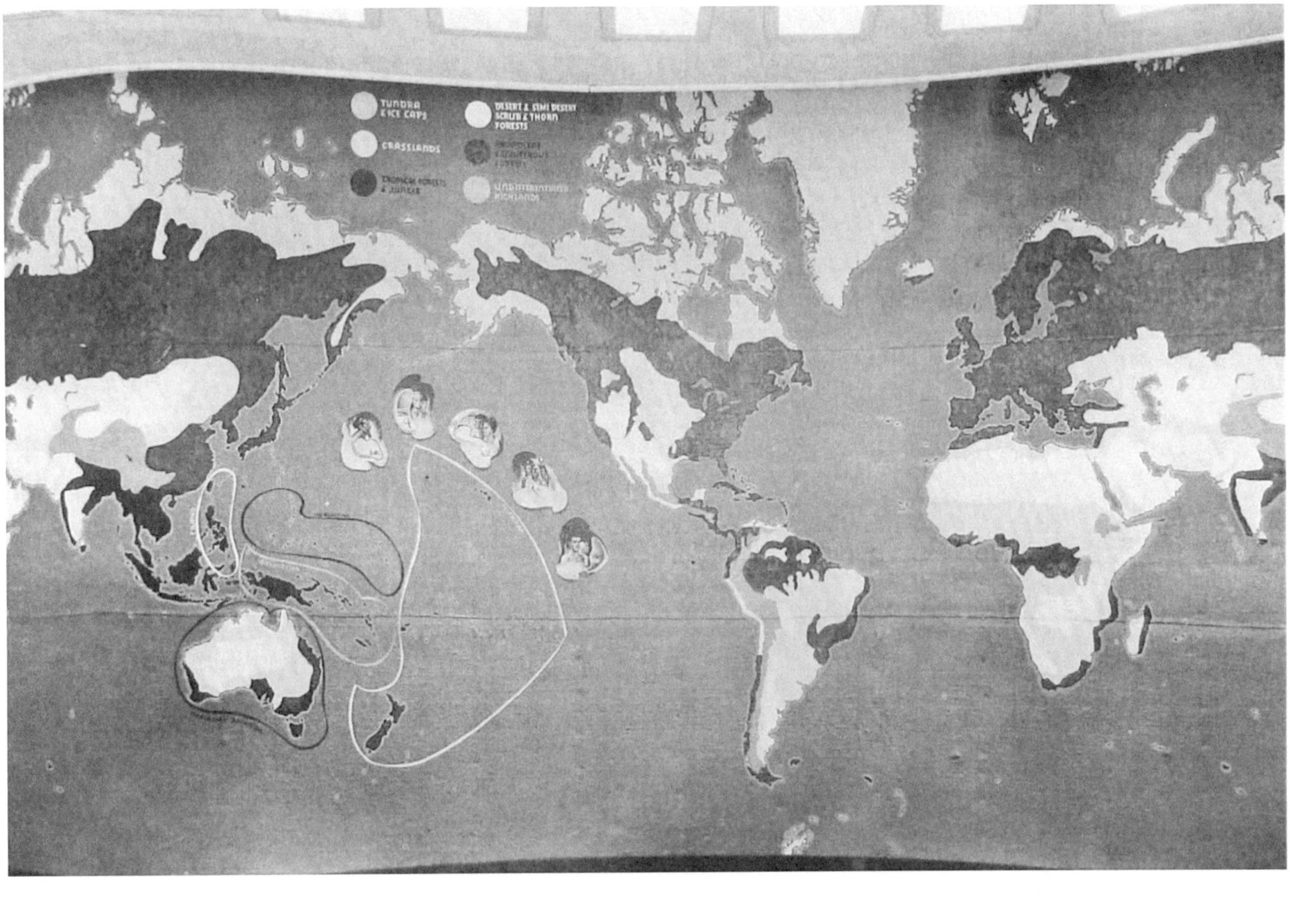

(Above left) The Pensacola Survival Exhibit also included a Chute Room exhibiting diagrams and parachute procedures. (National Archives)

(Above right) The Float Room consisted of life rafts and associated equipment, and along the walls were other emergency items, diagrams and ocean related articles. (National Archives)

(Above left) The Flight Room was constructed around the actual hull of a PBY aircraft and also included oxygen equipment. On the outside of the building was the Field Exhibit, further demonstrating such things as shelters, ovens, and snares to name a few orientated towards the tropics. (National Archives)

(Above) The Tropic Room was designed under the same principle. (National Archives)

(Left) The Arctic Room had several full-size dioramas illustrating principles of sub-zero survival. (National Archives)

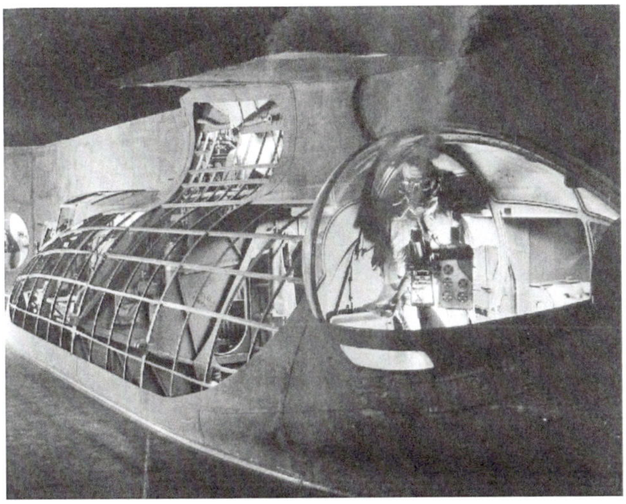

(Above, left and right) The Flight Room may have been the most unique feature of the exhibit. One half of the aircraft with the skin removed was inside where visitors could view the internal structure and placement of emergency equipment. The other half of the hull was outside complete with the skin. (National Archives)

(Right and below) The survival exhibit remained in existence until 1997 when it was finally closed and torn down. The PBY remained in storage until 2001 when it was transferred to the National Naval Aviation Museum's restoration facility, where it was completed, restored, and placed on permanent exhibit in the museum. (NNAM)

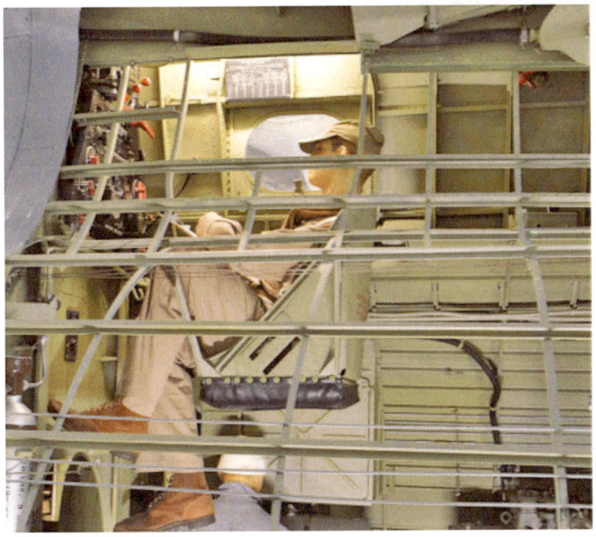

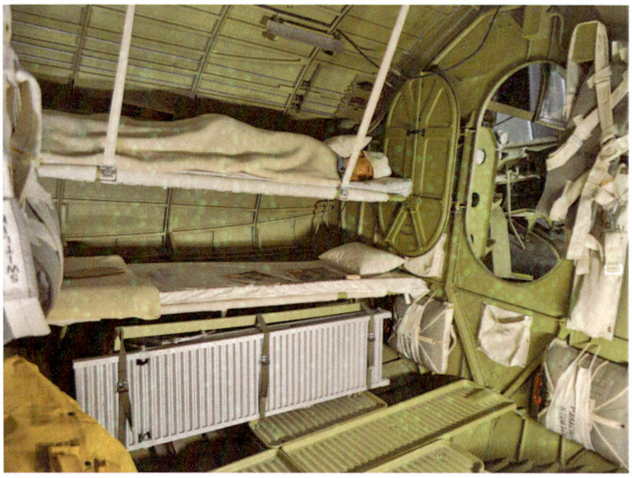

(Above left and right) Crew quarters on the PBY exhibit at the National Naval Aviation Museum. (NNAM)

(Below) Survival training at NAS Jacksonville. Aircrew would spend 90 days in one of the branches at a Technical Training School across the country for an enlisted specialist rate. After graduation, they could volunteer as air gunners, and those that were accepted took a two month course at one of three Naval air gunnery schools at NAS Jacksonville, Florida, Hollywood Beach, California or Purcell, Oklahoma. Inaugurated at NAS Jacksonville by John and Frank Craighead was a land survival training course where crewmen would receive classroom lectures and demonstrations on survival, paralleling the pilot program. Two hours were devoted to instruction and practical work in a demonstration area. The exhibit orientated the enlisted men to the pertinent information relative to tropical, seashore and arctic survival. Supplementing the program would be water survival techniques conducted in a swimming pool. (National Archives)

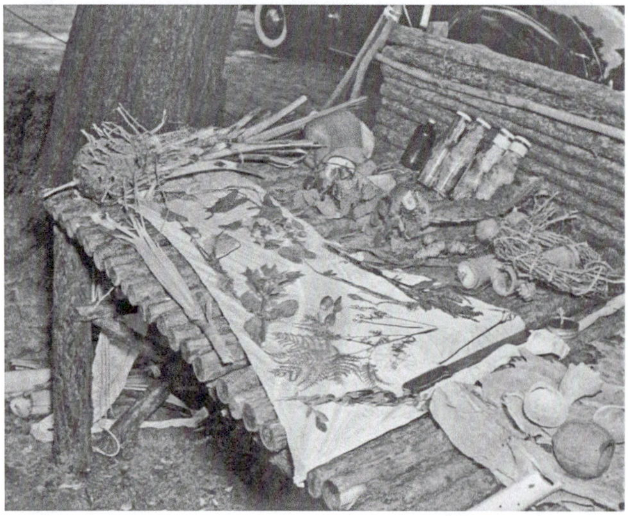

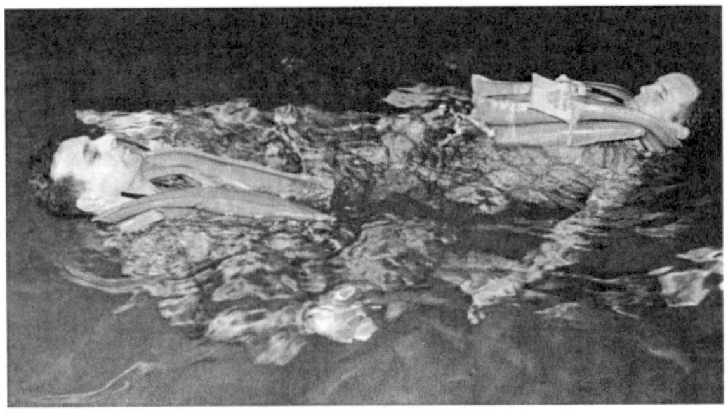

(This page) Of the physical training programs, swimming was the most strongly emphasized, starting with the simulation of battle conditions. Pilots and aircrew were indoctrinated in abandon ship techniques and water safety instructions. The program became progressively more difficult through each phase, maintaining minimum requirements from pre-flight to primary to intermediate. Testing was required every two months in operational training and fleet stations. Knowledge of emergency equipment started in the primary phase, such as learning to swim with a life preserver and boarding, or helping others, aboard a life raft. Primary training began with a formal introduction to pneumatic life rafts, back pad kits and other emergency kits. Shown are simulations of bailing out over water and other parachuting techniques. (National Archives)

(Above) During the intermediate stage of training, pilots would be introduced to a simulator to get a taste of what it feels like to bring a plane down in water, nicknamed "Dilbert Dunker". This training device was constructed from the cockpit of a SNJ-3. With the cadet strapped in, it was catapulted at a 45-degree angle traveling at 20 mph. Upon striking the water, the cockpit automatically flipped over, placing the trainee under water. He then had to release his harness and exit the cockpit, swim to the surface, then inflate and board his rubber raft as part of the exercise. (National Archives)

(Right and below) Aircrew would undergo aquatic training as standard for Navy personnel of abandon ship in their trade and gunnery schools. It wasn't until 1944 that survival training courses were introduced as part of the program. In November 1944, ditching drills were introduced. Using the hull of a PBY, pilots and aircrew were instructed to land and live under emergency conditions.

(Below) The crew assumes crash positions.

(Right) The plane is pushed down a ramp from dry land into the water. Holes in the hull allowed water to enter and rise inside.

(Below right) Upon the signal to abandon ship the crew began the procedure to gather all emergency provisions and exit the aircraft, they then inflated and boarded the pneumatic raft. (National Archives)

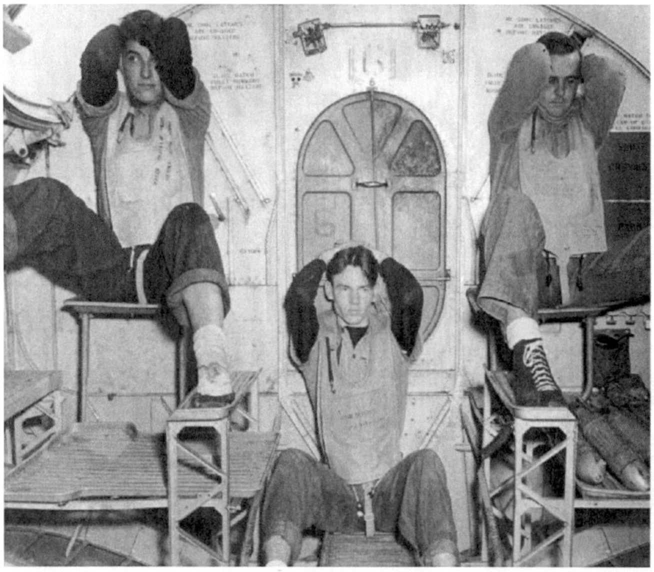

Due to the continually expanding survival training program and the institution of the equipment officer, a survival instructor's course was initiated in the fall of 1943. The course followed that of the pre-flight, primary, intermediate and operational training, but with additional reading assignments, field work, and methods of training.

Instructor trainees would be orientated to the pre-flight program at Chapel Hill for 12 days. During the first six days they would follow a pattern of lectures, films, slides and reading assignments each morning. In the afternoons they would be taken on field trips for practical application. The second week of instruction consisted of a six-day field trip through the countryside with a minimum amount of equipment, living off the land.

This introductory training would be followed by further indoctrination for nine days at NAS Pensacola, where training included advanced lectures on various fields such as flight safety, medicine and parachutes, as well as survival at sea, seashore and in the arctic. It was this course that greatly influenced the erection of the Survival Training Exhibit mentioned previously. Intensive instruction was given in air sea rescue training, use of emergency equipment and how to stow it in planes. Lectures also included parachutes and their uses – principles of jumping, descending and landing, ditching procedures and distress communications. After concluding the course, trainees learned air sea rescue organization and control between survivors and rescuers. This phase of the training included a live exercise using PBYs, fighter planes and crash boats to illustrate how the various types of rescue units were coordinated to affect a rescue.

The instructor's program concluded with a week of tropical survival methods at NAS Opalocka near Miami. This course consisted wholly of practical field training in tropical survival, erecting shelters, snares, identification of vegetation, fishing, swimming, swamp navigation and air sea rescue techniques.

(Right and below) Emergency equipment exhibits were eventually maintained at all naval air stations, typically consisting of basic displays along the walls in a classroom. It was in these classrooms where lectures and demonstrations would be outlined, followed by practical use in a swimming pool lead by an AESO. The same concept applied to fleet activities, but would often be in the ready room of the aircraft carrier, conducted by the equipment officer, flight surgeon, or intelligence officer. Continued education was maintained from pre-flight to the combat zone, a policy that paid dividends during emergencies.

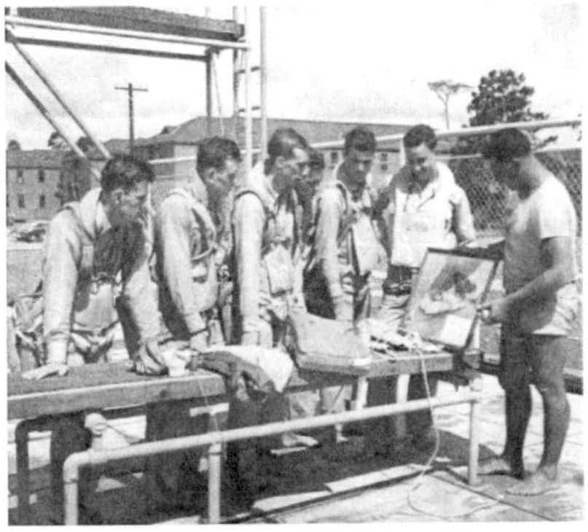

The USAAF training programs were very much like that of the US Navy but more diverse and peculiar to their needs. Cadets would enter pre-flight or college preparatory training for a four to nine week course of academics in aspects of math, physics, map reading, and meteorology, with preliminary flight training by CAA instructors. Upon graduation, they would enter primary training at one of the three regional sub commands of the Flying Training Command for 12 weeks. Here they would become familiar with aircraft, flying with an instructor, then move on to solo flying. Those who qualified would move on to basic flight training for nine to ten weeks, flying more powerful aircraft and receiving advanced instruction averaging about 70 hours of flight time.

From basic flight training, the selection process began for pilots of single or twin engine aircraft, and those who did not qualify were transferred to schools for bombardiers, navigators, gunnery or ground training. Cadets were then transferred for twin or single engine advanced flight training for 60-70 hours receiving further technical training.

It was within the transitional flight training that the newly-commissioned officers would begin specific training in fighters, medium bombardment, heavy bombardment or twin engine transport aircraft for ten weeks. Aircrew technical training would be conducted at various schools across the regions which typically consisted of four to six week courses. Pilots, navigators, bombardiers and aircrew would join in the operational training period as newly-activated groups (designated as Operational Training Units) for 30 days of simulated flying missions and academic studies of their combat functions.

The USAAF operated with a standard physical fitness program for all trainees of one hour a day, seven days a week, consisting of calisthenics, swimming and cross country running, plus about four hours a week in team sports beginning in pre-flight training and basic training for aircrew. The hours required gradually increased from 32 to 60 hours by May 1944, principally due to the increased length of the overall training course. Physical training would taper off in all subsequent training programs where emphasis was orientated towards flight, ground and technical training.

While the Navy placed emphasis on swimming and physical conditioning for the basis of survival, the USAAF developed a high altitude training program placing their emphasis on the technical and practical aspects of survival at high altitudes. As part the Altitude Training Program, a course for unit oxygen officers was instituted in July 1942 at the School of Aviation Medicine. This course was originally only five days, consisting of lectures on the physiological aspects of high altitude flying, demonstrations of oxygen equipment and altitude chamber flights. As the needs in the combat theater increased, oxygen officers, later re-designated as Personal Equipment Officers, were inaugurated in the combat zone.

As the war progressed through 1942, significant deficiencies were noted on practical survival knowledge in the event of emergencies. This was initially addressed by the establishment of the PEO to indoctrinate pilots and aircrew on bail-out and ditching procedures, emergency kits, life rafts and parachutes.

As a result of reports from the Pacific indicating a greater need for aquatic training, swimming pools began to be constructed at training centers. Curriculums in basic and pre-flight training integrated instruction on the inflation of clothing, swimming through burning oil, and exercises in artificial respiration, establishing proficiency standards.

By the end of 1943, trainees were indoctrinated in aquatic training, parachute landing training, and a more rigorous physical fitness emphasis. Additional subjects were added, in May 1944, for 6 hours of medical training and 9 hours of altitude training in pre-flight schools. The medical training was geared toward preparing individuals to live in lands infected with various disease-carrying insects, food and emergency first aid treatment.

USAAF personnel were required to have training in survival aquatics as outlined in AAF Letter 50-57, dated 12 October 1944, with additional instruction in life rafts, life preservers, abandon ship, defense against strafing, and escape from the parachute harness. Specialized survival training was integrated into the pre-flight and basic training curriculums to bring pilots and enlisted men to a physical and knowledgeable standard. They then entered technical academic training schools after which they were activated in operational training units. Operational training was conducted at the AAF School of Applied Tactics in Orlando, Florida, activated in October 1942.

AAFSAT was the successor to the Air Corps Tactical School at Maxwell Field, Alabama. During the war it was the principle center for development of air indoctrination and instruction. For consolidation of fighter command, bombardment and other air support aspects, the school was re-designated AAF Tactical Center, in October 1943, a school for postgraduate combat unit training, which also conducted research in all fields of air warfare. Here, special courses were prepared for survival and air sea rescue as part of the practical training syllabus, before departing to the combat theaters.

In March 1944, the PEO course was transferred from the School of Aviation Medicine to the AAFTC, and while at the station they would train combat aircrews in addition to their schooling.

AAFTC was again re-designated as the AAF Center, in June 1945, and comprised the AAF Proving Ground Command at Eglin Field, Florida plus the AAF School, AAF Board and the Arctic, Desert and Tropic Branch.

(This page) Aquatic training centers gradually began to be integrated into stateside training centers and overseas stations for practical survival training. Personnel would undergo abandon ship drills of basic swimming principles and orientation of life rafts and associated emergency equipment. Continued education courses would be available and represented in this series of images from AAFTC stations in the Philippines, England and Panama. (National Archives)

(Above) Practicing crash positions and escape hatch drills. Early training practices lacked essential equipment, making it difficult to give adequate instruction, often relying on training films. Due to the lack of facilities the students had no opportunity to see or practice stowage of equipment in fuselages, or other air sea rescue procedures.

(Top right and right) In late 1943 PEO and AAFTC schools requested salvaged fuselages so that actual demonstrations could be conducted. In early 1944 mock-ups became available allowing full demonstrations. Trainees would learn crash positions and other emergency procedure protocols. They would be orientated towards all emergency equipment such as oxygen installations, fire extinguishers, escape hatch drills, and dinghy drills.

(Below) Some installations did not have access to ponds, having the mock-ups on dry land, but the overall basic principles of training can be applied. (National Archives)

(Above) Practical mock-ups were provided in the overseas theater, principally in the ETO. In August 1942, the Provisional Medical Field Service School was officially opened at High Wycombe, England. In part it concentrated on high altitude bombing and airplane ditching. As the school expanded it provided indoctrination to ditching and emergency equipment, which ultimately saved many lives. In addition, the AAF utilized the RAF School of Air Sea Rescue in Blackpool, England, which was established in May 1943. (National Archives)

(Right) First Aid training. (National Archives)

(Below) The school at High Wycombe was eventually renamed the First Central Medical Establishment, with a second established in the Pacific, and a third in North Africa. (National Archives)

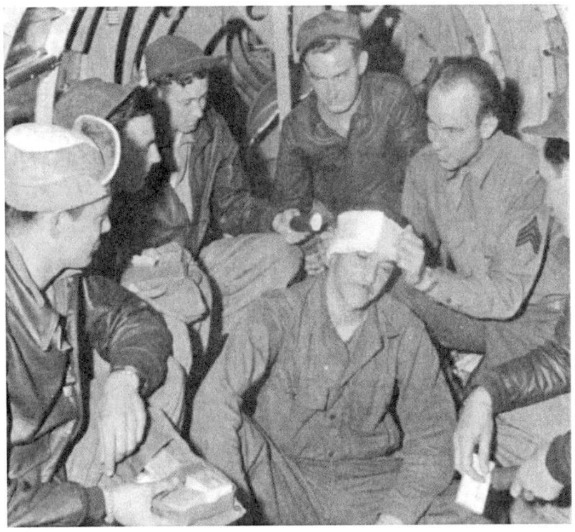

(This page) Lectures and continued education became part of the standard operating procedure for the USAAF for the duration of the war. Orientation classrooms with exhibits were available at most air fields, typically headed by the personal equipment officer. Images shown are from demonstration AAFTC demonstration rooms in Panama and the ETO. (National Archives)

(This page) Images from the Bernice P. Bishop Museum survival training school, Honolulu, Hawaii. In the early stages of the war, teaching of survival principles was left up to local commands. Some were unofficially assigned special designations such as Land Rescue Officer or Jungle Training Officer. Under direction of the Intelligence Center, POA in Hawaii, Kenneth P. Emory prepared a syllabus and literature for a jungle survival school. Established at the Bernice P. Bishop Museum in Honolulu, the school was headed by Emory. Like the school established by John and Frank Craighead, the curriculum comprised lectures and demonstrations on jungle survival. It was open to all services, becoming part of a standard rotation for military personal coming into theater and those in reorganization. An additional jungle training school was established in the Kahuna Valley, Oahu. Marine organizations established advanced survival training camps at Espiritu Santo and a jungle school in the New Hebrides. Brig. Gen. Walter G. Farrell, USMC commander of the Third Marine Aircraft Wing, developed a four-day advanced survival training camp in Hawaii for airmen entering the theater. The coarse mirrored the curriculum of that from Pensacola. These courses in the combat zones (both the PTO and ETO) were established to meet an immediate demand for education for the preservation of lives. Stateside training was amended to address the deficiencies noted by the overseas commands and, as time went by, pilots and aircrew entering the theater would be thoroughly trained for emergency situations. The schools already established would augment that training upon their arrival. By 1945, survival orientation was so intense that there were complaints from commanders that too much time was devoted to the subject, detracting from tactical application.

(Left) OSS personal of the 1st Detachment headed a jungle survival school in Taro, Burma. Here they familiarize personnel with the terrain, edible foods and native peoples. (National Archives)

(Below and right) AAF personnel participate in a jungle survival school on New Guinea headed by instructors from Australia and New Zealand. (National Archives)

Afterword

Visual signals contributed greatly to keeping survivors in sight and locating them when they were lost. The types of signal used most successfully, in order of their value, were: sea dye marker, smoke signals, signaling mirror and tracer ammunition.

The SCR-578 transmitter (Gibson Girl) was used successfully, and had a high favorability rating, but due to its bulk, its use was limited. Development of smaller transmitters was initiated during the conflict, which may have helped in the initial preservation and attracted rescue, but supplemental items were necessary to further sustain life. Personal aid kits could support the health of a survivor for a limited time.

Today, Air Sea Rescue is considered an integral facet of all military operations. Emergencies requiring survival and rescue became routine occurrences in World War II, especially in combat. Complete survival instruction is an integral part of the training of aviation personnel, as well as key to the preparation of the aviator's equipment.

Acknowledgments

This work could not have reached its potential without contributions and support from Robin Bartel and Paul Faulkner. Robin Bartel has been a friend and a peer for more than 25 years, and has contributed many artifacts for reference as well as solid support for the project, consistently urging me on. Paul Faulkner is a long-time friend who shares the same interests, and his knowledge of Naval aviation is unequalled. I would also like to thank Kriss Usherwood, Charlie Flick, Frank Trzaska, Bob Baldwin, Mathieu Bianchi, Remi Bouche, Jeff Klein, Jerry Kolakowski, Ben Major, Steve Milnthorpe, George Petersen, Gregory Pons, Geoff Pringle and Mick Prodger for their contributions and graciously shared information. Significant assistance was provided by Brett Stolle of the National Museum of the United States Air Force, Bob Thomas of the National Naval Aviation Museum and Maranda Gilmore of the Air Force Historical Research Agency.

Author's Note

I have maintained an interest in WWII for as far back as I can recall, playing army in the woods in my home state of Oregon. It was on my 13th birthday, back in 1987, that I established myself as a collector. My gift was an M1 Carbine. Rummaging through my trunk of field gear, I separated the WWII items from everything else, and those, together with the rifle, formed the basis of my collection.

My interest only grew, accumulating more and more items until today, 30 years later, all aspects of United States WWII memorabilia remain a passion of mine. Some 20 years ago I obtained an empty US Navy M-592 emergency back pad kit – my first piece of WWII survival equipment – and so began my transition solely into that area. This initiated a quest to find out and obtain what went in the kit. I quickly found myself frustrated at the lack of information available, gathering more questions than answers. As the investigation moved forward I became more and more interested in the overall subject, accumulating artifacts but still operating with a very limited reference library. Influenced by peers in my local collecting community, I eagerly joined in conversations about their adventures across the country, networking and traveling to various archival holdings. It soon became apparent that I would need make those pilgrimages in order to find the answers to questions that I, and many others, shared. After a relatively short time I decided that if I was going to make that effort, I should put my findings in a reference book for anyone who may be interested.

My initial research efforts were at local university libraries then inter-library loans, until the environment allowed me to visit the National Archives. I have since been back to the NARA multiple times, accumulating documents and images. Not every answer to every question was found, though it was certainly through no lack of effort. Those answers simply may not exist.

It wasn't mandatory for government agencies or facilities to retain all documents, once they were determined obsolete or irrelevant, often ending in their destruction. What survived post-war downsizing and finished up in repositories was at the grace and courtesy of individuals at these government installations, for which I am eternally grateful! However, many records are incomplete, and it can be viewed as a stroke of luck that they still even exist. Information from various sources was sometimes contradictory but through continued accumulation of data, a compilation of redundancy pointed towards the facts. I have attempted to detail each genealogy as clearly as possible.

The overall experience has been very educational for me, and hopefully those who read these studies will find them helpful.

Dustin Clingenpeel
May, 2019

Index

Index

www.ingramcontent.com/pod-product-compliance
Lightning Source LLC
LaVergne TN
LVRC101547220926
848486LV00041B/392

* 9 7 8 1 9 4 3 4 9 2 6 1 9 *